Electrical Power Engineering

Other McGraw-Hill Electrical Power Engineering Books of Interest

FINK, BEATY • *Standard Handbook for Electrical Engineers*
KURTZ, SHOEMAKER • *The Lineman's and Cableman's Handbook*
KUSKO • *Emergency/Standby Power Systems*
LINDEN • *Handbook of Batteries and Fuel Cells*
LUNDQUIST • *On-Line Electrical Troubleshooting*
SEIDMAN, MAHROUS, HICKS • *Handbook of Electric Power Calculations*
SMEATON • *Switchgear and Control Handbook*
WANG, MACDONALD • *Modern Power System Planning*

Electrical Power Engineering

H. W. Kabisama

California State Polytechnic University, Pomona
Formerly of the University of Dar Es Salaam

McGraw-Hill, Inc.
New York St. Louis San Francisco Auckland Bogotá
Caracas Lisbon London Madrid Mexico City Milan
Montreal New Delhi San Juan Singapore
Sydney Tokyo Toronto

Library of Congress Cataloging-in-Publication Data

Kabisama, H. W. (Henslay, William)
Electrical power engineering / H. W. Kabisama.
p. cm.
Includes bibliographical references and index.
ISBN 0-07-033157-X
1. Electric power systems. I. Title.
TK1005.K32 1993
621.31—dc20 93-9697
CIP

2 3 4 5 6 7 8 9 10 11 12 13 14 BKMBKM 9 9 8 7 6 5

ISBN 0-07-033157-X

The sponsoring editor for this book was Harold B. Crawford, the editing supervisor was David E. Fogarty, and the production supervisor was Pamela A. Pelton. This book was set in Century Schoolbook by Techna Type.

This book is dedicated to the memory of my late son, my late daughter, and my late parents:
Percival H. Kabisama,
Rabecca T. M. H. Kabisama,
William Percival Kabisama, and
Kelena Gilbert Kabisama.
May their souls rest in peace.

Contents

ABOUT THE AUTHOR

Dr. Henslay William Kabisama is a Professor in the Electrical and Computer Engineering Department of the California State Polytechnic University in Pomona. He teaches and conducts research in power systems engineering, electric machines, power plants, and electromagnetics. In addition, he works as a consultant for companies in the United States and abroad. Dr. Kabisama has designed and worked on rural electrification in developing countries. He is the author of numerous articles on various electrical power engineering topics. His previous books include *Design of Electrical Power Transmission Lines for Students and Practicing Engineers.*

Preface

This book is intended for electrical engineering professionals, as well as for students. The material presented has been developed in such a way that all readers will be able to read and understand basic concepts.

The book has eight chapters and an appendix. The first chapter covers different types of electric power plants. The plants include steam, hydro-, and nuclear plants. Other energy sources include cogeneration, geothermal, and solar power.

Chapter 2 focuses on magnetic fields and magnetic circuits. It is assumed here that readers have taken an electrical network theory course, an introductory course in electromagnetic fields, and a course in differential equations.

Chapter 3 covers ideal power transformers, real (exact) transformers, and three-phase transformers. Chapters 4, 5, and 6 cover synchronous machines, dc machines, and induction machines.

Chapter 7 covers electric power systems networks. The first part of this chapter covers the per-unit system, and the second part gives the mechanical and electrical design of power transmission lines. The final chapter covers power systems fault calculations. The chapter is subdivided into two major divisions: The first part discusses symmetrical three-phase faults and the second part discusses unsymmetrical faults with the use of symmetrical components.

Complex numbers, phasors, impedances, and polyphase circuits are included in the Appendix. The three-phase circuits are here included to highlight some of the simpler aspects of circuit theory that are important to the understanding of electric machines and transmission lines.

The material of this book is the result of development to meet the needs of the author's classes taught at the University of Dar-es-Salaam, Tanzania, and at the California State Polytechnic University, Pomona, California, over the past 22 years.

In conclusion, I would particularly like to acknowledge the reading of and suggestions to the manuscript by Mr. Athuman Mftukamba, a

Ph.D. student at Stanford University, California. I also want to express appreciation to my wife Masauko Kabisama for typing the text. A special thanks to my children Tayamika, Chimwemwe, Chifuno, and Mwawi for their patience and support.

H. W. Kabisama

Chapter

1

Energy Sources and Electric Power

1.1 Introduction

The great demand in the world for electric energy, in residential, commercial, and industrial applications, has led most developed countries such as the United States, Canada, and the countries of Western Europe, which consume a large portion of the world's energy, to think of developing alternative types of energy sources, rather than depending upon fossil fuels. In this chapter different types of energy sources and power plants will be discussed in brief.

1.2 Electric Power Plants

1.2.1 Conventional steam power plants

In many countries in the world where water power is not abundant or is unavailable, steam power plants are the main supply of electricity. The thermal efficiency of a steam power plant is mainly influenced by the choice of the steam cycle. The actual cycle of a steam power plant is shown in Fig. 1.1. The principal equipment and capabilities of a steam power plant are the boiler, superheating, feedwater heating, steam reheating, turbine, and generator. The performance of the steam power plant is as follows. The steam coming out of the boiler is superheated and enters the turbine at the same temperature T_1. In the turbine the steam expands adiabatically (isentropically) from i_1 to i_2.

In order to improve the thermal efficiency, the mean cycle temperature is increased and the steam is superheated. The condensation of steam occurs at a uniform temperature T_2. In the condenser a large

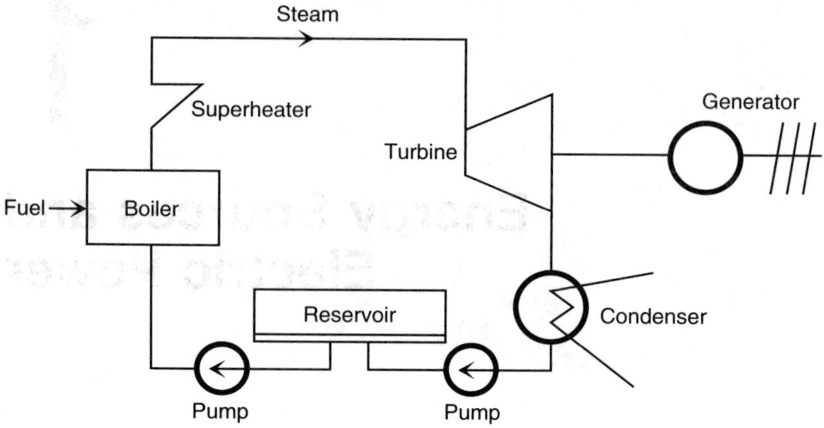

Figure 1.1 A diagram of the major components of a steam generating plant.

thermal loss as the latent heat of steam $(i_2 - i_w)$ is lost in the condenser cooling water. The increase in pressure occurs at increased entropy (Fig. 1.2).

In engineering calculations an enthalpy-entropy or Mollier (i-s) diagram is used rather than T-s. The Mollier diagram is used to extract energy at various temperatures and pressures. The Mollier diagram is reproduced in Fig. 1.3. In this diagram the steam reaches a temperature of 100°C, the enthalpy is measured in kcal/kg, while the entropy is measured in kcal/(kg · degree).

The theoretical thermal efficiency assumes an ideal turbine, condenser, boiler, and pumps, all working at 100 percent. The internal

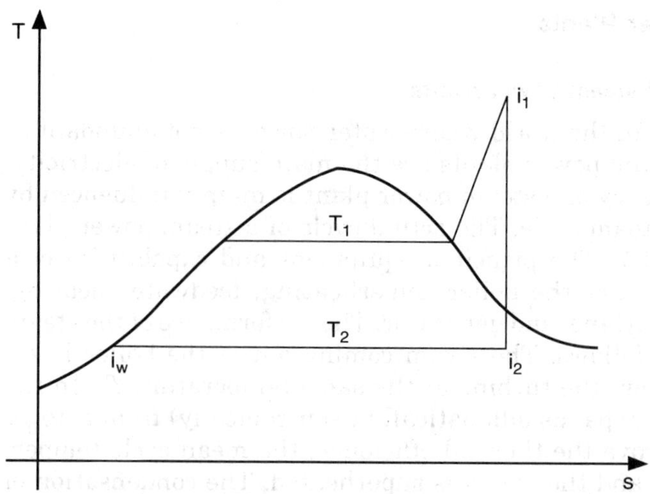

Figure 1.2 Temperature-entropy (T-s) diagram.

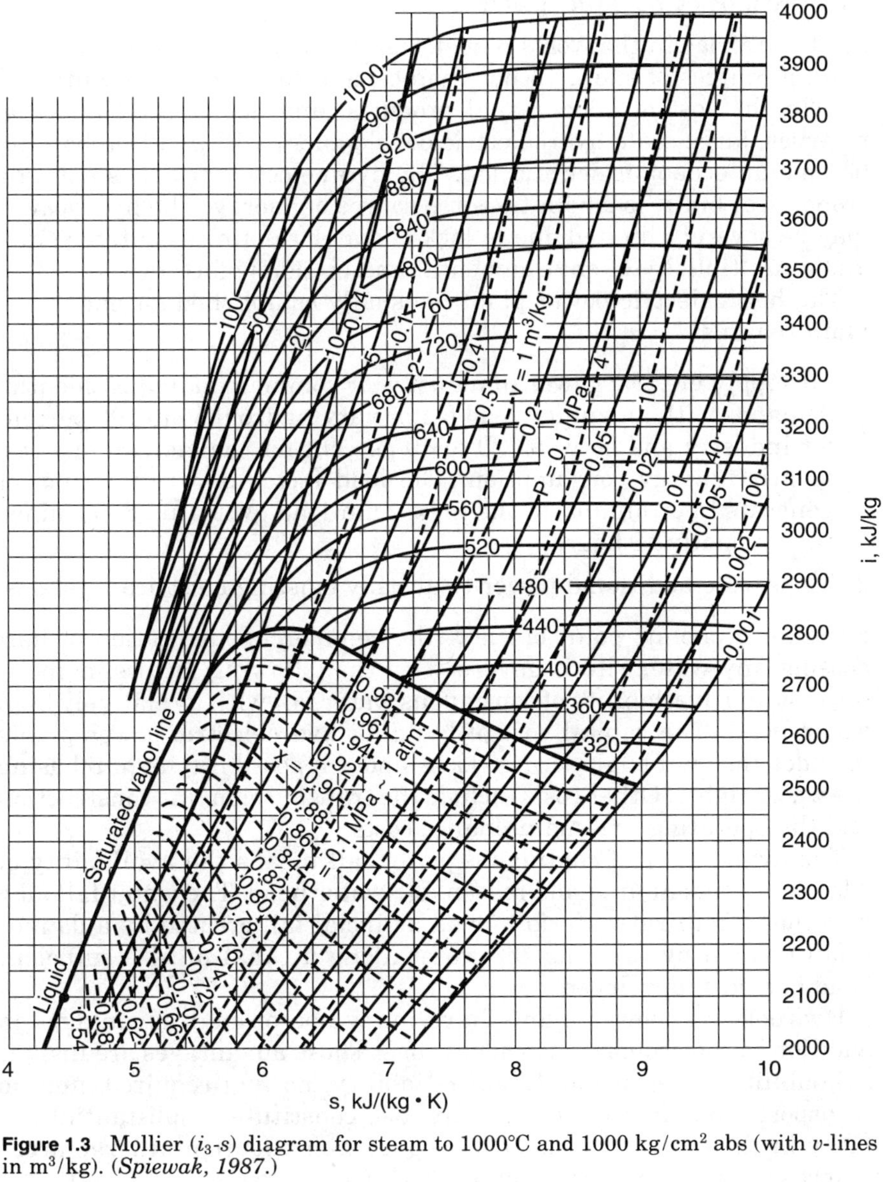

Figure 1.3 Mollier (i_3-s) diagram for steam to 1000°C and 1000 kg/cm² abs (with v-lines in m³/kg). (*Spiewak, 1987.*)

efficiency is

$$\eta_{\text{thermal}} = \frac{\text{energy output of turbine}}{\text{energy input of boilers}} = \frac{i_1 - i_2}{i_1 - i_w} \tag{1.1}$$

A typical value for η_{thermal} is 0.40.

1.2.2 Hydroelectric power plants

Most countries in the world have focused considerable attention upon finding sources of energy other than oil and coal. With the continued rise in the cost of oil and coal products, and depleted oil and coal reserves, large-scale hydroelectric development will in all probability become an economic speculation. The hydroelectric power is not only economical but it is a pollution-free source of energy. The hydropower may be severely limited due to lack of water resources, suitable sites, and a multiplicity of environmental and economic factors.

The hydroelectric power plant is usually established for one of two main reasons:

1. To supply electric power to industry or commercial and residential demands with an annual load factor varying from 80 to 100 percent for industry and between 20 to 40 percent for commercial and residential energy consumption. The load factor is a ratio of power generated per annum to the maximum load over a predetermined time, usually ½ hour.
2. To provide additional power to already existing electrical systems.

From an economic point of view, the commercial utilization of harnessing any water site is governed by the following factors, namely, the cost and availability of continuous water fall with a relatively high water head. The capital cost of building hydroelectric power plants includes the civil engineering works and the possibility of purchasing the water rights. Other costs include the transmission line construction and the operating and maintenance costs.

The availability of continuous water fall means the availability of a large catchment area and reasonable water head. The water fall may be either a natural waterfall or an artificial fall created by a dam or a fall created by carrying the water along a canal until a sufficient head has been developed.

Hydroelectric power plants have many operational advantages in addition to the economic considerations. These advantages are instant availability of service and high reliability, no fuel required, and no transportation charges. Dams may also constitute a substantial liability in the event of a major earthquake, enundation of large areas of fertile arable land, and water-borne diseases.

Operation of hydroelectric power plants. The general mode of operation of hydroelectric power plants includes the study of types of water heads and types of turbines. The water head is defined as the vertical difference between the upper reservoir and the level of the turbine. The water falling through this head gains kinetic energy which is imparted to the turbine blades. There are three types of heads. For each head, there is a particular type of turbine employed. Below is a summary of different heights of water heads and their corresponding turbines:

1. *High head.* The estimated height is about 184 to 1840 m (600 to 6000 ft). With such heights Pelton turbines are used.
2. *Medium head.* The height for a medium head is 37 to 490 m (120 to 1600 ft). Francis turbines are used for medium heights.
3. *Low head.* The height for low heads is up to 60 m (220 ft). A practical minimum height is about 3 m. Kaplan turbines with axial rotors and variable pitch blades are used.

Figure 1.4 shows a general flow diagram of a hydroelectric power plant from the catchment area to the powerhouse.

Efficiency of turbines. Turbines with large output and low heads operate at slow speeds. There is a critical speed at which any turbine operates under the most economical conditions, and if this speed is exceeded without decrease in diameter of the wheel, the efficiency is reduced. The turbines' efficiency curves are shown in Fig. 1.5.

The power available from hydroelectric power is given by

$$P = \rho g Q H = 9.81 QH \qquad \text{kW} \tag{1.2}$$

where ρ = density (1000 kg/m^3)
g = gravitational force (9.81 m/s^2)
Q = rate of flow of water through the turbine (in m^3/s)
H = height of upper water level above the lower (in meters)

Power and energy output of hydroelectric plant. The energy available depends on the catchment area, and is obtained in two ways; either from flow or from storage.

1. From the flow the energy is calculated as follows:

$$\text{kWh} = \frac{\text{ft}^3/\text{s} \times \text{head (ft)}}{15}$$

2. From storage, the energy is calculated as follows:

$$\text{kWh} = \frac{\text{thousands cubic feet stored}}{56} \times \text{head (ft)}$$

or $$\text{kWh} = \text{million cubic feet stored} \times \text{head (ft)} \times 17$$

where 1 in of rain = 3600 ft^3 per cubic area = 2.33 million ft^3/mi^2

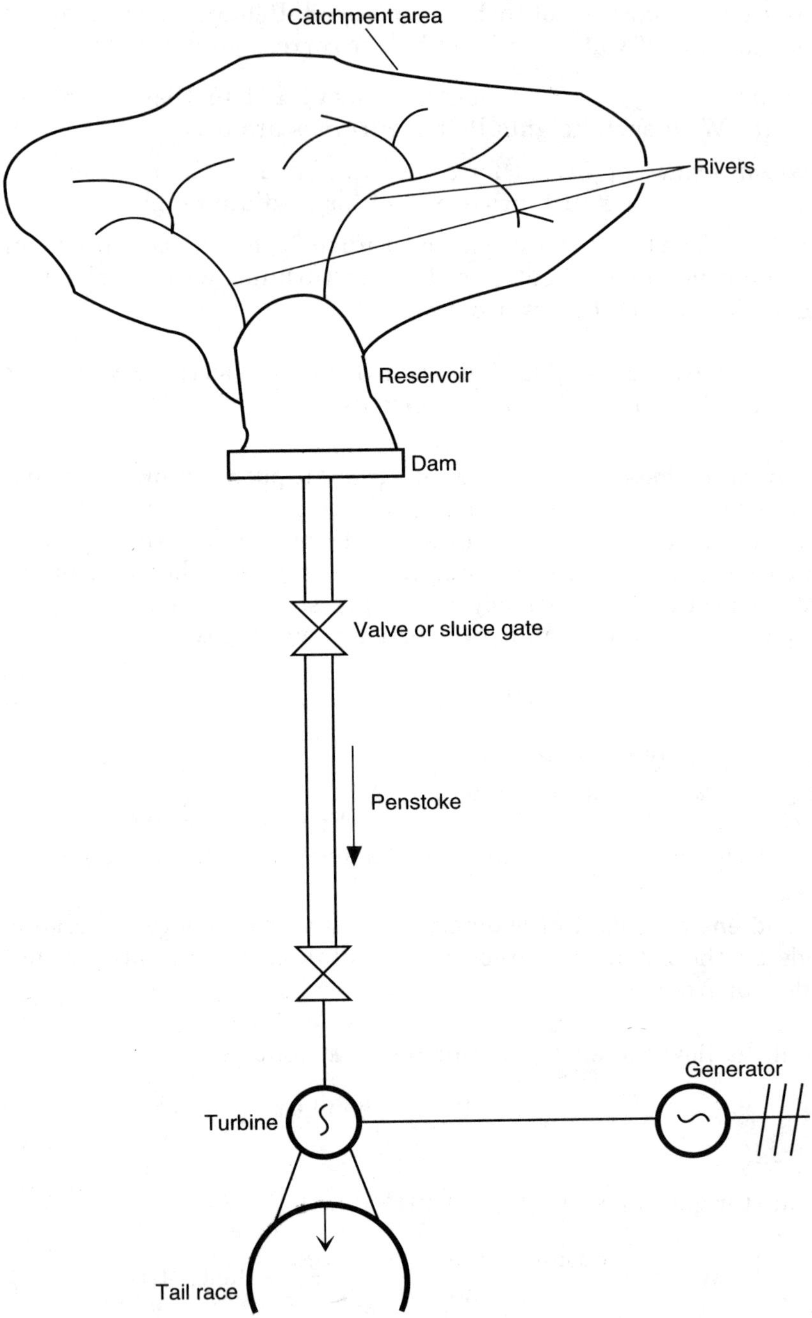

Figure 1.4 Flow diagram of hydroelectric power plant.

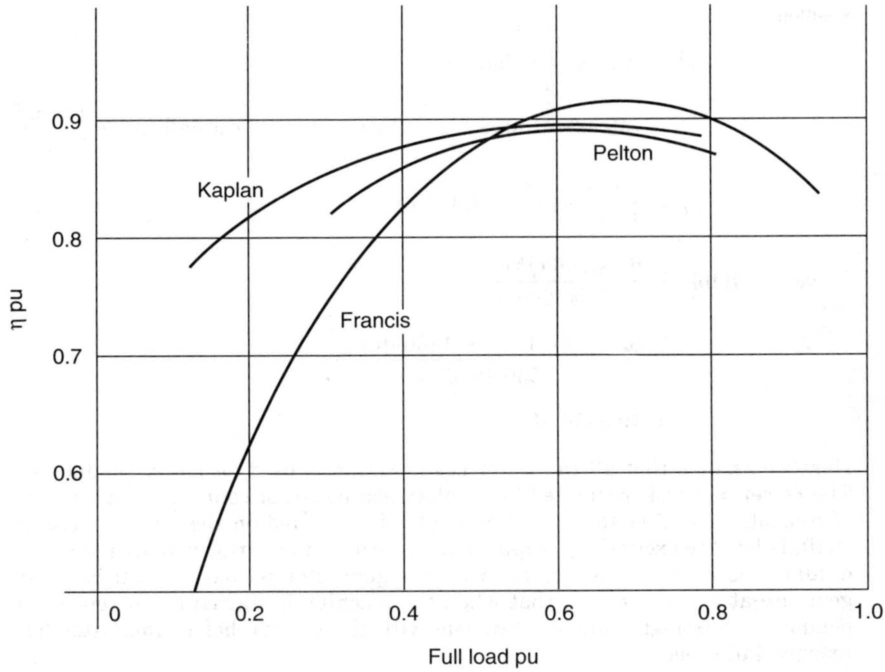

Figure 1.5 Efficiencies of turbines.

In calculating output it is useful to remember that 1 ft^3/s = 1.13 $m^2 \cdot ft$ per annum, that is, 1 ft of water over an area of 1 mi^2 gives a flow of 1 ft^3/s for 1 year.

The theoretical water horsepower available from a supply of Q ft^3/s water, a net head of h ft is equal to 62.4 $Qh/550$, with the turbine efficiency of η [brake horsepower (BHP)] at the turbine shaft is

$$\text{BHP} = \frac{62.4Qh_2\eta}{550} \qquad \text{or} \qquad \text{kW} = \frac{62.4(746)Qh\eta}{550} \tag{1.3}$$

If the efficiency of the electric generator is η' then the overall efficiency or EHP at the switchboard is $62.4Qh\eta\eta'/550$ or

$$\text{kW} = \frac{62.4(746)Q\eta\eta'}{550(1000)}$$

The following example will illustrate the method of estimating the power available under given conditions.

Example A hydroelectric plant has a catchment area of 40 mi^2 and the run-off is 75 percent. The average rainfall is 55 in and the head is 1400 ft. Calculate the power available if the overall efficiency of the plant is 85 percent.

solution

$$Q = \text{cu. sec. available}$$

$$= \text{catchment area (sq. miles)} \times \text{annual rainfall (ft)} \times \frac{\text{run-off}}{1.13}$$

$$\therefore Q = \frac{40(55)}{1.13(12)}(0.75) = 121.68$$

$$\text{Power available} = \frac{62.4(746)Qh\eta\eta'}{550(1000)}$$

$$= \frac{62.4(746)(121.68)(1400)(0.85)}{550(1000)}$$

$$= 12.2556 \text{ MW}$$

Here we assume that all the water is stored and is used for power production. The power output is regulated by an electrical governor control. In one system of control, a small (1 to 5 kWA) generator is mounted on the main generator shaft, below the exciter, and supplies the power to a motor-driven actuator pendulum. The characteristic demand of this generator is that it shall begin to generate at a low speed, so that when the machine is started up the governor pendulum is brought into synchronism with the turbine before the latter has reached full speed.

1.2.3 Peaking power units

Introduction. The demand for electric energy is usually called the load of the system, and is usually measured in kilowatts or megawatts. The demand for electric energy varies from hour to hour, from day to day, and from season to season. Figure 1.6 shows the variation of electric power demand of the City of Los Angeles during one day (Wednesday, June 19, 1991).

In order to meet the power demand the utility companies install peaking power plants which will help to meet the peak power demands. Some of the types of peaking power units will be briefly discussed here. These are the pumped storage and gas turbines.

Pumped-storage electric power plant. The pumped-storage power plant is a variation of the hydroelectric power plant. The pumped-storage power plant consists of an upper and a lower reservoir and turbine-generators which can be used as motor-pump. Figure 1.7 shows a block diagram of a pumped-storage power plant.

The mode of operation of the pumped-storage power plant is as follows: During peak power demand, the water in the upper reservoir is released to generate the extra power that the steam plant is unable to

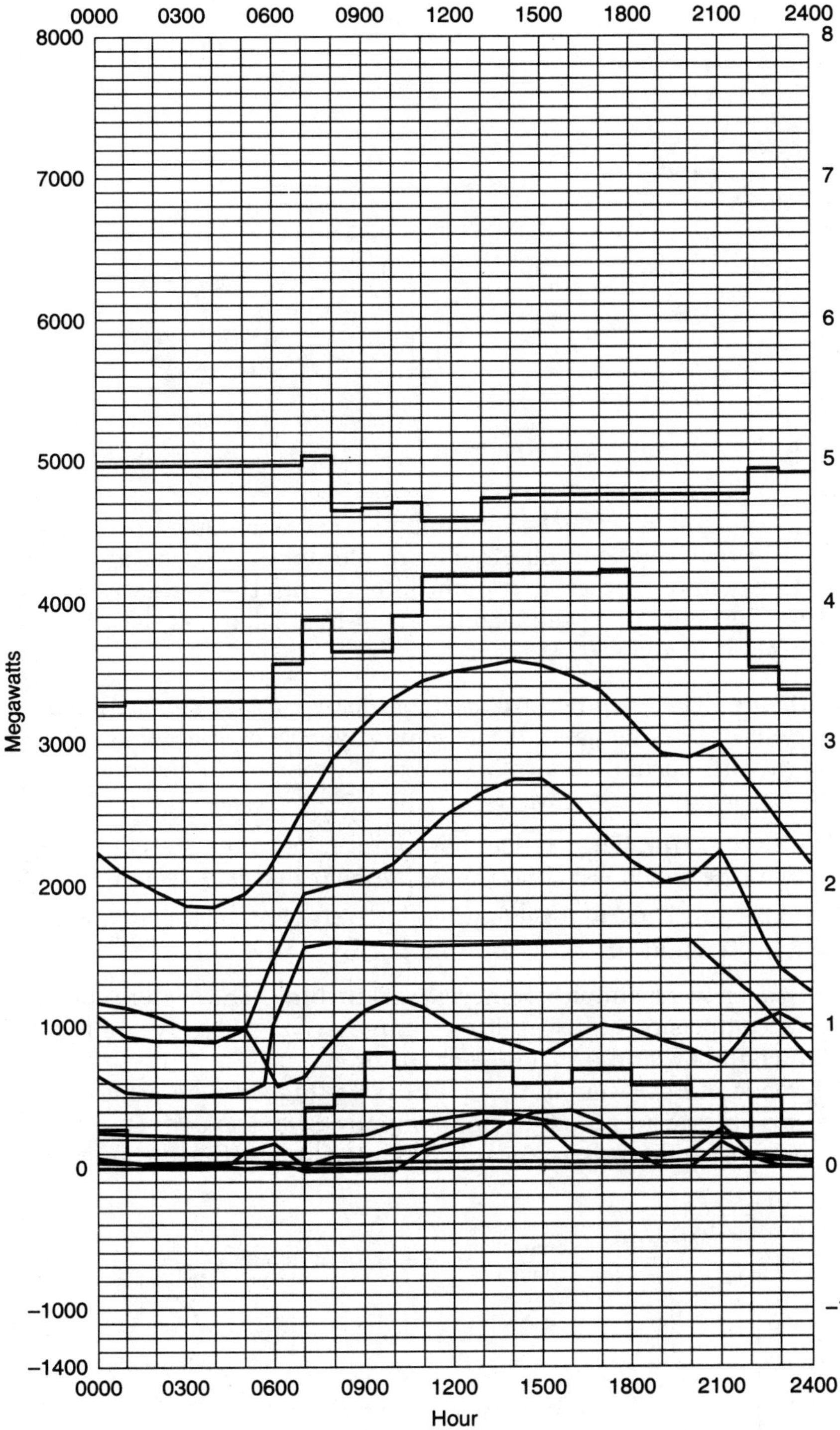

Figure 1.6 Typical daily load curve of the City of Los Angeles. (*Courtesy of DWP LA.*)

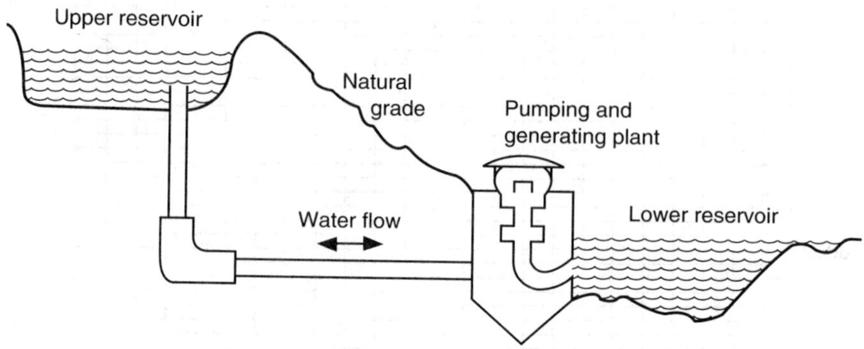

Figure 1.7 Pumped-storage system.

generate since it has reached its maximum value. During the night when only the base load plants (steam plants) are in operation, the water in the lower reservoir is pumped back into the higher reservoir ready for the next day's peak load. The overall efficiency of the pumped-storage plant is about 67 percent.

Gas turbines. Gas turbines are used for electric power generation during periods of peak demand. It has some advantages over steam plants although with normal running it is less economical to operate.

Two major advantages are

1. Its inherent simplicity. It does not contain the boiler or steam supply system, the condenser, and the waste heat disposal system.
2. Its ability to start and take a load quickly. For instance, a simple open-cycle plant, its start-up time is only 2 to 8 min.

Figure 1.8 shows a simple open-cycle gas turbine. The system consists of a unit comprising an axial flow compressor, a relatively small combustion chamber, and a turbine-generator set.

The mode of operation of this plant is as follows: The atmospheric air is compressed in an axial flow compressor and is delivered to the combustion chamber where the fuel is burnt along with the compressed

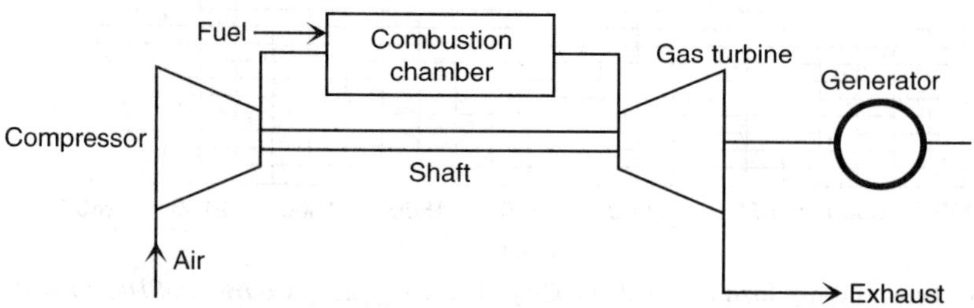

Figure 1.8 Simple open-cycle plant.

air. The products of combustion expand in the turbine, and the exhaust gases are released to the air after passing through the turbine. The overall efficiency of a simple open-cycle plant 15 to 25 percent and that of open-cycle plants is between 21 to 29 percent. When the plant has two- and three-shaft engines, the overall efficiency is between 33 to 35 percent. The size of these units is typically 10 to 50 MW which makes them at least an order of magnitude smaller than typical central station generators.

A combustion gas turbine power unit as shown in Fig. 1.8 must be operated above 1000°F in order to achieve very high efficiencies in its turbine and compressor components. At 1200°F for each 1 kW at the generator terminals 4 kW must be produced by the turbine, the balance being used to compress the gas. At 1000°F the gas cycle efficiency increases approximately three times as fast as the steam cycle efficiency for a given top temperature increase. In view of the low boiling pressure of the gas turbine, the energy per pound of gas is small. For instance, a 5 MW, 3600 rpm simple cycle unit, the gas flow with the inlet pressure and temperature of 88 psi and 1200°F would be about 510,000 lb/h. Figure 1.9 gives a comparison between steam and closed-cycle turbines.

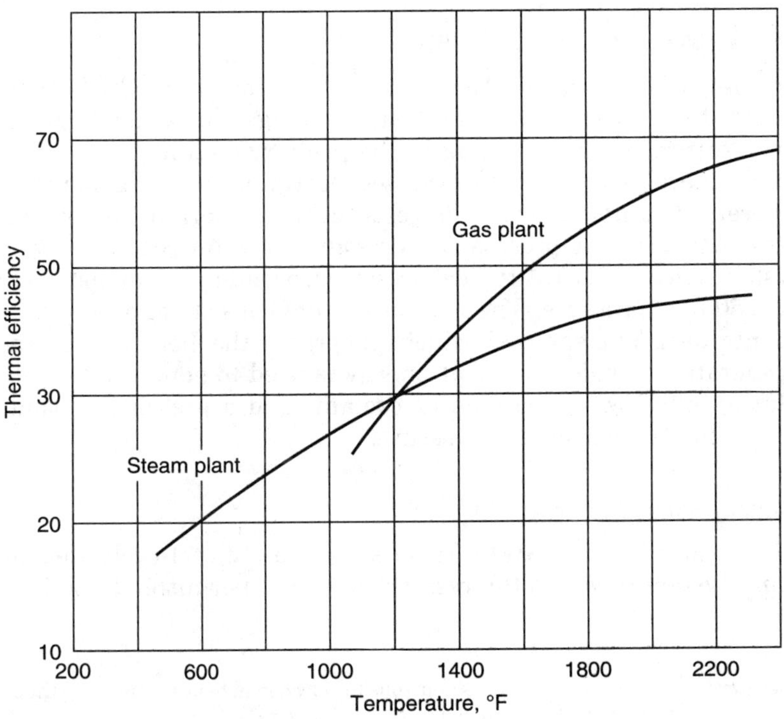

Figure 1.9 Steam and gas cycle turbines.

1.3 Nuclear Power Plants

1.3.1 Introduction

In a nuclear power plant, nuclear fission takes place when a free neutron strikes the nucleus of a fissile material, such as ^{235}U. On impact the nucleus splits into several particles: two major fission fragments such as radioactive krypton and cesium, and 2 to 5 new free neutrons releasing energy which is manifested as heat. In this process some of the new neutrons released collide with each other fissile nuclei which also split, and a chain reaction is set up.

The plant containing the fissile material is called a reactor or a pile. The reactor produces heat which must be converted into electric energy via a heat exchanger, turbine, and generator. There are many different types of reactors with different coolants and moderators,* and different types of fissile fuel. Here only two types of reactors will be discussed briefly:

1. Pressurized water reactor (PWR) where water acts as a coolant and as a moderator
2. A boiling water reactor (BWR)

1.3.2 Pressurized water reactor (PWR)

A pressurized water reactor (PWR) is a power reactor in which heat is transferred from the core to the heat exchanger by water kept at high temperatures without boiling in the primary circuit.

The heat energy transferred to the second water system generates steam therein. The steam in turn drives a turbine generator set, which produces electricity. The process of a pressurized water reactor power plant is shown in Fig. 1.10. A typical pressurized water reactor operates at about 2500 psi pressure. The reactor system consists of a pressure vessel containing nuclear fuel which generates the heat energy, a steam generator in which the heat energy is used to generate steam, a circulating pump which circulates coolant, and a pressurizer that maintains and controls system pressure.

1.3.3 Boiling water reactor (BWR)

The boiling water reactor (BWR) power system is a direct-cycle steam-generating system in which the reactor coolant is permitted to boil in

*A moderator is a substance that causes neutrons to slow down, hence increasing their probability of interacting with fissile nuclei.

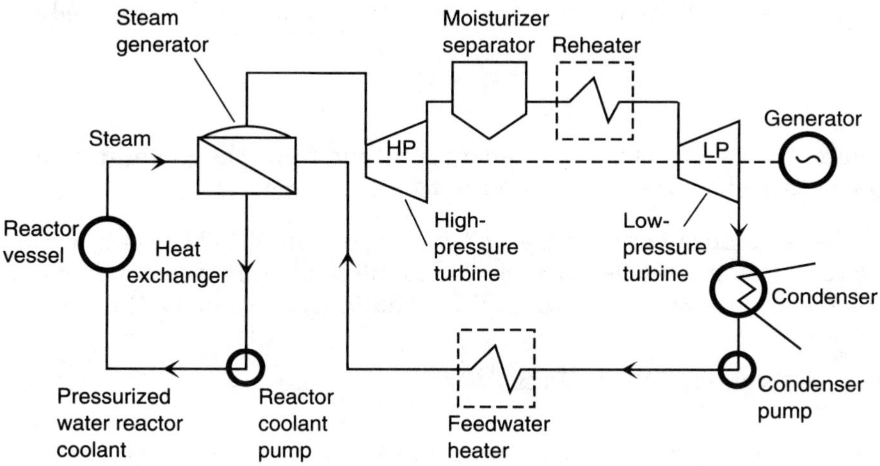

Figure 1.10 Pressurized water reactor (PWR) power plant.

the core, and the resulting steam flows directly to a turbine via moisture separators and steam dryers. After passing through a suitable heat cycle, the steam is condensed. From the condenser, the condensate is pumped through full-flow polishing demineralizers, feedwater heaters, and then back to the reactor. Figure 1.11 shows the diagram of the boiling water reactor power plant.

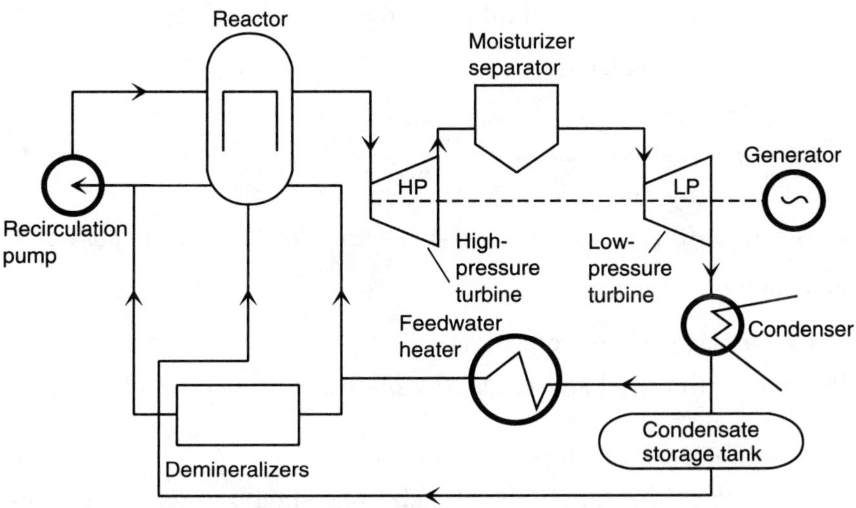

Figure 1.11 Schematic diagram of the boiling water reactor plant (BWR).

The overall efficiency of a nuclear power plant can be expressed by

$$\eta = \frac{W}{Q_R} \tag{1.4}$$

where W = rate of electric energy output from the generator and Q_R = rate of thermal energy output from the reactor.

The difference between Q_R and W is the sum of heat losses in the overall system plus the heat rejected to the condenser. Typical thermal efficiencies are 30 percent for PWR and 35 percent for BWR designs.

1.4 Emergency Power Units

1.4.1 Introduction

The diesel engine acts as a prime mover for the drive of electric generators but use of the diesel engines has been curtailed by the development of steam turbines. The diesel engine plays the following two important roles in the fields of power generation:

1. First, as a stand-by set especially for start-up of auxiliaries in steam and gas turbine power plants, and also for emergency supply to hospitals, factories, airports, large hotels, etc.
2. Second, as a continuous power generation out of plants where the power output is not exceeding 4 MW

The diesel engine has advantages which make it very suitable for small and medium output power plants. Some of the advantages are

1. High operating efficiency.
2. No stand-by losses.
3. Needs very little cooling water.
4. Quick starting and quick load pick-up is possible.
5. Easier handling of fuel storage space and no refuse to dispose of.

Some disadvantages are

1. Emissions degrade air quality.
2. In areas with diesel shortages, the plant would not operate.

1.4.2 Selection of engine type and size

When selecting the engine type and size one should take into consideration the possible engine cycles and the factors influencing the en-

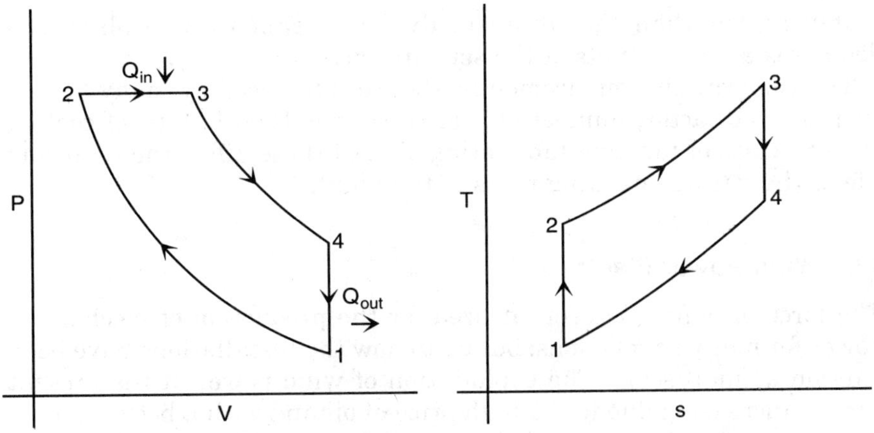

Figure 1.12 The theoretical diesel engine cycle in p-V and T-s diagrams.

gine, such as plant output, thermal efficiency, investments, maintenance costs, and operational availability.

The diesel cycle is characterized by compression of intake air to pressures of about 30 to 100 atmospheres (atm) and consequently high temperatures and injection of the fuel into the hot air for ignition and combustion. Figure 1.12 shows the resulting theoretical pressure versus volume (p-v) and temperature versus entropy (T-s) diagrams of the cycle of the reciprocating engine. Air at 1 is compressed and heated to 2. Fuel is injected at 2. The fuel-air mixture is burned at constant pressure from 2 to 3 expands from 3 to 4 and is exhausted at 4. The theoretical thermal efficiency of an ideal diesel engine is

$$\eta_{th} = \frac{Q_{in} - Q_{out}}{Q_{in}} = 1 - \frac{C_v(T_4 - T_1)}{C_p(T_3 - T_2)} = 1 - \frac{T_4 - T_1}{K(T_3 - T_2)}$$

where Q_{in} = heat energy supplied to cycle (fuel energy)

Q_{out} = energy of exhaust gases

C_p = specific heat at constant pressure

C_v = specific heat at constant volume

K = C_p/C_v (where for air $K = 1.40$)

1.4.3 Costs of diesel electric plants

Diesel engine plants are generally only used for isolated stations of comparatively small outputs, except in those countries where fuel oil is a local product. Most of these diesel engine plants are peak load stations and operate on a one- or two-shift basis. The capital cost per kilowatt of plant installed (engine, generator, and exciter) is considerably higher than that of an equivalent steam turbine plant and decreases as the outputs of the sets are increased.

The renewal and replacement costs are influenced by many factors such as load factor, number of cold starts per 1000 h, type of cooling system, class of fuel and lubricating oil. All these affect the wear and life of the various working parts of the plant.

1.5 Wind Power Plants

The force of wind has been utilized for the production of mechanical power for many generations; but up to now the installations have been only on a small scale. The exploitation of wind power at the present time is increasing due to the high price of oil and with a better knowledge of possible depleting of oil and coal deposits. The great drawback on the utilization of wind power arises from both the intermittence and unreliability of the source of energy. The design of windmills has almost been restricted to small sets and to duties for which continuous service is not essential. Both in degree of variation and the length of periods of deficiency, wind power probably has an advantage over water power which may be affected not just by dry season, but also by dry years.

The largest sets of windmills may have towers approximating 150 m (500 ft) in height. At such height the wind energy is approximately three times that at the ground level.

The great advantage of windmills is that they may be installed in any locality where the topographical and metrological conditions are suitable and require no outside supplies for its operation except for a stand-by battery and lubricating oil. The choice of site of windmills depends on the following factors:

1. Windy area
2. Scattered population
3. Economic advantage in fuel saving, e.g., coal and oil
4. Automatic operation

The output power of the windmill is given by

$$P_{\text{out}} = 2.14Ad\left(\frac{v^3}{10^3}\right) \qquad \text{kW}$$

where A = area swept by the windmill
d = wind density of air
v = wind velocity in miles per hour

Further investigation shows that a modern $^{11}/_{55}$-kW wind turbine can be connected to the grid and contribute approximately 100 MWh annually without causing appreciable changes to the voltage quality. The disadvantages of windmills are

1. Towers may blow down in a hurricane or typhoon.
2. Acoustic noise level is high.
3. The metal blades cause noticeable electromagnetic interference.

1.6 Combined Cycle and Heat Recovery, Cogeneration

Cogeneration is a process of capturing waste heat from manufacturing, industrial process or heating and cooling systems, and using it to generate electric power. The electric power generated by the cogeneration can be used to run one's own equipment and can be sold to the utility company.

Two cogeneration system arrangements are possible; namely, topping and bottoming cycles.

1. In topping cycles, the input energy is first used to produce power and the exhaust heat from the power-producing prime movers is used to generate steam or used directly in a heating process.

2. Bottoming cycles are the reverse, utilizing waste heat from a heating process, such as an industrial furnace, to produce electricity. Commonly the waste heat is used to produce steam in a heat recovery boiler and the steam is expanded through a steam turbine connected to a generator. An additional system arrangement is the combined cycle, in which the prime mover drives a generator, and steam produced in a heat recovery boiler is expanded through a steam turbine generator.

Cogeneration systems simultaneously produce steam and electricity with an overall efficiency of 80 to 85 percent. One of the more promising cogeneration technologies is the gas turbine topping. Several gas turbines are available in a wide range of sizes that provide good overall performance as cogeneration units. For instance, a gas turbine topping cogeneration unit producing 3 MW of electric power also produces about 20,000 lb/h of 150-psi steam.

Figure 1.13*a* and *b* shows two industrial processes: (*a*) an industrial process without cogeneration, and (*b*) an industrial process with cogeneration.

Some of the companies in the United States which use cogeneration systems to produce their own electric power are Dow Chemicals, Garden State Paper, Kerr-McGee, Loma Linda University, etc. The Kerr-

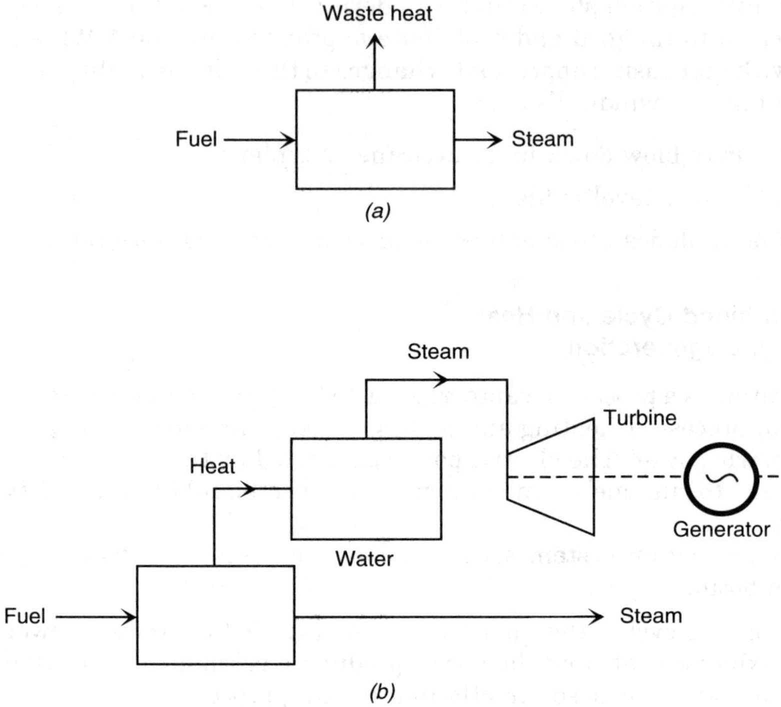

Figure 1.13 Schematic diagram of (*a*) an industrial process without cogeneration and (*b*) an industrial process with cogeneration.

McGee chemical plant for instance generates about 80 MW through cogeneration. When the company does not need all of its electric power, it sells the excess to Southern California Edison Company. Other cogeneration methods are from the saw mills, sugar cane plants (biomass), and waste-water treatment plants, etc.

1.7 Geothermal Power

Geothermal power is defined as the transfer of heat from the hot interior of the earth to the cooler exterior. The geothermal heat originates from molten rock in the earth's crust. Geothermal steam or hot water is obtained by drilling down into porous rock. Water is then pumped to the surface by deep well pumps. The water gives up its heat in a heat exchanger to heat a liquid to a boiling point and then it is superheated. The steam is injected in the turbine. The condensate from the turbine is changed into water. The cooled water is then injected into the well, nearby. Figure 1.14 shows the block diagram of a geothermal power plant.

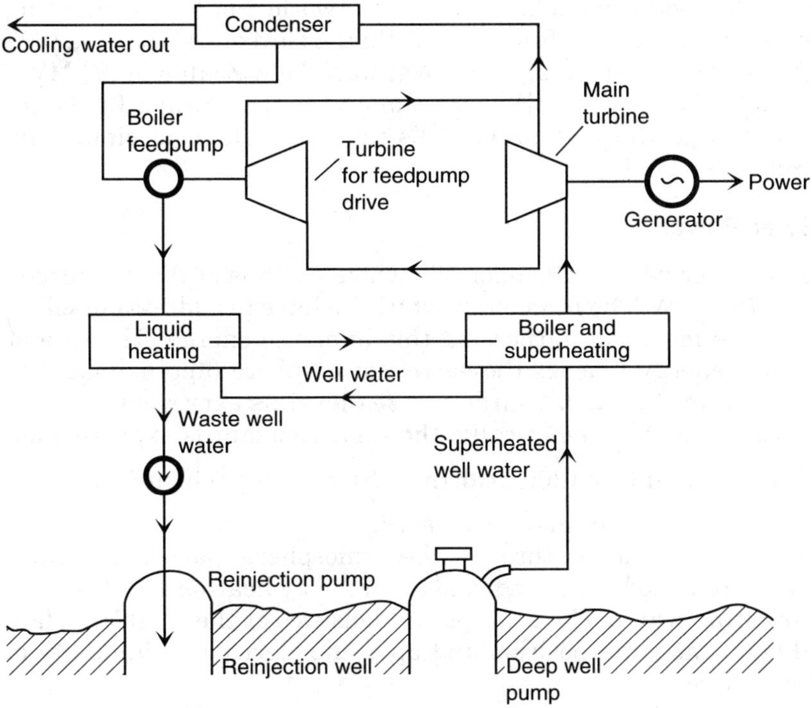

Figure 1.14 Block diagram of a geothermal power plant.

The geothermal power is divided into two divisions according to the heat production—the high heat energy (enthalpy) systems, above 200 cal/g, and low energy (enthalpy) systems, above 100 cal/g. The high enthalpy geothermal systems are found in the young volcanic crust of rifting, such as the Rift Valley of East Africa and the Middle East, the Mediterranean basin of Europe and North Africa, across Asia to the Pacific (Alaska to South America) and the island groups of the mid-Atlantic. The lower-enthalpy geothermal systems are found in the Gulf Coast of the United States, Western Siberia, and Central Europe.

Since heat dissipates rapidly, geothermal fluids cannot be transported long distances without losses. A length of about 5000 ft has been used to transport geothermal fluids at Wairakei in New Zealand. The geothermal power/geothermal efficiency of a perfect transfer of geothermal heat from the hot interior of the earth to cooler exterior is given by

$$\eta = \frac{T_1 - T_2}{T_1}$$

where T_1 is the absolute temperature of the hot body and T_2 is the absolute temperature of the cold body.

Some of the geothermal plants around the world and their generating capacities are Larderollo, Italy (406 MW); the Geysers, California (608 MW); Matsukawa, Japan (20 MW); Wairakei, New Zealand (192 MW); Cerro Prieto, Mexico (75 MW); etc. Impure and corrosive fluids are involved with geothermal power. This causes problems of piping and heat exchanger fouling.

1.8 Solar Power

The solar power which falls upon the whole earth is of the magnitude of 1.77 × 1014 kW. The problem to which we have to address ourselves is finding the means of harnessing this immense energy. The rate at which solar energy reaches the earth's atmosphere (approximated to be 150 km from the earth's surface at sea level) is expressed in terms of solar constant. Mathematically, the solar constant is expressed as

$$\text{Solar constant} = 1200 \text{ kcal}/(\text{m}^2 \cdot \text{h}) = 1.39 \text{ kWh}/(\text{m}^2 \cdot \text{h})$$

where m^2 is the area in meters squared.

Where the sun passes through the atmosphere, part of the sun's radiation is reflected, scattered, and absorbed by heating air, dust, and by evaporating water. The solar power densities at the earth's surface depend upon the sun's position and upon the clarity and humidity of the atmosphere.

There are two distinct applications of solar power. These are

1. Space and water heating on a domestic scale; e.g., photovoltaic collectors
2. Large-scale heat collection used for production of electric power

Figure 1.15 shows a schematic diagram of a solar water heater. The solar collector is facing south when it is installed in the northern

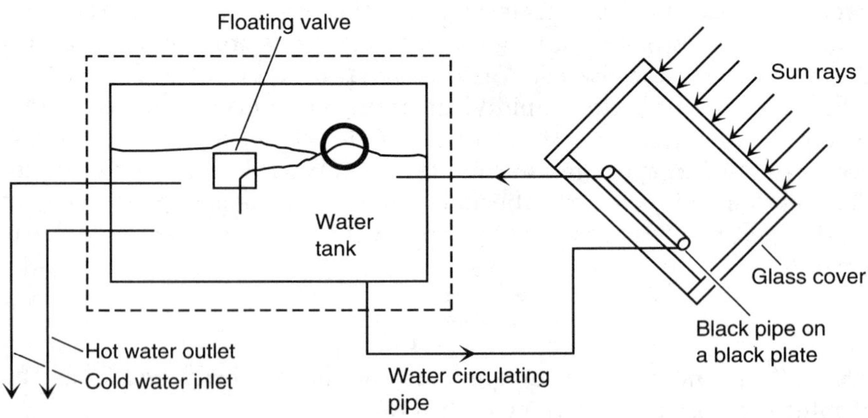

Figure 1.15 Schematic diagram of a solar water heater.

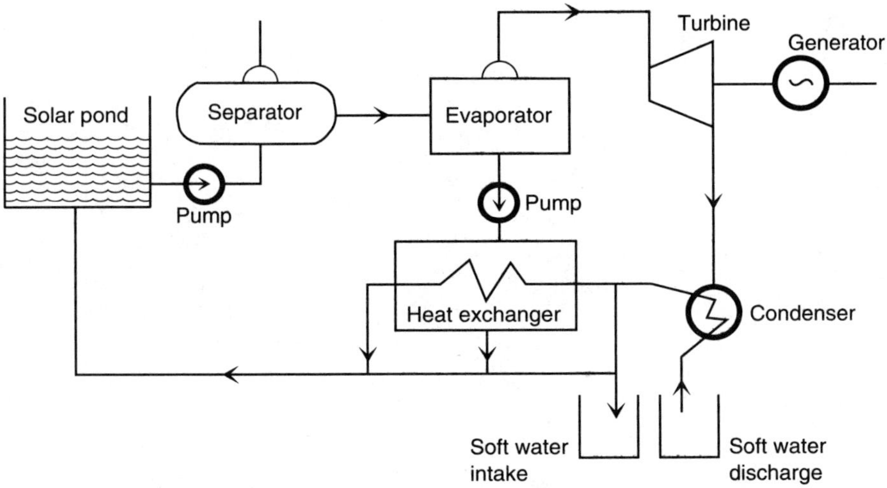

Figure 1.16 Schematic diagram of a steam power plant using a solar pond.

hemisphere and faces north when installed in the southern hemisphere. As the temperature of the solar-collecting surface rises, it radiates heat (infrared). The cover plate over the collecting surface could be designed in such a way that it would pass nearly all the solar radiation and reflect all the infrared. As the cold water circulates in the black pipe, it is heated up. The solar collectors can attain a temperature of about 150°C.

For large-scale heat collection used for the generation of electric power, the sun's rays may be concentrated by lenses or mirrors to the solar pond with a proper mechanism for steering the lenses or mirrors to follow the motion of the sun (tracking system). Figure 1.16 shows the steam power plant with the steam generated by hot water from a solar pond.

Bibliography

Aschen, F. S., *Planning Fundamentals of Thermal Power Plants,* Wiley, New York, 1978.

IEEE Transactions on Power Apparatus and Systems, Discussion of Sacramento Municipal Utilities District's 100 MWE Photovoltaic Solar Farm, September 1984.

Kruger, P., and C. Otte, *Geothermal Energy,* Stanford University Press, Stanford, Calif., 1973.

Payne, W. F. (compiled and edited), *Cogeneration Sourcebook,* Fairmont Press Inc., Atlanta, Ga., 1985.

Pedersen, E. S., *Nuclear Power,* vol. 1, Ann Arbor Science Publishers Inc., Ann Arbor, Mich., 1979.

Spiewak, S. A., *Cogeneration and Small Power Production Manual,* Fairmont Press, Englewood Cliffs, N.J., 1987.

Weedy, B. M., *Electric Power Systems,* 3d ed., Wiley, New York, 1979.

Chapter

2

Magnetic Fields and Magnetic Circuits

2.1 Introduction

In this chapter a review of the basic concepts of field theory will be presented. A representation of some aspects of magnetism of value to the design of electromagnetic devices will also be discussed. It is assumed that the reader is familiar with elementary field theory which covers the following three most important magnetic theories:

1. Faraday's law
2. Ampere's law
3. The magnetic force

1. Faraday's law

$$\text{emf} = -N\frac{d\phi}{dt} = -\frac{d\lambda}{dt} \tag{2.1}$$

where

$$\phi = \int B\, dA \tag{2.2}$$

$$\lambda = N\phi \tag{2.3}$$

The minus sign in Eq. (2.1) means that the flux due to the induced current is in the direction to oppose the increasing flux that induced the current (Lenz's law).

ϕ is the magnetic flux in webers (Wb).

B is the magnetic flux density in webers per meter squared (Wb/m^2).

λ is the flux linkage in webers-turns and is given as the product of the number of turns and the magnetic flux linking the number of turns.

A is the cross-sectional area in meters squared (m^2).

Equation (2.1) shows a time-varying flux linking a stationary coil, yielding a time-varying voltage. If a coil is moving through a magnetic field, an induced voltage will be generated and will be given by

$$\text{emf} = \int (\mathbf{v} \times \mathbf{B}) \cdot dl \tag{2.4}$$

where $\mathbf{v}$ is the velocity of the conductor, and l is the length of the moving conductor perpendicular to the flux density $\mathbf{B}$ and a constant velocity $\mathbf{v}$.

2. Ampere's law states that the line integral of the magnetic field intensity H, about a closed path, is exactly equal to the direct current enclosed by that path.

$$\oint \mathbf{H}\, d\mathbf{l} = I_{\text{enclosed}} \tag{2.5}$$

The $\oint \mathbf{H}\, d\mathbf{l}$ is called the magnetomotive force (mmf). It is measured in ampere-turns. The turn in ampere-turns is dimensionless and the fundamental unit of mmf is only amperes (A).

3. *The magnetic force* From magnetic theory we know that the electric field causes a force to be exerted on a charge which may be either stationary or in motion. The magnetic field, on the other hand, is capable of exerting a force only on a moving charge. It must be mentioned here that the magnetic field cannot arise from stationary charges. Focusing attention to the forces on current-carrying conductors, it is found that a force at a differential current filament is given by

$$d\mathbf{F} = I\, d\mathbf{l} \times \mathbf{B} \tag{2.6}$$

Integrating Eq. (2.6) over a closed path we get

$$\mathbf{F} = \int I\, d\mathbf{l} \times \mathbf{B} \tag{2.7}$$

A result to Eq. (2.6) can be obtained by applying it to a straight conductor in a uniform magnetic field, thus

$$\mathbf{F} = I\mathbf{l} \times \mathbf{B} \tag{2.8}$$

The magnitude of the force is given by

$$F = Bli \sin \alpha \tag{2.9}$$

where α is the angle between the vectors. It must be noted here that Eqs. (2.7) and (2.8) apply only to a portion of the closed circuit.

2.2 Magnetic Circuits

In this section the fundamental techniques involved in solving magnetic problems known as magnetic circuits will be discussed. As a convenient starting point first let us enumerate main magnetic field equations. The first relationship is between the magnetomotive force (mmf) and the magnetic field intensity:

$$\mathbf{F} = \int_b^a \mathbf{H}\, d\mathbf{l} \tag{2.10}$$

where a and b are the limits of integration.

The magnetic flux density and the magnetic field intensity are related through the relationship:

$$\mathbf{B} = \mu\mathbf{H} \tag{2.11}$$

where μ = permeability ($= \mu_r \mu_o$)

μ_r = relative permeability and a dimensionless constant

μ_o = permeability of free space ($= 4\pi \times 10^{-7}$ H/m)

The relative permeabilities of some magnetic materials are tabulated in Table 2.1.

Consider an atom in which the small magnetic fields produced by the motion of the electrons in their orbits and those produced by the electron spin combine to produce a zero net field. In other words, it can be stated that the electron spin moments oppose each other as well as the orbiting electron moments, to produce zero net permanent magnetic moment in the absence of an external magnetic field.

When an external field is applied there is a small torque in each atomic moment, and these moments align themselves with the external field. The alignment acts to increase the value of the flux density B. If the net result is a decrease in flux density the material is called

TABLE 2.1 The Values of Relative Permeabilities for Some Ferromagnetic Materials

Materials	Type	μ_r
Bismuth	Diamagnetic	0.9999834
Silver	Diamagnetic	0.0999980
Copper	Diamagnetic	0.9999910
Vacuum	Nonmagnetic	1.0000000
Aluminum	Paramagnetic	1.0000000
Nickel chloride	Paramagnetic	1.0000000
Cobalt	Ferromagnetic	250.0000000
Nickel	Ferromagnetic	600.0000000
Mild-steel	Ferromagnetic	2,000.0000000
Iron	Ferromagnetic	5,000.0000000
Mumetal	Ferromagnetic	100,000.0000000
Superalloy	Ferromagnetic	800,000.0000000

diamagnetic. If, however, there is an increase in the flux density the material is called *paramagnetic*.

To determine the total magnetic flux flowing through the cross section of a magnetic circuit, it is necessary to integrate the flux over a cross-sectional area.

$$\phi = \int B \, dA \tag{2.12}$$

Finally let us consider the reluctance of a flux path by taking a toroid of magnetic material carrying N uniformly spaced turns, as shown in

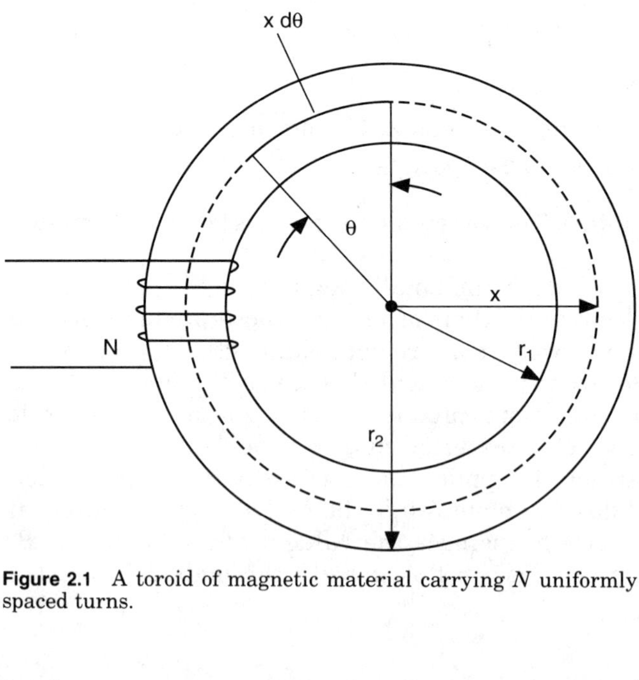

Figure 2.1 A toroid of magnetic material carrying N uniformly spaced turns.

Fig. 2.1. The reluctance is proportional to the length of the conductor and inversely proportional to the cross-sectional area.

Let the dotted line be the mean toroid radius (X). If the relative permeability is greater than one ($\mu_r > 1$), it is assumed that the line of flux will remain inside the toroid and will form concentric circles. Considering a flux line at the mean radius X and integrating mean field intensity around this line we get

$$F = \int H\,dl = H\int X\,d\theta = HX\int_0^{2\pi} d\theta = (H)(2\pi)(X) = NI \qquad (2.13)$$

Therefore,

$$H = \frac{IN}{2\pi X} \qquad (2.14)$$

and IN is the magnetomotive force (mmf) measured in ampere-turns (At). If the radius r_1 of the toroid section is less than X ($r_1 < X$), then the magnetic field intensity is within the toroid. Since the magnetic field is constant over the section, then

$$\phi = BA = \mu_r\mu_o HA = \mu_r\mu_o\frac{INA}{2\pi X} \quad \text{Wb} \qquad (2.15)$$

Therefore,

$$IN = \frac{2\pi X\phi}{\mu_r\mu_o A} = R\phi \qquad (2.16)$$

where R is the reluctance of the flux path, and is given by

$$R = \frac{2\pi X}{\mu_r\mu_o A} = \frac{l}{\mu A} \quad \frac{\text{At}}{\text{Wb}} \qquad (2.17)$$

For uniform permeability and the cross-sectional area A, l stands for the corresponding portion of the length of the magnetic circuit. The reciprocal of the reluctance is the permeance (P). If the ferromagnetic systems contain air gaps the permeability becomes infinity and the reluctance at the air gap is given by

$$R_g = \frac{l_g}{\mu_o A} \quad \frac{\text{At}}{\text{Wb}} \qquad (2.18)$$

Figure 2.2 shows the magnetization curves of some of the ferromagnetic materials.

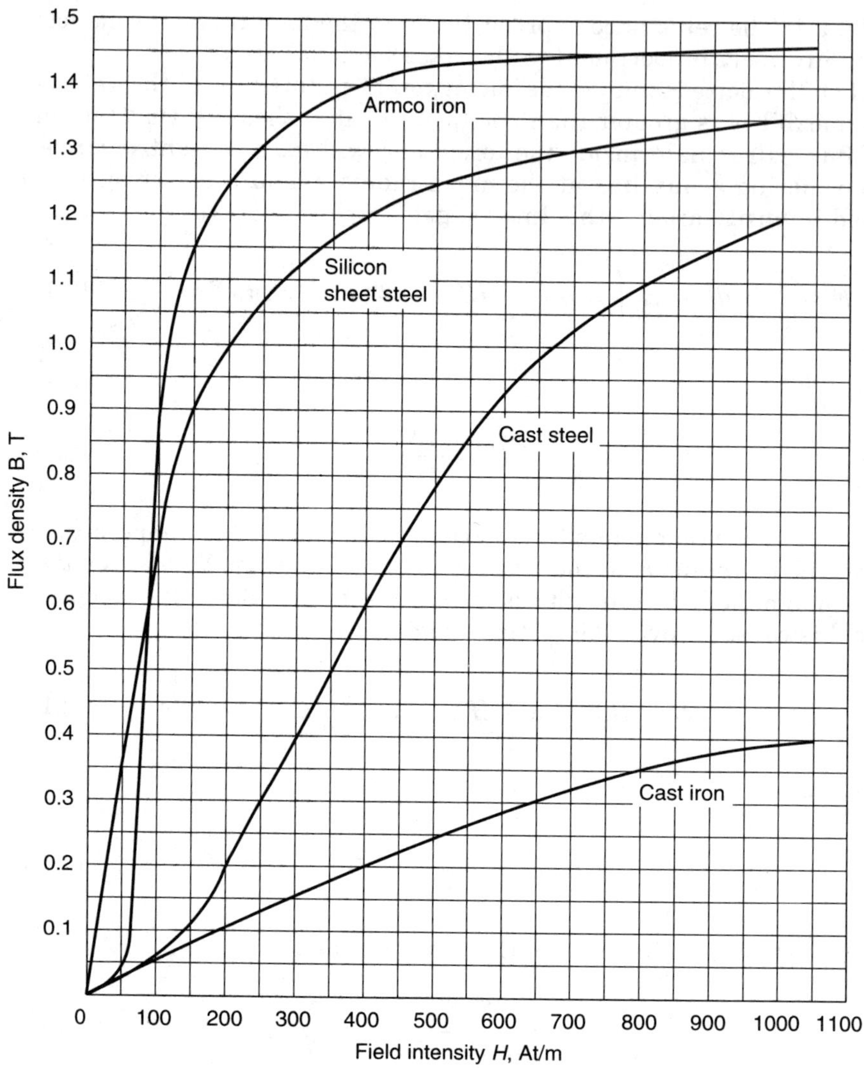

Figure 2.2 Magnetization curves.

Example 2.1 A ferromagnetic material has a relative permeability of $\mu_r = 3000$. The core cross-sectional area is $A = 4$ cm^2 and the mean core length is $l = 15$ cm as shown by Fig. 2.3. If the coil has 300 turns and $e(t) = 100 \cos 1000t$ V, find

1. the expression for the steady-state flux
2. the steady-state current $i(t)$

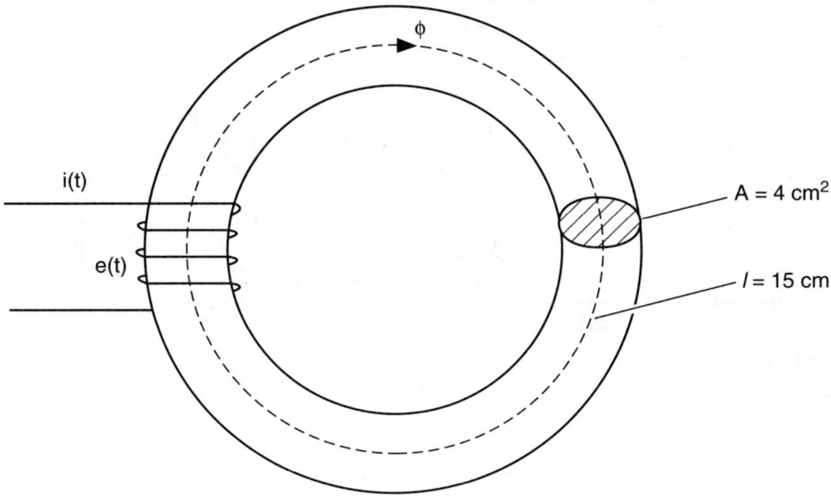

Figure 2.3 Example 2.1.

solution

1. $e(t) = -N \dfrac{d\phi}{dt}$

$$\frac{d\phi}{dt} = -\left(\frac{l}{N}\right) e(t)$$

$$\phi = -\int \frac{l}{N} e(t)\, dt = \frac{100 \sin 1000t}{300(1000)} = 3.3 \times 10^{-4} \sin 1000t \qquad \text{Wb}$$

$$= \frac{Ni(t)}{R}$$

2. $i(t) = \dfrac{\phi R}{N} = 0.109418 \sin 1000t \qquad \text{A}$

Example 2.2 Figure 2.4 shows a cast steel core with all its dimensions. Calculate the magnetomotive force (mmf) required to establish an air-gap flux of 0.7×10^{-3} Wb. The relative permeability of cast steel is $\mu_r = 620$ at a flux density of $B = 1.2$ T. The cross-sectional area of the cast steel is 6 cm².

solution Air-gap flux density

$$B_g = \frac{\phi_g}{A} = \frac{0.7 \times 10^{-3}}{6 \times 10^{-4}} = 1.166 \text{ T}$$

The magnetic field intensity at the air gap is

$$H_g = \frac{B_g}{\mu_o} = 928{,}669.09 \text{ At/m}$$

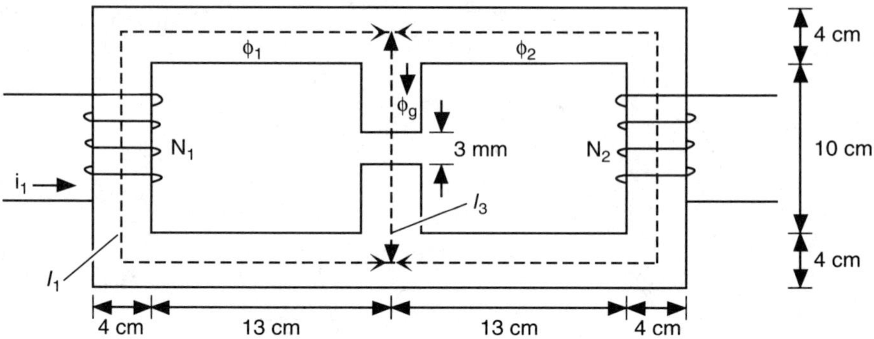

Figure 2.4 Example 2.2.

The mmf at the air gap is

$$F_g = H_g l_g = 928{,}669.09 \times 0.003 = 2786.0 \text{ At}$$

From Fig. 2.2 for $B = 1.2$ T, $H = 1000$ At/m

$$F_1 = H_1 l_1 = 1000 \times 0.44 = 440.0 \text{ At}$$

$$F_2 = H_2 l_2 = 1000 \times 0.44 = 440.0 \text{ At}$$

$$F_3 = H_3 l_3 = 1000 \times 0.137 = 137.0 \text{ At}$$

$$F_T = F_1 + F_2 + F_3 = 1017.0 \text{ At}$$

Example 2.3 The magnetic circuit shown in Fig. 2.5 has the following dimensions: cross-sectional area, $A = 4$ cm^2, the mean length, $l = 35$ cm, the number of turns, $N = 200$ turns, relative permeability of the core is $\mu_r = 10^3$ and $\mu_o = 4\pi \times 10^{-7}$ H/m. Neglecting the leakage flux and winding resistance, the steady-state current is $i(t) = 7 \cos 1000t$ A, calculate

1. The flux of the core
2. The flux leakages $l(t)$
3. The voltage $e(t)$

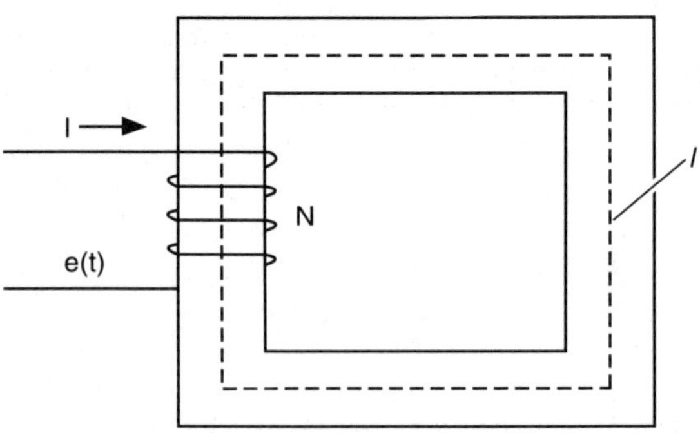

Figure 2.5 Example 2.3.

solution The magnetic flux is

$$\phi = \frac{Ni}{R}$$

and the reluctance is

$$R = \frac{l}{\mu A} = 0.007 \times 10 = 7 \times 10^{-2}$$

where $\mu = \mu_r \mu_o$.

1. $\phi = \dfrac{Ni}{R} = 0.002 \cos 1000t \qquad \text{Wb}$

Using Eqs. (2.1) and (2.3) we have

2. $\lambda = N\phi = 200 \times 0.002 \cos 1000t = 0.4 \cos 1000t \qquad \text{Wb}$

$$L = \frac{\lambda}{i} = \frac{N\phi}{i} = 0.057 \text{ H}$$

3. $e(t) = -\dfrac{d\lambda}{dt} = -\dfrac{N\,d\phi}{dt} = -200 \times 0.002 \times 1000 \sin 1000t$

$$= 400 \sin 1000t \qquad \text{V}$$

Example 2.4 Figure 2.6 shows a ferromagnetic material with an air gap of 0.8 mm, and a cross-sectional area, including fringing, of 0.8×10^{-4} m^2 cut in it. A 5000-turn coil carries a current $i = 30$ mA. A cross-sectional area of the iron core is 0.6×10^{-4} m^2 and the mean length of the flux path in the iron is 0.22 m. The relative permeability of the material $\mu_r = 700$. Calculate

1. The air-gap flux
2. The energy stored in the air-gap magnetic field

solution Total reluctance of the core is given by summing up the two reluctances:

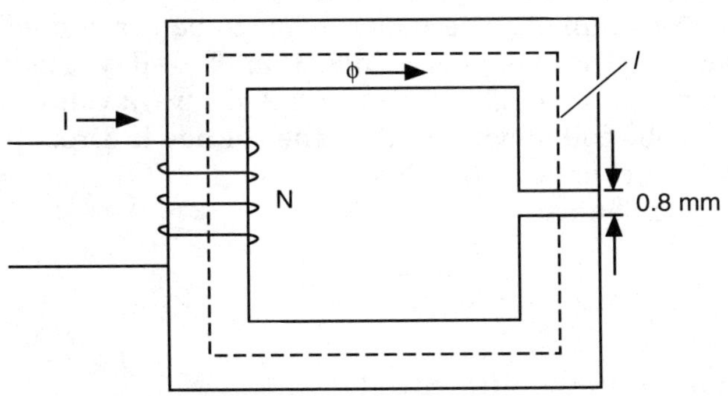

Figure 2.6 Example 2.4.

1. Reluctance of the core:

$$R_c = \frac{l_c}{\mu A_c} = \frac{0.22}{(700)(4\pi \times 10^{-7})(0.6 \times 10^{-4})} = 4.168 \times 10^6 \text{ At/Wb}$$

2. Reluctance of the gap:

$$R_g = \frac{l_g}{\mu_o A_g} = \frac{0.0008}{(4\pi \times 10^{-7})(0.8 \times 10^{-4})} = 7.958 \times 10^6 \text{ At/Wb}$$

Total reluctance:

$$R_T = R_c + R_g = \frac{l_c}{\mu A_c} + \frac{l_g}{\mu_o A_g}$$

$$= 4.168 \times 10^6 + 7.958 \times 10^6 = 12.126 \times 10^6 \text{ At/Wb}$$

$$\phi_g = \phi = \frac{Ni}{R_T} = \frac{(5000)(0.030)}{12.126 \times 10^6} = 12.37 \times 10^{-6} \text{ Wb}$$

$$B_g = \frac{\phi}{A_g} = \frac{12.37 \times 10^{-6}}{0.8 \times 10^{-4}} = 0.1546 \text{ T}$$

The flux density B and magnetic field intensity H are uniformly distributed in the air gap, and hence the stored energy in the air gap is

$$W = \tfrac{1}{2} BH \times \text{air-gap volume}$$

$$= \tfrac{1}{2} R\phi^2 = \frac{B_g^2}{\mu_o} \times \text{volume}$$

$$\text{Stored energy} = \tfrac{1}{2} (0.1546)^2 \left(\frac{1}{4\pi \times 10^{-7}}\right)$$

$$\times (0.8 \times 10^{-4})(0.0008) = 6.09 \times 10^{-4} \text{ J}$$

2.3 Self- and Mutual Inductances

In this section the various ways in which the magnetic field can affect its surroundings will be examined. The major effect to be considered here is Faraday's law of electromagnetic induction. Faraday's law states that when a flux passes through a turn of a coil of wire, a voltage is induced in the turn of the coil of wire, that is, the voltage is directly proportional to the rate of change of the flux.

In equation form it is given as

$$e_{\text{ind}} = -N \frac{d\phi}{dt} \tag{2.19}$$

where e = voltage induced in the turn of coil
N = number of turns
ϕ = flux passing through coil

The minus sign in Eq. (2.19) is an indication that the induced voltage is in such a direction as to produce a current whose flux if added to the original flux would reduce the magnitude of the induced voltage. This statement is called Lenz's law as mentioned in Sec. 2.1. To illustrate this theory let us examine Fig. 2.7. If no other magnetic flux passes the coil in Fig. 2.7 than that of the current in the coil itself, then the magnetic flux is proportional to the current and the number of turns or

$$\phi = KNi \tag{2.20}$$

From Eq. (2.19) the induced voltage is given by

$$e_{\text{ind}} = -N\frac{d\phi}{dt} = -N\frac{d(KNi)}{dt} = -KN^2\frac{di}{dt} = -L\frac{di}{dt} \tag{2.21}$$

where the self-inductance is given by

$$L = KN^2 \tag{2.22}$$

Note in Eq. (2.22) that the inductance is proportional to the square of the number of turns.

Substituting Eq. (2.20) into Eq. (2.22) we have

$$L = N\frac{\phi}{i} = \frac{\lambda}{i} \tag{2.23}$$

where $\lambda = N\phi$ flux linkage. The self-inductance is directly proportional to the flux and inversely proportional to the current through the coil.

The self-inductance in a coaxial cable can be determined by applying Ampere's law about a circular path of radius r.

$$\oint \mathbf{H} \cdot d\mathbf{l} = \int_0^{2\pi} Hr\, d\theta = H(2\pi r) = I_{\text{enclosed}} = Ni \tag{2.24}$$

Therefore, the magnetic field intensity is given by

$$H = \frac{Ni}{2\pi r} \tag{2.25}$$

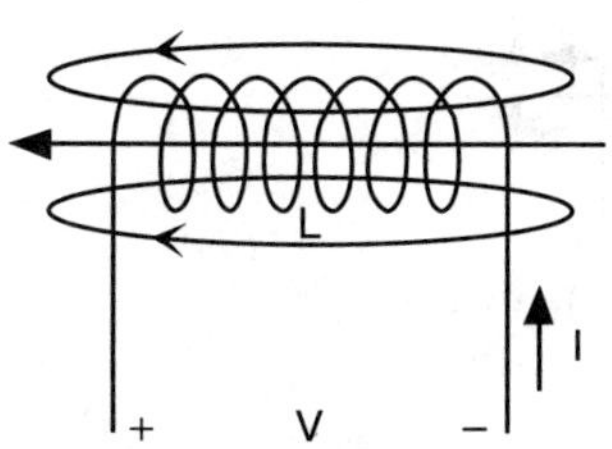

Figure 2.7 Self-inductance.

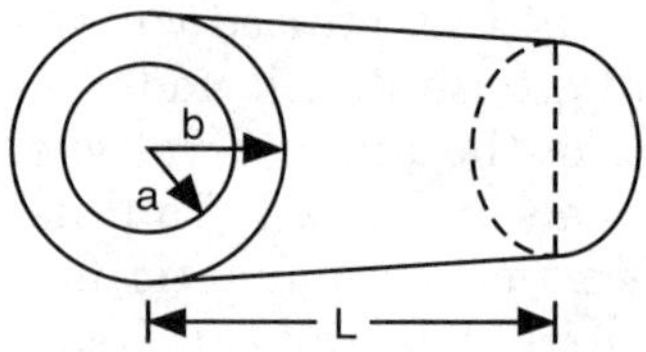

Figure 2.8 Coaxial cable.

but

$$B = \mu_o H = \frac{\mu_o N i}{2\pi r} \tag{2.26}$$

The magnetic flux of a coaxial cable of a length l is the flux crossing any radial plane extending from $r = a$ to $r = B$ and from say $z = l$ to $z = 0$ as shown in Fig. 2.8.

$$\phi = \int B\, dA = \int_0^L \int_a^b B\left(\frac{\mu_0 l}{2\pi r}\right) drdz = \frac{\mu_o N i L}{2\pi} \ln \frac{b}{a} \tag{2.27}$$

From Eq. (2.26) the self-inductance of the coaxial cable is

$$L = \frac{\phi}{i} = \frac{N\mu_o L}{2\pi} \ln \frac{b}{a} \tag{2.28}$$

To determine the self- and magnetic circuit mutual inductance consider two coils, 1 and 2, in Fig. 2.9 which are partially linked by a mutual flux; the two coils can have two possibilities, namely, aiding or bucking fluxes. In this case, however, only the aiding case will be considered.

The sum of two partial fluxes ϕ_1 due to the current I_1 will remain unchanged even if coil 2 is removed, and ϕ_{21} due to current I_2 which passes through the windings of coil 1 will remain unaltered even if I_1 becomes zero. In both cases the flux is the total linkage flux and is

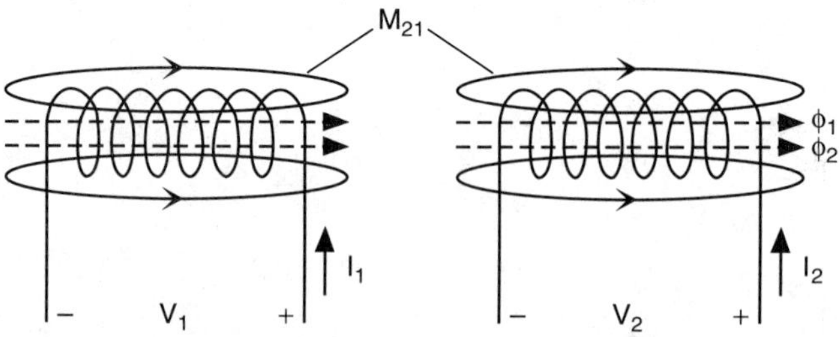

Figure 2.9 Mutual inductance.

measured in weber-turns. The counterelectromotive force in circuit 1 is given by $d(\phi_1 + \phi_2)/dt$ and according to Fig. 2.9 for the aiding case we have

$$V = \frac{\delta\phi}{\delta t} + RI \qquad (2.29)$$

Therefore the equation for the first coil is given by

$$V_1 = R_1 I_1 + \frac{d\phi_1}{dt} + \frac{d\phi_{21}}{dt} \qquad (2.30)$$

From Eq. (2.23) we have $\phi_1 = L_1 I_1$ where L_1 is the inductance of coil 1. Flux ϕ_{21} is proportional to current I_2, therefore,

$$\phi_{21} = M_{21} I_2 \qquad (2.31)$$

where M_{21} is the mutual inductance between coils 1 and 2. Substituting Eq. (2.31) into Eq. (2.30) we have

$$V_1 + R_1 I_1 + L_1 \frac{dI_1}{dt} + M_{21} \frac{dI_2}{dt} \qquad (2.32)$$

Similarly for the second circuit we have

$$V_2 = R_2 I_2 + L_2 \frac{dI_2}{dt} + M_{12} \frac{dI_1}{dt} \qquad (2.33)$$

where the mutual inductances in both expressions are equal to $M_{21} = M_{12} = M$. Multiplying Eq. (2.32) by I_1 and Eq. (2.33) by I_2, we obtain power in the two magnetic circuits:

$$V_1 I_1 = R_1 I_1^2 + L_1 I_1 \frac{dI_1}{dt} + M I_1 \frac{dI_2}{dt} \quad \text{W} \qquad (2.34)$$

$$V_2 I_2 = R_2 I_2^2 + L_2 I_2 \frac{dI_2}{dt} + M I_2 \frac{dI_1}{dt} \quad \text{W} \qquad (2.35)$$

From Eqs. (2.34) and (2.35) the following can be said:

1. Expressions $V_1 I_1$ and $V_2 I_2$ are powers delivered by the two coils 1 and 2, respectively.
2. The terms $R_1 I_1^2$ and $R_2 I_2^2$ represent the ohmic losses in the windings. The sum of the remaining terms represents the rate of change of

magnetic energy in the two coils; therefore,

$$\frac{d(W_m)}{dt} = L_1 I_1 \frac{dI_1}{dt} + L_2 I_2 \frac{dI_2}{dt} + M\left(I_1 \frac{dI_2}{dt} + I_2 \frac{dI_1}{dt}\right) \tag{2.36}$$

Integrating Eq. (2.36) we get the magnetic energy stored in the coils.

$$W_m = \tfrac{1}{2} L_1 I_1^2 + \tfrac{1}{2} L_2 I_2^2 + M I_1 I_2 \tag{2.37}$$

Example 2.5 Consider two coupled windings (Fig. 2.9) with the following parameter values: $L_1 = 10$ mH, $L_2 = 16$ mH, and $M = 3$ mH. The currents applied to winding 1 and winding 2 are $i_1 = 20$ A and $i_2 = 10$ A, respectively.

1. Find the stored energy in the system.
2. If the current in the second winding is zero, $i_2 = 0$, and a 10-V rms, 1000-Hz sinusoidal voltage is applied to the first winding, find the current through the first winding.

solution Using Fig. 2.9 and Eq. (2.37)

1. $W_m = \tfrac{1}{2} L_1 I_1^2 + \tfrac{1}{2} L_2 I_2^2 + M I_1 I_2 = 3.4$ J
2. $I_1 = L_1 i_1 + M i_2 + L_1 i_1$

$$\frac{d\lambda_1}{dt} = L_1 \frac{di_1}{dt} = e_1$$

But

$$E_1 = 10 \angle 0°$$

$$= j\omega L_1 I_1$$

$$I_1\left(\frac{E_1}{j\omega L_1}\right) = 0.1592 \angle 90°$$

$$|I_1| = 0.1592 \text{ A}$$

2.4 Magnetization Curves of Ferromagnetic Materials

The electromagnetic devices have cores of ferromagnetic materials which have very high permeabilities. The relation between the magnetic field intensity H and the magnetic flux density B is obtained empirically. Figure 2.10 shows the B/H relationships of most common ferromagnetic materials used in the building of electromagnetic devices, such as transformers, ac and dc machines. The B/H relationships are called the magnetization curves. These curves are different for different ferromagnetic materials. The curves are linear until the flux density passes beyond the "knee." Beyond this point the flux density

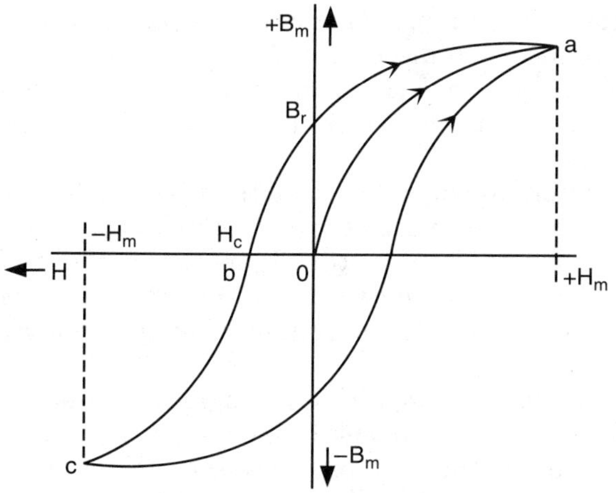

Figure 2.10 A hysteresis loop.

becomes insensitive for any more change of the field intensity, such a state is known as saturation and the magnetization curve is also known as the saturation curve.

Since there is no single value of the permeability which can be obtained from the magnetization curve beyond the "knee," point *a* in Fig. 2.10, the permeability is not constant. This leads to a nonlinear flux-reluctance relationship and thereby complicates the solution of magnetic circuit problems. From Fig. 2.10 when magnetic field intensity H is zero, there is a residual flux density B_r. To reduce the flux density to zero, from B_r, requires demagnetizing coercive force H_c. Coercive force H_c is the magnitude of the demagnetizing force corresponding to zero flux density and gives a measure of the material's resistance to demagnetization. When the magnetic field intensity is increased to positive H and then reduced to negative H, as shown in Fig. 2.10, it is found that the curve does not retrace along the magnetization curve *a*0 but rather along *abc*. In Fig. 2.10 we observe that there is a tendency of the flux density to lag behind the field intensity when the ferromagnetic material undergoes symmetrical cyclical magnetization; this phenomenon is called hysteresis, which means lagging behind.

2.5 Eddy-Current Losses in Ferromagnetic Materials

In order to conclude the discussion on magnetic circuits, it is important at this point to examine losses occurring in ferromagnetic materials when a current passes in conductors wound on the ferromagnetic materials. The nonlinear properties of magnetic materials, as characterized by the hysteresis loop, has been discussed in Sec. 2.3. Numerous electromagnetic devices such as transformers and self-excited dc generators depend upon the nonlinear characteristics of their magnetic circuits for operation.

The process of magnetization and demagnetization of ferromagnetic materials involves a storage and a release of energy which is not completely reversible as shown in Fig. 2.11. In this figure, the amount of the magnetic energy stored during magnetization of the material *abea* at half cycle is greater than the amount of magnetic energy during release *abf*.

The time-varying fluxes produce losses in ferromagnetic materials known as core losses. The core losses consist of hysteresis and eddy-current losses. In order to compute the hysteresis power loss, Steinmetz obtained an empirical formula for hysteresis power loss. The formula is

$$P_h = vfK_h B_m^n \qquad \text{W} \tag{2.38}$$

where B_m = maximum flux density in hysteresis and eddy-current losses, teslas

f = frequency, Hz

v = speed, m/s

K_h = constant, depends on the material

n = integer, lies between 1.5 and 2.5

The eddy-current loss is associated with the circulating currents which are found to exist in closed paths within the body of the ferromagnetic

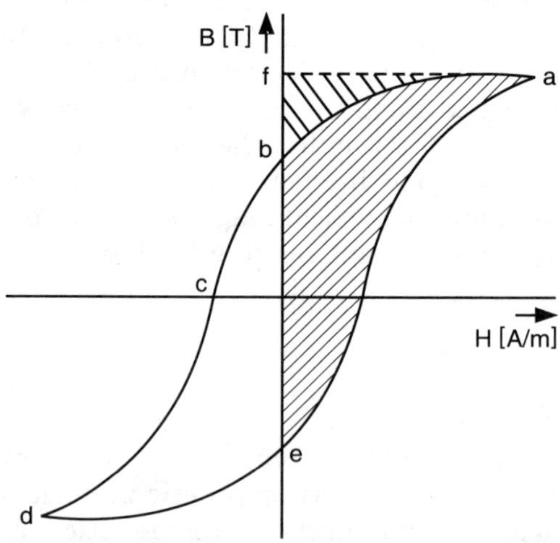

Figure 2.11 Hysteresis loop and magnetic energy relationship.

materials causing an appreciable heat loss. The circulating currents are the result of the differences in potentials which exist in the entire body of the ferromagnetic materials, owing to the change of flux. To reduce the losses, the change of the electromagnetic devices, their magnetic circuits are made by thin laminations. The empirical equation for the eddy-current loss is

$$P_e = K_e f^2 B_m^2 \tau^2 V \qquad \text{W} \tag{2.39}$$

where K_e = constant, depends on the material
B_m = maximum flux density, tesla
f = frequency, Hz
t = lamination thickness
V = total volume of the material

The two equations, (2.38) and (2.39), reveal that the hysteresis loss varies directly with the frequency while the eddy-current loss varies as the square of the frequency.

2.6 Stored Magnetic Energy

The magnetic energy stored in the magnetic field accounts largely for the common use of electromagnetic devices for electromechanical energy conversion. The total energy stored in a steady magnetic field in which B is linearly related to H is

$$W_H = \frac{1}{2}\int_{\text{vol}} \mathbf{B} \cdot \mathbf{H}\, dV \tag{2.40}$$

where dV is the elementary volume.

Let $B = \mu H$, Eq. (2.40) takes the following forms:

$$W_H = \frac{1}{2}\int_{\text{vol}} \mu H^2\, dV \tag{2.41}$$

or

$$W_H = \frac{1}{2}\int_{\text{vol}} \frac{B^2}{\mu}\, dV \tag{2.42}$$

It is convenient to think of this energy as being distributed through the volume with an energy density of $\frac{1}{2}BH$ J/M. At present there is no mathematical justification to this expression. In the linear region of the magnetization curve, such as at the air gap, the magnetic energy stored in the magnetic field is expressed by

$$W_{m,\text{air}} = \frac{1}{2}\left(\frac{B_g^2}{\mu_o}\right)(\text{volume}) = \frac{1}{2}F\phi = \frac{1}{2}R\phi^2 = \frac{1}{2}i\lambda \tag{2.43}$$

Equation (2.43) gives the energy stored upon increasing flux density from zero to air-gap density. With nonlinearity and hysteresis neglected, the reluctance and permeance are constant. The flux and mmf are directly proportional, as shown in Eq. (2.43).

The principles of magnetic circuits can be applied to the lifting of weights by means of magnetic cranes. Consider the magnetic circuit of Figure 2.12, consisting of two parts of the same magnetic material with two air gaps.

Here we can determine the weight that can be lifted by the crane for a given set of parameters. The magnetic energy stored in a differential volume of space is given by Eq. (2.40), or

$$dW_H = \tfrac{1}{2}(\mathbf{B} \cdot \mathbf{H})\, dV \tag{2.44}$$

Due to the symmetry, working with one pole of the magnetic circuit, the incremental change in volume becomes

$$dV = A\, dl_g \tag{2.45}$$

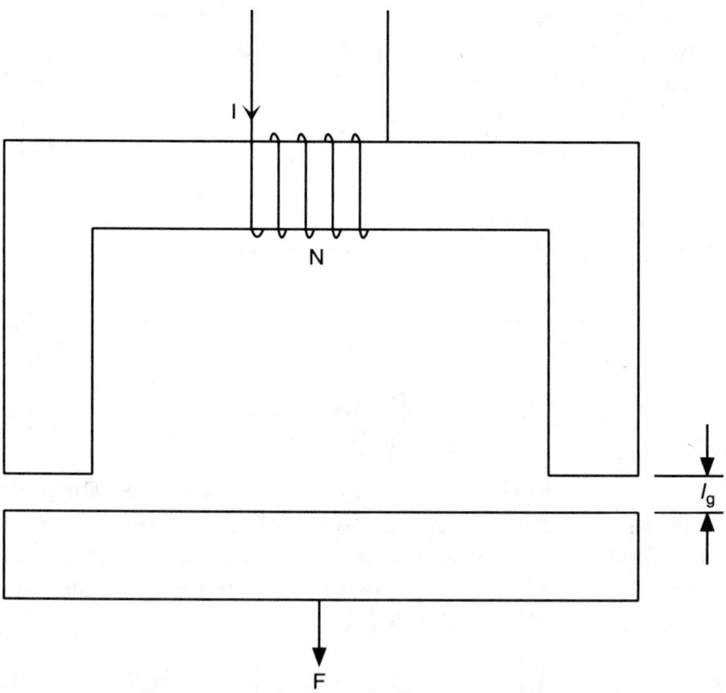

Figure 2.12 Magnetic circuit for weight lifting.

Neglecting leakage and fringing we get

$$dW_H = \frac{1}{2} BH\, dV = \frac{1}{2} BHA\, dl_g = \frac{1}{2}\left(\frac{B^2}{\mu_o}\right) A\, dl_g \qquad (2.46a)$$

Work is defined as

$$dW = \mathbf{F} \cdot d\mathbf{L} = F\, dl_g \qquad (2.46b)$$

While a magnetic pull is exerted upon the bar, an energy dW equal to the magnetic energy W_H stored in the magnetic field is expended. Therefore,

$$dW_H = dW \qquad (2.47)$$

$$\frac{1}{2}\left(\frac{B^2}{\mu_o}\right) A\, dl_g = F\, dl_g \qquad (2.48)$$

$$F = \frac{1}{2}\left(\frac{B^2}{\mu_o}\right) A \qquad (2.49)$$

Equation (2.49) gives the pulling force per pole on the bar. The total pulling force on the bar becomes

$$F_{\text{total}} = 2\left(\frac{1}{2}\right)\left(\frac{B^2}{\mu_o}\right) A = \left(\frac{B^2}{\mu_o}\right) A \quad \text{N} \qquad (2.50)$$

2.7 Instrument Transformers

It is important at this stage to introduce instrument transformers such as current transformers, also called CT, and potential transformers, or PT, which are very important to electric power engineers.

Instrument potential transformers are used in voltage measurements and instrument current transformers are used in current measurements. Both types also serve to insulate the low-voltage instruments from the high-voltage system. Instrument transformers are also employed to operate relays used for power system protection such as overcurrent relays, undervoltage relays, impedance relays, etc. Figure 2.13 shows a current transformer CT circuit connection.

The instrument current transformers are connected in series with the line to step high values of current down to a rated value of about 5 A for ammeters and voltmeters. Potential transformers are usually connected line-to-neutral in the three-phase installations, although line-to-line connections are not uncommon. The secondaries of instrument transformers are grounded for reasons of safety, to eliminate the

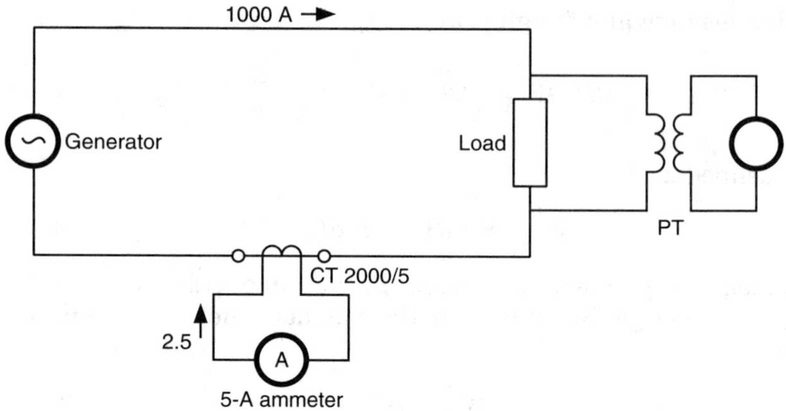

Figure 2.13 Current transformer circuit connection.

hazard of raising the secondary winding to a potential through the capacitance coupling with the primary winding. Operating the secondary circuit under load, there would then be no mmf to oppose the primary mmf, and all the primary current would become exciting current, thus inducing a very high voltage in the secondary. Therefore, before disconnecting instruments and relays from a current transformer, the secondary must be shorted.

The core losses and the nonlinearity of the exciting current caused by nonlinearity in the ferromagnetic core present the true ratio of primary to secondary current from being exactly equal to the inverse of the turns ratio and also causes a small phase shift. The instrument error is compensated by the manufacturers of current transformers by making the turns ratio slightly different from the nameplate ratio. The relative instantaneous polarities of instrument transformer terminals are indicated by ± markings or H_1 and X_1 markings. Letter H_1 denotes the high-side polarity terminal and X_1 the corresponding low-side polarity terminal.

Apparatus wattmeters, voltmeters, power factor meters, synchronoscopes, relays, etc. all require specific polarity connections. Wrong polarity connections might cause errors in instrument readings and may cause extensive damage to ac generators and distribution systems that utilize protective relays. Therefore, polarity tests should always be conducted on rebuilt, repaired, and newly installed instrument transformers to make sure that the polarity markings are correct. The normal current rating of a current transformer secondary has been standardized at 5 A with a second standard of 1 A being used in Europe. For a short period of time, the rating of CT secondary windings can be exceeded without damaging the windings. Current of more than 10

TABLE 2.2 Standard Current Transformer Ratios

Current ratio	Current ratio	Current ratio
50:5	300:5	800:5
100:5	400:5	900:5
150:5	450:5	1000:5
200:5	500:5	1200:5
250:5	600:5	

or 20 times are often encountered in current transformer windings during short circuits on the power system. Standard current transformer current ratios have been established and some of these are shown in Table 2.2.

To illustrate the usage of a current transformer, let the alternating flux produced by current in the power line conductor induce a current in the closed secondary circuit that is approximately proportional to the primary current. The actual current transformer ratio as stamped on the nameplate is generally expressed as a current ratio with respect to a 5-A secondary. Figure 2.13 shows the circuit connection for metering a 1000-A load with a 2000/5-A current transformer. The current to the ammeter is

$$I_{\text{meter}} = 1000\left(\frac{5}{2000}\right) = 2.5 \text{ A}$$

The secondary winding of a current transformer, as mentioned earlier, must be short-circuited at its terminals.

Problems

1 A ferromagnetic material core, shown in Fig. 2.14, has the following parameters: $l_1 = l_2 = 20$ cm, $l_3 = 5$ cm, $A_1 = A_2 = 150$ cm^2, $A_3 = 20$ cm^2, $\mu_{r1} = \mu_{r2} = 200$, $\mu_{r3} = 1000$, $N = 30$. Calculate the flux densities in the three branches of the circuit, when the current through the coil is 0.8 A.

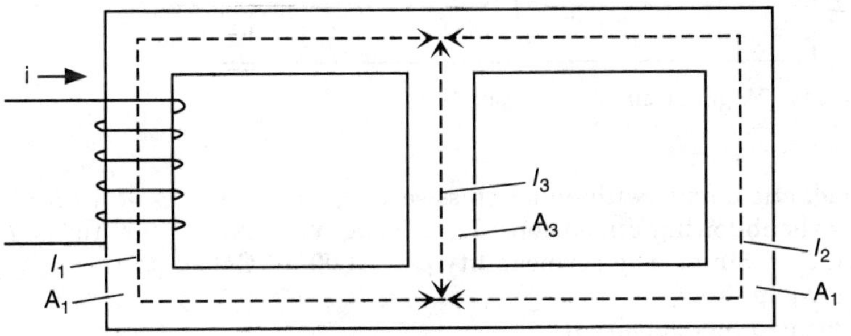

Figure 2.14 Magnetic circuit for Prob. 1.

2 An iron core is made of sheet steel laminations. The core has two parts with the following dimensions: $A_1 = 0.0038\ \text{m}^2$, $l_1 = 0.2$ m, $A_2 = 0.004\ \text{m}^2$, and $l_2 = 0.6$ m. A coil of 100 turns is energized from a 60-Hz voltage source. The flux is $\phi(t) = 0.0055 \sin wt$ Wb, using Fig. 2.15. The permeability of the core is $\mu_r = 1000$. Calculate

a The induced voltage
b The effective current

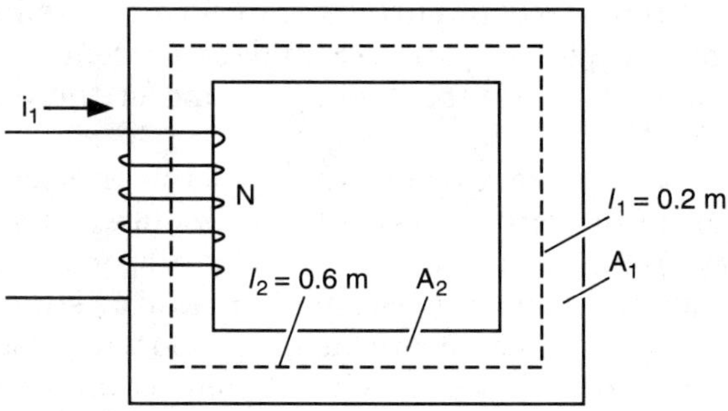

Figure 2.15 Magnetic circuit for Prob. 2.

3 The magnetic circuit of Fig. 2.16 has the following dimensions: $I_1 = 36$ cm, $I_2 = 4.1$ cm, $I_3 = 12$ cm, $I_g = 3$ mm, $A_1 = 0.0024\ \text{m}^2$, $A_g = 0.0020\ \text{m}^2$. The coil has 8600 At. The air-gap flux is 0.003 Wb. Determine the cross-sectional area of segment two (A_2). $\mu_r = 200$.

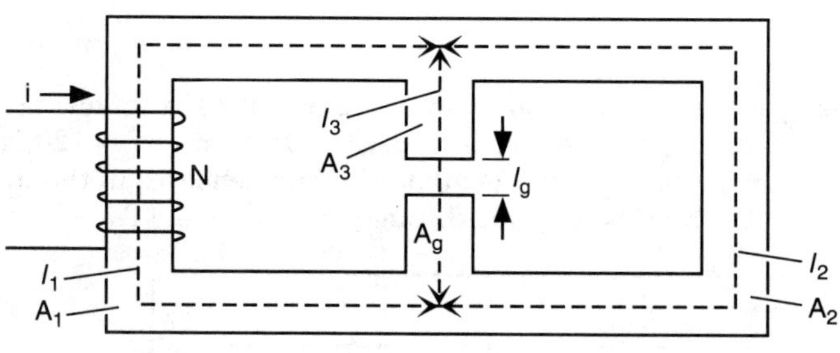

Figure 2.16 Magnetic circuit for Prob. 3.

4 A magnetic circuit with equal cross-sectional area $A = 4 \times 10\ \text{m}^2$ (Fig. 2.17) has the following dimensions: $I = 20$ cm, $N_1 = N_2 = 1000$ turns, $I_1 = I_2 = 4$ A, $l_g = 5$ mm. The permeability $\mu = 1000\ \mu_o$. Calculate

a The air-gap flux f_g
b The air-gap flux density B_g
c The air-gap magnetic field intensity H_g

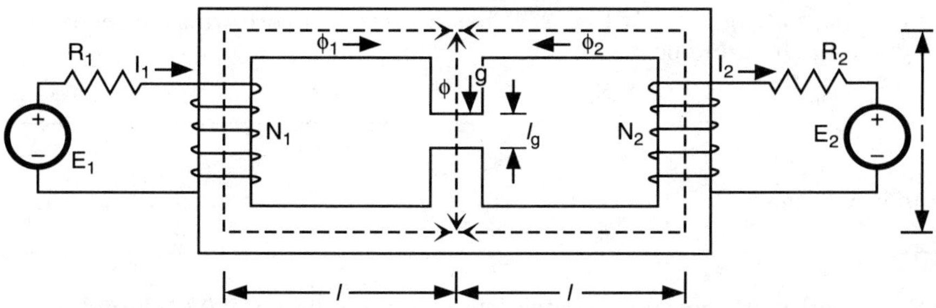

Figure 2.17 Magnetic circuit for Prob. 4.

5 A magnetic crane is used to lift a weight as shown in Fig. 2.18. The magnetic circuit has the following parameters: $N = 1000$ turns, $I = 50$ A, $l_1 = 3.5$ m, $l_g = 4$ mm, $l_2 = 1$ m; $A = 3 \times 10^2$ m^2 and $\mu_r = 10^6$. Find the force that can be lifted by the crane.

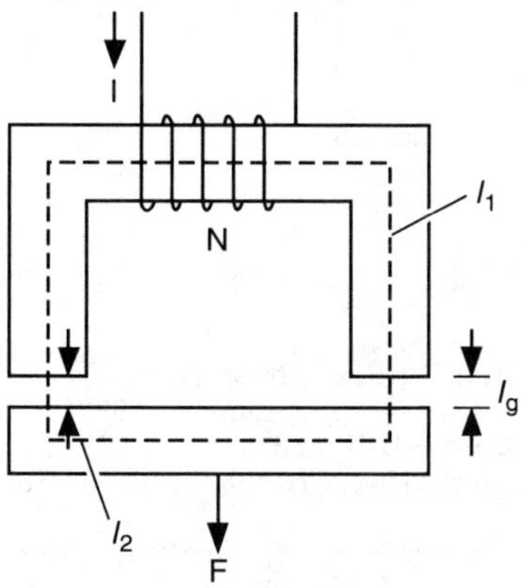

Figure 2.18 The magnetic crane circuit for Prob. 5.

6 The dimensions of the core and armature of a relay constructed of electric sheet steel are given in Fig. 2.19. Calculate the number of ampere-turns required to produce a flux density at the air gap of 1.6 Wb.

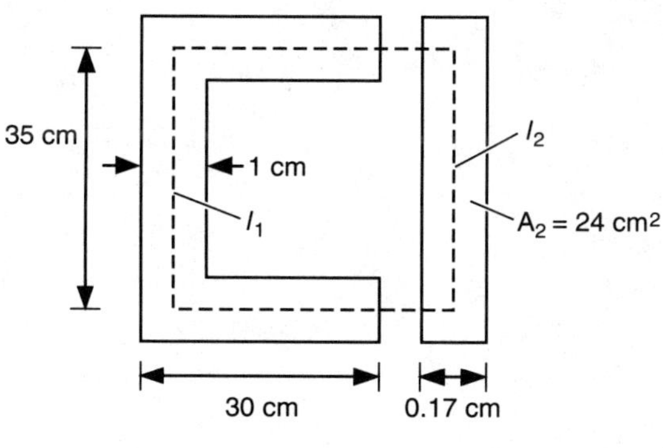

Figure 2.19 Magnetic circuit for Prob. 6.

7 The magnetic circuit of Fig. 2.20 has a cast steel core whose dimensions are given in the table below.

	ab	*bc*	*cd*	*ad*	*dea*
Mean length, cm	10	0.01	10	8	15
Area, cm^2	4	4.04	4	2	6

Find the coil mmf required to establish an air-gap flux of 0.02 Wb and μ_r = 620.

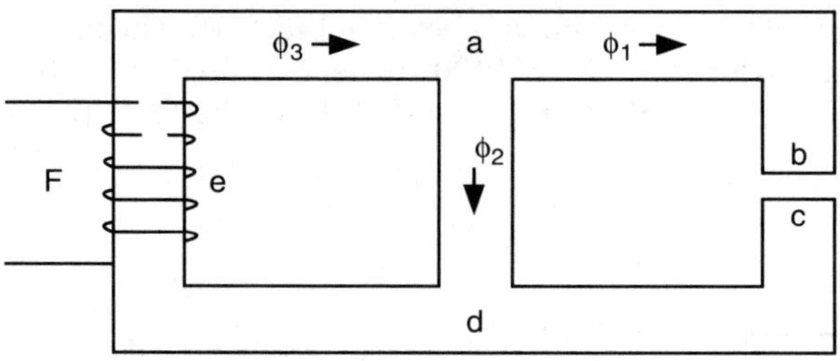

Figure 2.20 Magnetic circuit for Prob. 7.

Bibliography

Adams, A. T., *Electromagnetic for Engineers,* The Ronald Press, New York, 1971.
Elgerd, O. I., *Basic Electric Power Engineering,* Addison-Wesley, London, 1977.
Hayt, W. H. Jr., *Engineering Electromagnetics,* McGraw-Hill, 1984.
Nasar, S. A., and J. J. Cathey, *Basic Electrical Engineering, Schaum's Outline Series,* McGraw-Hill, 1984.
Skitek, G. G., and S V. Marchall, *Electromagnetic Concepts and Applications,* Prentice-Hall, Englewood Clifts, N.J., 1982.

Chapter

3

The Power Transformer

3.1 Introduction

A transformer is an electromagnetic device which comprises a highly permeable metallic core, a primary winding with N_1 turns, and a secondary winding with N_2 turns to which the load is connected. The range of applications of transformers in electrical systems extend from low voltage and low power used in electronics and control circuits, to high-voltage power used in transmission lines and distribution lines.

3.2 Ideal Transformer

An *ideal transformer* is defined as one that has no power losses, that is, the core has no hysteresis or eddy currents and no leakage flux. This means that all the flux in the core couples both windings, and the resistance of the transformer winding is zero. The transformer action is based on Faraday's law of electromagnetic induction. From Fig. 3.1, the magnitude of the induced voltages e_1 and e_2 are given by Faraday's law as

$$e_1 = \frac{d\lambda_1}{dt} \quad \text{and} \quad e_2 = \frac{d\lambda_2}{dt} \tag{3.1}$$

where λ_1 and λ_2 are flux linkages.

The two windings have N_1 and N_2 turns, R_1 and R_2 resistances, L_{11} and L_{22} self-inductances, and $L_{21} = L_{12}$ mutual inductances, respectively. Using Kirchhoff's voltage law for each loop gives the terminal voltages as

$$V_1(t) = L_1R_1 + \frac{d\lambda_1}{dt} \tag{3.2}$$

$$V_2(t) = L_2R_2 + \frac{d\lambda_2}{dt} \tag{3.3}$$

Assuming the resistance of the transformer winding to be zero, Eqs. (3.2) and (3.3) become

$$V_1(t) = \frac{d\lambda_1}{dt} = N_1 \frac{d\phi_1}{dt} = e_1(t) \tag{3.4}$$

$$V_2(t) = \frac{d\lambda_2}{dt} = N_2 \frac{d\phi_1}{dt} = e_2(t) \tag{3.5}$$

From Fig. 3.1 the flux $\phi_1 = \phi_2 = \phi$. Therefore, Eqs. (3.4) and (3.5) become

$$\frac{V_1}{V_2} = \frac{e_1}{e_2} = \frac{N_1}{N_2} \tag{3.6}$$

$$\frac{N_1}{N_2} = a \tag{3.7}$$

where a is the turns ratio.

The current ratio of the transformer and its dot conversion is discussed by using Fig. 3.1. When the transformer's secondary side is loaded, as shown in Fig. 3.1, the dots placed in Fig. 3.1 help to determine the polarity of the voltages and currents in the core. When a current flows into the dotted end of a winding, a positive magnetomotive force (mmf) F is produced. Likewise, a current flowing into an undotted end of a winding produces a negative magnetomotive force (mmf). If two currents are both flowing into the dotted ends of their respective windings their magnetomotive forces will add together. If one current flows into the dotted end of a winding and the other flows out of the dotted end, the magnetomotive forces will subtract from each other.

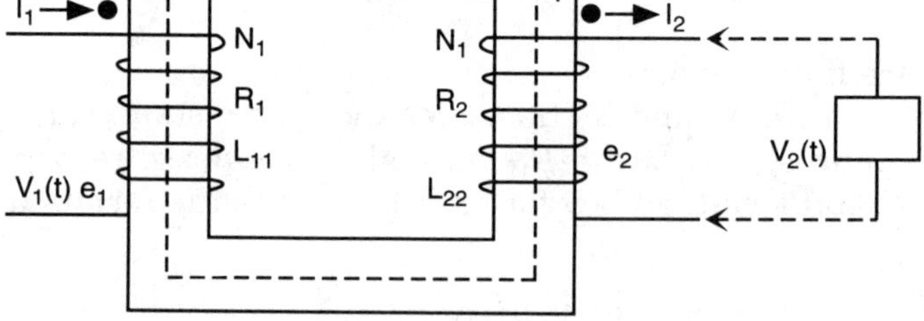

Figure 3.1 Schematic diagram of an ideal transformer.

From Fig. 3.1 the primary current produces a positive magnetomotive force, $F_1 = N_1 i_1$, and the secondary current produces a negative magnetomotive force, $F_2 = -N_2 i_2$. From the above statements the net magnetomotive force is given by

$$F_{\text{net}} = N_1 i_1 - N_2 i_2 = \phi R \tag{3.8}$$

where R is the reluctance of the transformer core and ϕ is the magnetic flux in the transformer core.

Since the reluctance of the transformer is very small or zero, Eq. (3.8) can be written as

$$F_{\text{net}} = N_1 i_1 - N_2 i_2 = 0 \tag{3.9}$$

When the applied voltage is a sinusoid of constant amplitude, Eq. (3.6) can be written as Eq. (3.10), where uppercase letters denote phasors. Therefore,

$$\frac{V_1}{V_2} = \frac{N_1}{N_2} \tag{3.10}$$

and Eq. (3.9) can be written as

$$N_1 i_1 = N_2 i_2$$

$$\frac{i_1}{i_2} = \frac{N_2}{N_1} = \frac{1}{a} \tag{3.11}$$

The last parameter of the transformer is the determination of the impedance. Using Eqs. (3.10) and (3.11) we get

$$V_1 = \frac{N_1}{N_2} V_2 \tag{3.12}$$

and

$$i_1 = \frac{N_2}{N_1} i_2 \tag{3.13}$$

Therefore,

$$Z_1 = \frac{V_1}{I_1} = \left(\frac{N_1}{N_2}\right)^2 \frac{V_2}{I_2} = a^2 Z_2 \tag{3.14}$$

Equation (3.14) gives the impedance of the secondary side referred to the primary. The term *primary* means that winding which is connected to the energy source, that is, coil 1.

3.2.1 Determination of electromagnetic equation of transformer

Since the emf varies sinusoidally [if $V_1(t)$ of Eq. (3.2) is sinusoidal], the flux also varies sinusoidally. Therefore,

$$\phi = \phi_{\max} \sin \omega t \tag{3.15}$$

Then from Eq. (3.4), the induced voltage is

$$e = N\frac{d\phi}{dt} = N\frac{d}{dt}(\phi_{\max} \sin \omega t) = N\omega\phi_{\max} \cos \omega t \tag{3.16}$$

From Eq. (3.16), the maximum value of the induced voltage is

$$E_{\max} = \omega N \phi_{\max} \tag{3.17}$$

or in rms value is

$$E_{\text{rms}} = \frac{\omega N \phi_{\max}}{\sqrt{2}} = 4.44 f N \phi_{\max} \tag{3.18}$$

where $\omega = 2\pi f$ rad/s. Equation (3.18) is called the electromotive force (emf) equation of the transformer.

Example 3.1 A single-phase transformer with $E = 200$ V, made of laminated silicon steel core, has a cross-sectional area of 0.005 m and a mean length of 0.6 m. The primary has 150 turns and the secondary winding has 400 turns. Determine

1. The primary current (I_1) with the secondary side open
2. The primary current (I_1) when the secondary side has a load resistance of 100 Ω (Fig. 3.2).

solution Maximum flux

$$\phi_m = \frac{\sqrt{2}E_1}{N_1\omega} = \frac{\sqrt{2}(200)}{150(2\pi)(60)} = 5 \times 10^{-3} \text{ Wb}$$

$$B_m = \frac{\phi_m}{A} = \frac{5 \times 10^{-3}}{0.005} = 1 \text{ T}$$

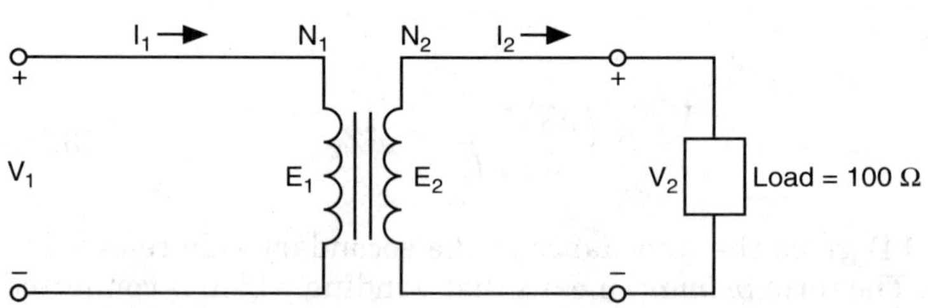

Figure 3.2 Example 3.1.

From Fig. 2.5

$$H = 340 \text{ At/m}$$

1. The magnetizing current (rms)

$$I_m = \frac{F}{\sqrt{2}N} = \frac{Hl}{\sqrt{2}N_1} = \frac{340(0.6)}{\sqrt{2}(150)} = 0.962 \text{ A}$$

$$V_2 = \frac{N_2}{N_1}V_1 = \frac{400(200)}{150} = 533.33 \text{ V}$$

2. With the load connected

$$I_2 = \frac{V_2}{R_2} = \frac{533.3}{100} = 5.33 \text{ A}$$

The primary current due to the load

$$I_{1L} = \frac{N_2}{N_1}I_2 = \frac{400(5.33)}{150} = 14.21 \text{ A}$$

where I_{1L} is the primary current with the load.

Note: The magnetizing current (I_m) is in quadrature with the load current so their vector sum has almost the same magnitude as the load current referred to the primary alone.

Example 3.2 A transformer with a magnetic core made of silicon steel laminations has a net cross-sectional area of 16 cm^2 and the voltage source of 120 V rms at 400 Hz. Calculate the number of primary winding turns N_1 required so that the maximum flux density at rated applied voltage is $B = 1.25$ T.

solution

$$\phi_{max} = BA = (1.25)(16 \times 10^{-4}) = 20.0 \times 10^{-4} \text{ Wb}$$

From Eq. (3.18) we have

$$E = 4.44fN\phi_{max} = (4.44)(400)(N)(20 \times 10^{-4}) \quad \text{V}$$

$$N = \frac{120}{(4.44)(400)(20 \times 10^{-4})} = 33.78 \text{ turns}$$

3.3 Equivalent Circuit of a Real Transformer

The real transformer differs from an ideal transformer in that the former takes into account the losses. The major losses in the transformer are the core losses (I^2R), these are resistive heating losses in the primary and secondary windings of the transformer; the eddy-current losses, these are resistive heating losses in the core of the transformer; and finally the hysteresis losses, these are associated with the rearrangement of magnetic domains in the core during each half cycle. Furthermore, the windings of the transformer are not perfect, that is, not all the fluxes link with the primary and secondary winding simultaneously because of leakage. These leakage fluxes ϕ_{l1} and ϕ_{l2} produce self-inductances in the primary and secondary coils, and the effects of the inductance in the two coils must be accounted for. Figure 3.3 shows the real transformer with flux leakages.

In order to draw an equivalent circuit of a real transformer all major losses and flux leakages must be taken into account. First let us discuss the effect of copper losses. These resistive losses are represented by placing a resistor R_1 in the primary and a resistor R_2 in the secondary of the transformer circuit, as shown in Fig. 3.4. As discussed in Sec. 3.2 the leakage flux in the primary winding ϕ_{l1} produces a voltage e_1 and the leakage flux in the secondary winding ϕ_{l2} produces a voltage e_2.

The two voltages can be expressed as

$$e_1 = N_1 \frac{d\phi_{l1}}{dt} \tag{3.19}$$

and

$$e_2 = N_2 \frac{d\phi_{l2}}{dt} \tag{3.20}$$

Since most of the leakage flux passes through the air, the reluctance of the air is much higher than that of the core. Therefore, it can be stated that the leakage fluxes are directly proportional to their respective currents; thus

$$\phi_{l1} = N_1 \frac{i_1}{R} \tag{3.21}$$

and

$$\phi_{l2} = N_2 \frac{i_2}{R} \tag{3.22}$$

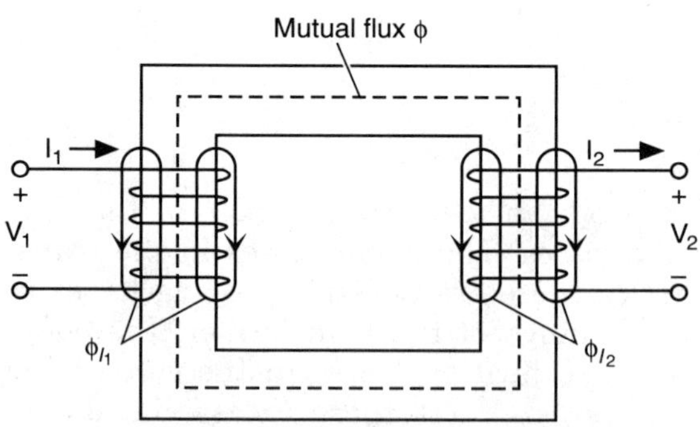

Figure 3.3 A real transformer showing flux leakages.

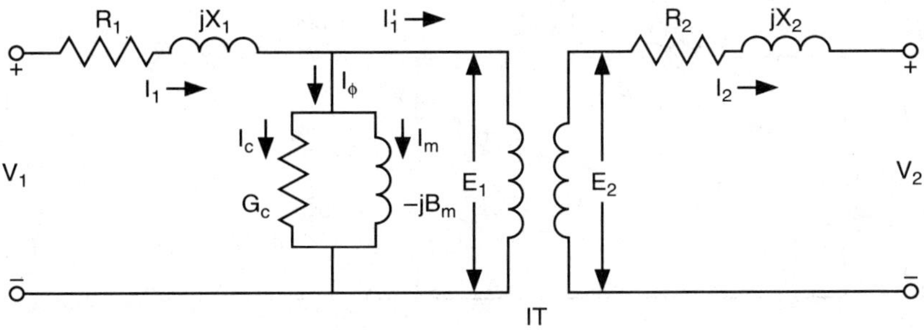

Figure 3.4 An exact equivalent circuit of a transformer.

where R is the reluctance of the flux path. Substituting Eqs. (3.21) and (3.22) into Eqs. (3.19) and (3.20) results in

$$e_1 = \frac{N_1^2}{R}\frac{di_1}{dt} = L_1\frac{di_1}{dt} \tag{3.23}$$

and

$$e_2 = \frac{N_2^2}{R}\frac{di_2}{dt} = L_2\frac{di_2}{dt} \tag{3.24}$$

Therefore, the inductances in the primary and secondary are given by

$$L_1 = \frac{N_1^2}{R} \tag{3.25}$$

and

$$L_2 = \frac{N_2^2}{R} \tag{3.26}$$

where L_1 and L_2 are the primary and secondary self-inductances, respectively. The leakage fluxes are therefore represented in Fig. 3.4 as inductive reactances in the primary and secondary of the transformer circuit. Finally the core excitation effect is represented by the magnetization reactance X_m placed across the primary voltage source and

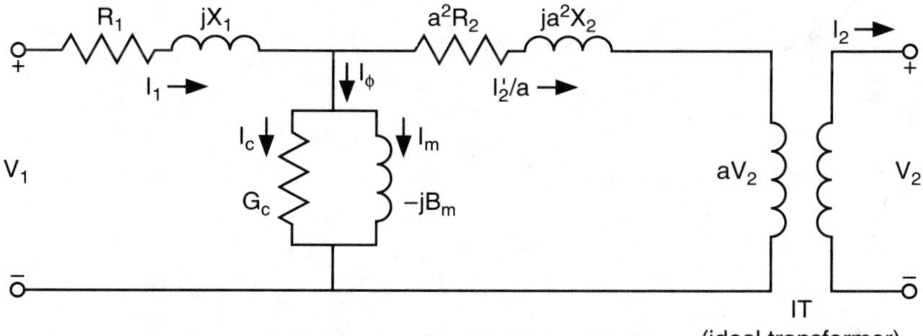

Figure 3.5 Transformer equivalent circuit referred to primary side of a transformer.

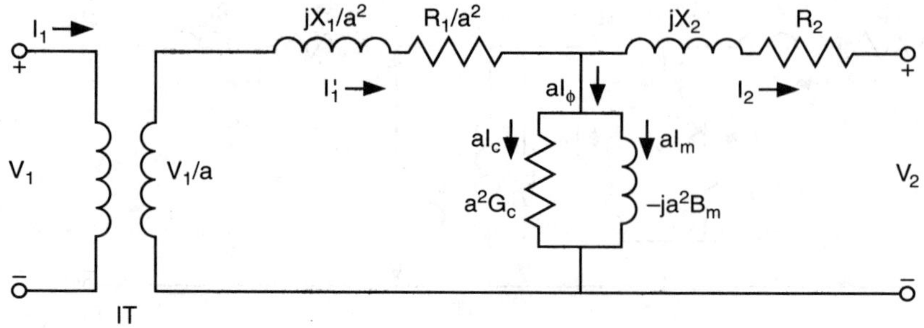

Figure 3.6 Transformer equivalent circuit referred to secondary side of a transformer.

in parallel with the core resistance R_c. The excitation current I_ϕ has two components, namely, the magnetization current I_m which lags the applied voltage by 90° and the core current I_c which is in phase with the applied voltage.

Figures 3.5 and 3.6 show the equivalent circuits of the transformer referred to the primary and secondary sides, respectively. Figure 3.7 shows the phasor diagram of the equivalent circuit of the transformer referred to the primary side.

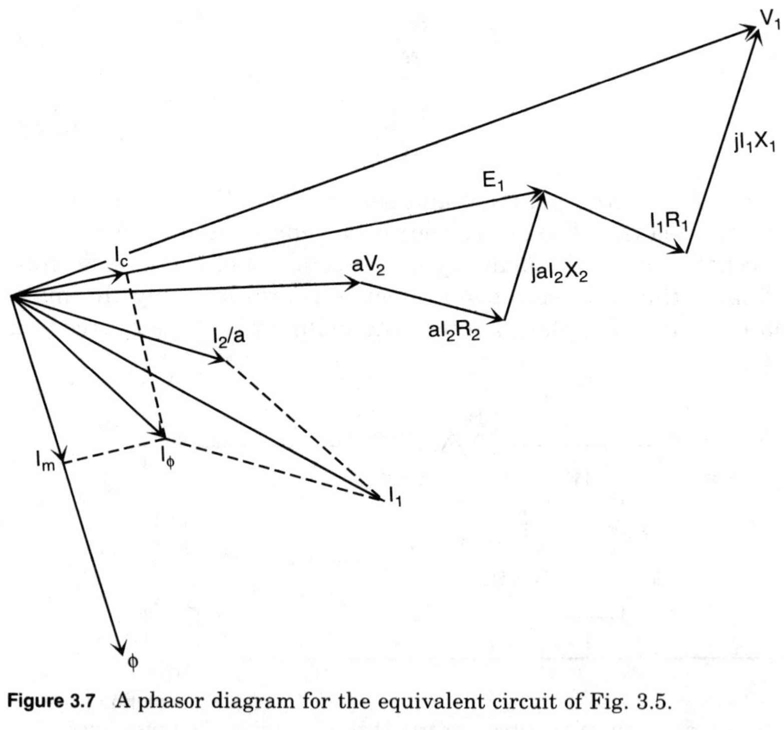

Figure 3.7 A phasor diagram for the equivalent circuit of Fig. 3.5.

3.3.1 Approximate circuits

The calculations of iron-core power transformers are simplified by using approximate circuits. A number of approximate circuits are shown in Fig. 3.8. Figure 3.8*a* is a modification of Fig. 3.5. This is effected by placing the exciting admittance across the primary terminals. The error due to this shift is small in the iron-core transformers since the exciting current I_ϕ is only about 3 to 5 percent of the full-load current I_1. The voltage drop across the primary leakage reactance X_1 is very

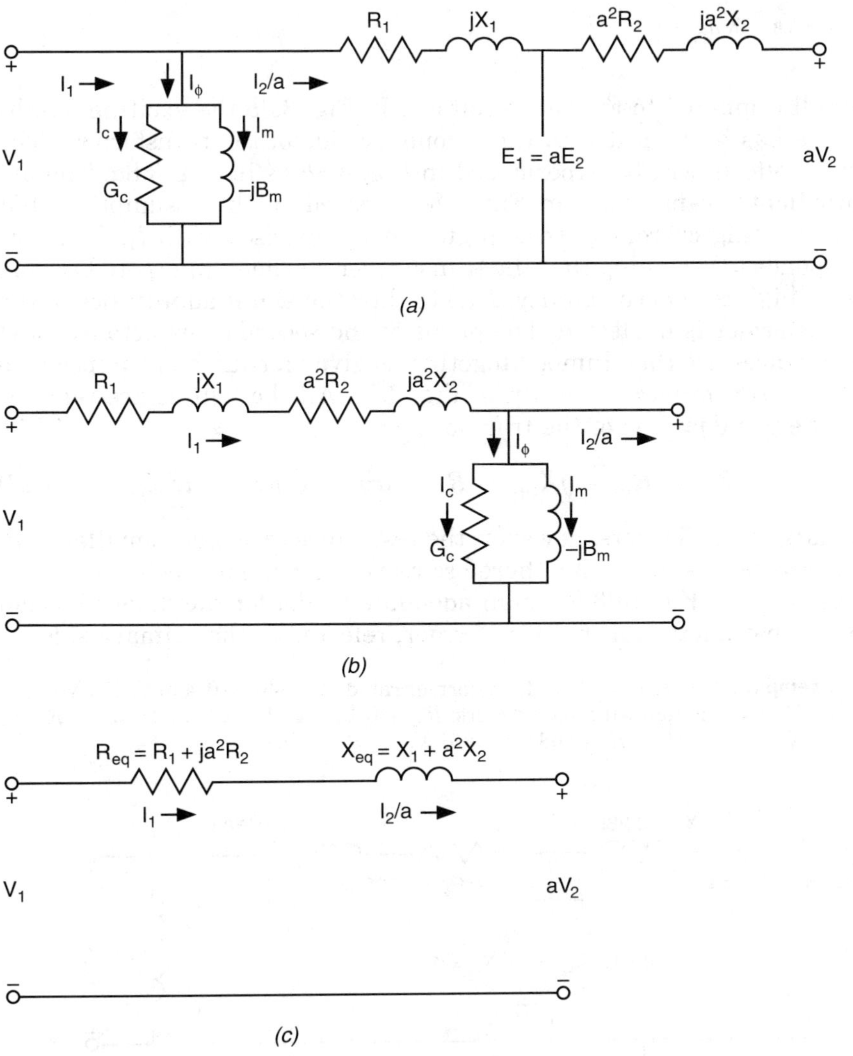

Figure 3.8 The approximate transformer circuits referred to the primary side.

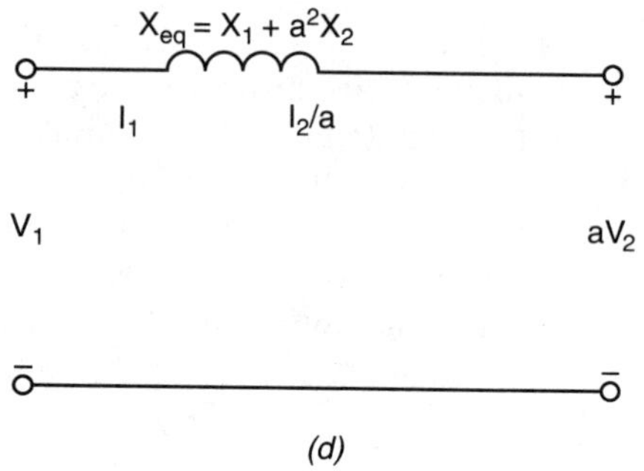

Figure 3.8 (*Continued*)

small compared to the rated voltage. In Fig. 3.8*b* the exciting admittance has been shifted to the secondary side of the transformer. The same effects will be experienced in Fig. 3.8*b* as in Fig. 3.8*a*. The approximate transformer in Fig. 3.8*c* is based on the assumption that the exciting current I_ϕ is negligible in comparison with I_1. This corresponds to assuming that the transformer has negligible core loss and very high core permeability. Due to this, the shunt admittance of the transformer is neglected. The primary and secondary resistances and reactances are then lumped together to give an equivalent impedance of the transformer as shown in Fig. 3.8*c*. All the figures are referred to the primary side of the transformer.

$$Z_{eq} = R_{eq} = jX_{eq} = (R_1 + a^2R_2) + j(X_1 + a^2X_2) \qquad (3.27)$$

In large transformers, however, the resistances are much smaller than the reactances ($R << X$), therefore resistances are neglected as shown in Fig. 3.8*d*. Figure 3.8*d* is an adequate model for the determination of the parameters of the transformer, referred to the primary side.

Example 3.3 A single-phase transformer rated 220/440, 220/440 V, 25 kVA, and 60 Hz has the following parameters: $R_1 = 0.155\ \Omega$, $R_2 = 0.039\ \Omega$, $X_1 = j0.285\ \Omega$, $X_2 = j0.0171\ \Omega$, $R_c = 68.6\ \Omega$, and $X_m = 23.15\ \Omega$.

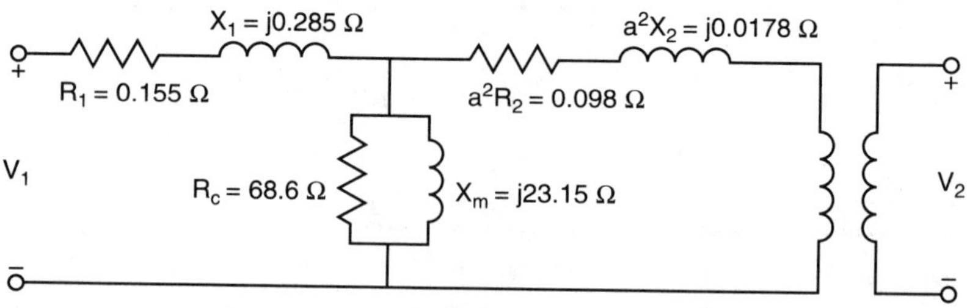

Figure 3.9 Solution for Example 3.3(1).

1. Refer all quantities to the primary side (low voltage).
2. Refer all quantities to the secondary side (high voltage).

Draw an equivalent circuit in each case.

solution Referring to the primary side of Fig. 3.5, we have

$$a = \frac{N_1}{N_2}\frac{V_1}{V_2} = \frac{220}{440} = 0.5$$

$$R_1' = a^2 R_2 = (0.5)^2(0.039) = 0.5$$

$$X_1' = a^2 X_2 = j(0.5)^2(0.0171) = j0.00427\ \Omega$$

$$aV_2 = 0.5 \times 440 = 220\ \text{V}$$

$$R_2' = \frac{R_1}{a^2} = \frac{0.155}{(0.5)^2} = 0.62\ \Omega$$

$$X_2' = \frac{X_1}{a^2} = \frac{0.285}{(0.5)^2} = j1.14\ \Omega$$

$$R_2' = \frac{R_c}{a^2} = \frac{68.6}{(0.5)^2} = 274.4\ \Omega$$

$$X_m = \frac{X_m}{a^2} = \frac{23.15}{(0.5)^2} = j92.6\ \Omega$$

See Figs. 3.9 and 3.10 for circuits for Example 3.3.

3.3.2 Transformer efficiency

The *efficiency of a device,* such as a motor, a generator, and a transformer, is defined by the following efficiency equation:

$$\eta = \frac{P_{\text{out}}}{P_{\text{in}}} \times 100\% = \frac{P_{\text{out}}}{P_{\text{out}} + P_{\text{loss}}} \times 100 \tag{3.28}$$

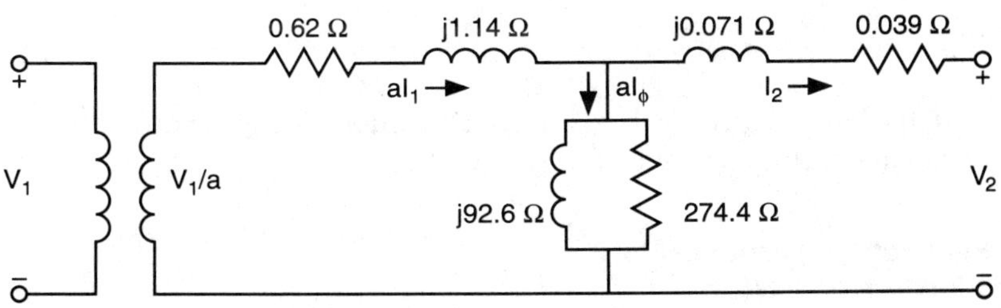

Figure 3.10 Solution for Example 3.3(2).

There are three types of losses present in a transformer; namely, the copper losses, the hysteresis losses, and eddy-current losses. The core loss depends upon the maximum flux density produced in the core and in turn depends upon the terminal potential difference. The efficiency of a transformer at a given load is given by

$$\eta = \frac{P_{\text{out}}}{P_{\text{out}} + P_{\text{Cu}} + P_{\text{core}}} \times 100\% = \frac{V_2 I_2 \cos\theta}{V_2 I_2 \cos\theta + P_{\text{Cu}} + P_{\text{core}}} \tag{3.29}$$

Using Fig. 3.8*a*, the core and copper losses are calculated as

$$P_{\text{core}} = \frac{V_1^2}{R_c} \quad \text{W} \tag{3.30}$$

$$P_{\text{Cu}} = R_{\text{eq}} \frac{I_2^2}{a} \quad \text{W} \tag{3.31}$$

3.3.3 Voltage regulation

The study of voltage regulation of power transformers is the study of the effects of imperfections. If the imperfections were negligibly small, the effective value of the primary potential difference V_1 would be a times the effective value of the secondary potential difference, that is, aV_2. Due to the imperfections represented by the equivalent impedance Z_{eq}, there is a difference between V_1 and aV_2. The difference depends largely on the operating conditions of the transformer.

The *voltage regulation* is therefore defined as the measure of the ability of the transformer to limit deviations of the output voltage within small limits of the rated value for a given input voltage and output power factor while the output current is varied from zero to full load. This means that on no-load, the secondary terminal voltage may vary from its rated value due to the effect of impedance on the transformers. The voltage regulation is then expressed as

$$\text{Percentage regulation} = \frac{V_{2,\text{NL}} - V_{2,\text{FL}}}{V_{2,\text{FL}}} \times 100 = \frac{V_1 - aV_2}{aV_2} \times 100\% \tag{3.32}$$

where, on no-load, $V_{2,\text{NL}} = V_1$. The better the voltage regulation is, the lower is its numerical value.

3.3.4 Parameter measurements of real transformer

The transformer parameters R_1, R_2, R_c, X_1, X_2, and X_m can be obtained by measuring the input voltage, current, and power (V_s, I_s, and P_s) to the primary, first with the secondary short-circuited and then with it open-circuited, as shown in Figs. 3.11 and 3.12, respectively.

Short-circuit test. With the secondary short-circuited, a voltage of about 2 to 12 percent of the rated primary voltage is applied in order to obtain full-load primary or secondary current. In this test a high-voltage side (HV) is taken as the primary. The low voltage impressed produces a very low magnetizing flux which means that the exciting current is negligible, $I_\phi = 0$, and the excitation impedance is infinite as shown in Fig. 3.11.

V_{sc}, I_{sc}, and P_{sc} are the applied primary voltage, current and power, then the short-circuit impedance Z_{sc}, its resistance R_{sc}, and reactance X_{sc} components referred to the primary are calculated by using Ohm's law. Since in conventional transformers the exciting current is small compared with the rated current it can therefore be neglected. The short-circuit impedance Z_{sc} is therefore assumed equal to the series equivalent impedance of the transformer, and we get

$$Z_{eq} = Z_{sc} = \frac{V_{sc}}{I_{sc}} \tag{3.33}$$

Since the core loss is negligible at low values of V_{sc} the equivalent resistance is

$$R_{eq} = R_{sc} = \frac{P_{sc}}{V_{sc}\, I_{sc}} \tag{3.34}$$

The equivalent leakage reactance and power factor are given by

$$X_{eq} = X_{sc} = \sqrt{Z_{sc}^2 - R_{sc}^2} \tag{3.35}$$

and

$$\cos \Psi_{eq} = \frac{P_{sc}}{V_{sc}\, I_{sc}} \tag{3.36}$$

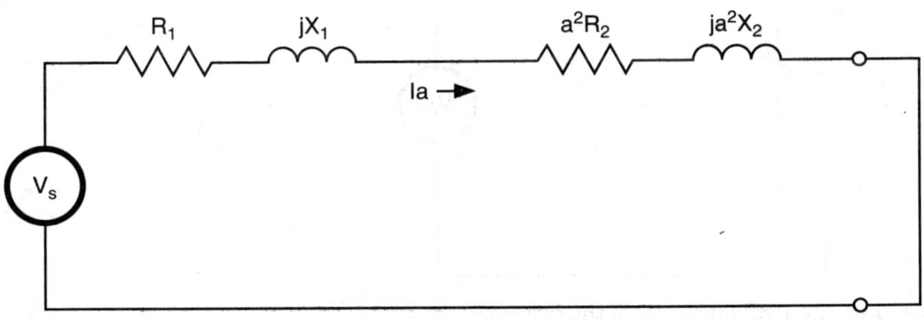

Figure 3.11 Equivalent circuit of the transformer under short-circuit test.

Open-circuit test. In an open-circuit test a voltage source is impressed, say, on the primary side, and the secondary side is open. In Fig. 3.12 an exciting current I_f is only about 2 to 6 percent of the full-load current. The low-voltage side is usually taken as the primary in this test. This small exciting current creates a negligible voltage drop across the primary impedance resulting in negligible primary copper losses. Therefore the impressed voltage appears across the exciting admittance (V) and the input power is very nearly equal to core losses P_c, as shown in Fig. 3.12.

If V_{oc}, I_{oc}, and P_{oc} are the applied voltage, current, and power, the exciting admittance Y_ϕ and its conductance G_c and susceptance B_m can be calculated by using Ohm's law, thus:

$$Y_\phi = \frac{I_{oc}}{V_{oc}} \tag{3.37}$$

The exciting conductance G_c is

$$G_c = \frac{P_{oc}}{V_{oc}^2} \tag{3.38}$$

The exciting susceptance B_m and power factor are given by

$$B_m = \sqrt{Y_\phi^2 - G_c^2} \tag{3.39}$$

$$\cos \Psi = \frac{P_{oc}}{V_{oc} I_{oc}} \tag{3.40}$$

The values obtained by Eqs. (3.37) to (3.39) are referred to the side which is used as the primary in this test.

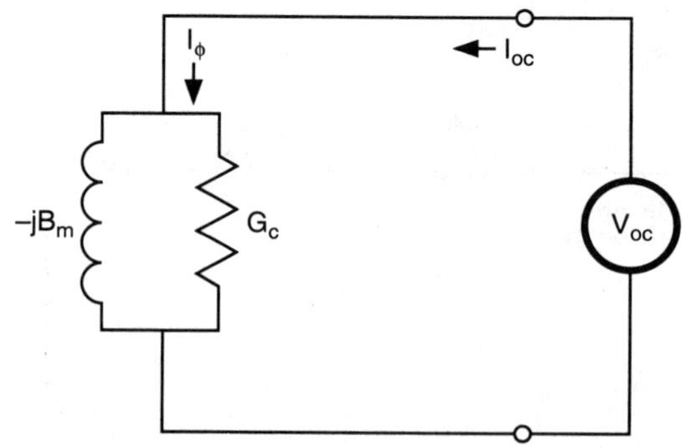

Figure 3.12 Equivalent circuit of the transformer under open-circuit test.

Example 3.4 A 40-kVA, 2400/240-V transformer feeds a load with a power factor 0.8 lagging. The transformer was under tests and the following measurements were recorded:

1. The short-circuit measurements on the high-potential side: $V_{sc} = 46$ V, $I_{sc} = 25.2$ A, and $P_{sc} = 650$ W
2. The open-circuit measurements on the low-potential side: $V_{oc} = 240$ V, $I_{oc} = 431$ A, and $P_{oc} = 195$ W

Determine

1. The parameters of the equivalent circuit of the transformer
2. The efficiency of the transformer
3. The voltage regulation of the transformer

solution

1. Using short-circuit parameters and Eqs. (3.33), (3.34), (3.35), and (3.36), we have

$$Z_{sc} = Z_{eq} = \frac{V_{sc}}{I_{sc}} = \frac{46}{25.2} = 1.825\ \Omega$$

$$R_{sc} = R_{eq} = \frac{P_{sc}}{I_{sc}^2} = \frac{650}{(25.2)^2} = 1.0235\ \Omega$$

$$X_{sc} = X_{eq} = \sqrt{(1.825)^2 - (1.0235)^2} = 1.5109\ \Omega$$

Therefore,

$$R_1 = \frac{1.0235}{2} = 0.512\ \Omega$$

$$R_2 = \frac{1.0235}{2} = 0.512\ \Omega$$

$$= \left(\frac{240}{2400}\right)^2 (0.512) = 0.00512\ \Omega$$

$$X_1 = \frac{j1.51}{2} = j0.755\ \Omega$$

$$X_2 = \frac{j1.51}{2} = j0.755\ \Omega$$

$$X_2' = \left(\frac{240}{2400}\right)^2 (j0.755) = j0.0076\ \Omega$$

2. Using open-circuit parameters and Eqs. (3.37), (3.38), and (3.39), we have

$$Y_\phi = \frac{I_{oc}}{V_{oc}} = \frac{4.31}{2400} = 1.8 \times 10^{-3}\ \text{mho}$$

$$G_c = \frac{P_{oc}}{V_{oc}^2} = \frac{195}{(2400)^2} = 0.339 \times 10^{-4}\ \text{mho}$$

$$B_m = \sqrt{Y_\phi^2 - G_c^2} = \sqrt{(18 \times 10^{-4})^2 - (0.339 \times 10^{-4})^2}$$
$$= 17.99 \times 10^{-4} \text{ S}$$

$$\cos \psi = \frac{P_{oc}}{V_{oc} I_{oc}} = \frac{195}{2400(4.31)} = 0.0189$$

Figure 3.13 gives the parameters calculated. The approximate calculations using the circuit parameters referred to the primary side:

$$I_2' = \frac{W_o}{V_2} = \frac{40{,}000}{2400} = 16.67 \text{ A}$$

$$P_T = R_{eq}|I_2'|^2 = 1.024(16.67)^2 = 284.56 \text{ W}$$

$$P_c = G_c|V_{sc}|^2 = (0.339 \times 10^{-4})(2400)^2 = 195.26 \text{ W}$$

$$P_o = |W_o| \cos \theta_L = 40{,}000(0.8) = 32{,}000 \text{ W}$$

$$P_1 = P_T + P_c + P_o = 284.56 + 195.26 + 32{,}000 = 32{,}479.76 \text{ W}$$

Efficiency:

$$\eta = \frac{P_{out}}{P_1}(100) = \frac{32{,}000 \times 100}{32{,}479.76} = 98.52\%$$

3. Voltage regulation equation (3.32)

$$\text{Voltage regulation (VR)} = \frac{V_1 - aV_2}{aV_2}$$

Using Fig. 3.14, we get

$$V_1 = I_2(1.024 + j1.51) + aV_2$$

But

$$I_2 = \frac{40{,}000}{240} = 166.67 \text{ A}$$

$$I_2' = \frac{I_2}{a} = \frac{166.67}{10} = 16.67 \text{ A}$$

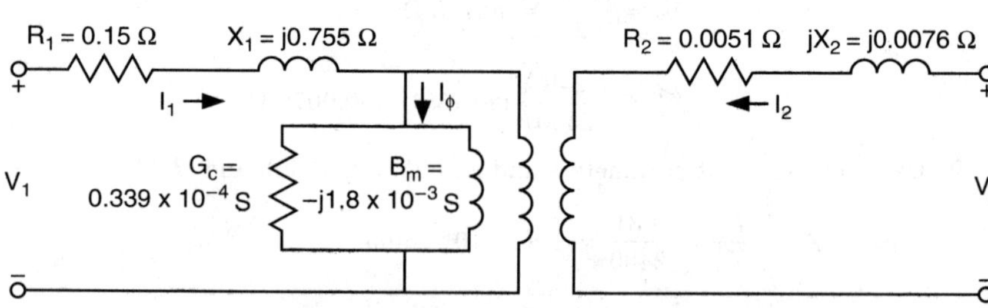

Figure 3.13 An impedance diagram for Example 3.4, as given in Fig. 3.4.

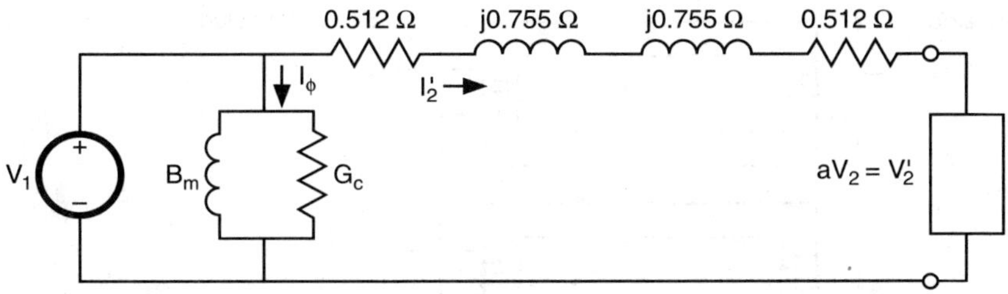

Figure 3.14 An equivalent circuit of a transformer referred to the primary side, for Example 3.4.

For 0.8 power factor* lagging, we have

$$I_2 = a(16.67\angle -36.8°) = 166.67\angle -36.8° \text{ A}$$

Therefore,

$$V_1 = (166.67\angle -36.8°)(1.024 + j1.51) + 2400$$

$$= 2647.3 + j99.47 = 2649.17\angle -2.15° \text{ V}$$

Therefore, voltage regulation is

$$VR = \frac{2649.17 - 2400}{2400} \times 100\% = 10.38\%$$

3.3.5 Three-phase transformer

The preceding transformer analysis was mainly centered on single-phase transformers that operate either over a range of frequencies, as in communication systems, or a single-phase frequency, as in single-phase power systems. Since electric energy is generated and transmitted in the form of three-phase voltages, it is necessary to discuss three-phase transformers. According to their construction, three-phase transformers are divided into two types:

1. Three separate single-phase transformers properly connected, as shown in Fig. 3.15
2. Single core on which three sets of windings are placed in the proper manner, as shown in Fig. 3.16

Three single-phase transformers can be connected into a three-phase bank. The primary and secondary windings may be connected either in wye (star) or delta. There are four common connections; these are wye-wye, wye-delta, delta-wye and delta-delta, and open-delta or V-V

*The reader who has no knowledge of power factor should refer to the Appendix.

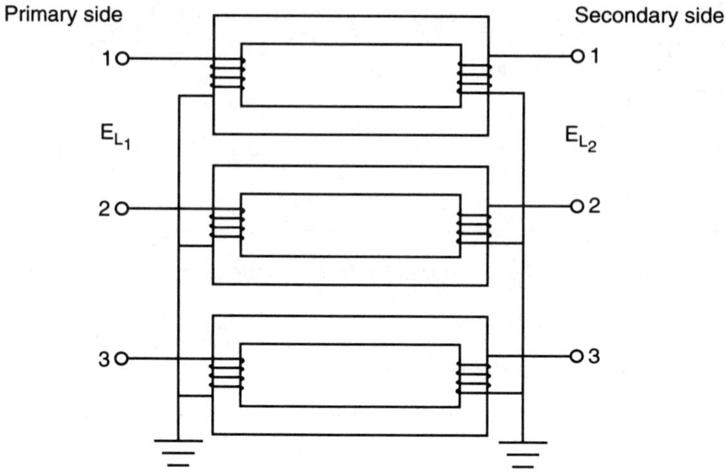

Figure 3.15 Three separate single-phase transformers connected into a three-phase bank.

connection. These five types of connections are each discussed in the following sections.

Wye-wye transformer connection. Figure 3.17 shows three single-phase transformers with their primaries and secondaries connected in wye. The primary voltage on each phase of the transformer is given by the following relation: $V_{\phi 1} = V_{L1}/\sqrt{3}$. The primary phase voltage is related to secondary phase voltage by the turns ratio, $N_1/N_2 = a$, of the transformer. Likewise the phase voltage of the secondary side of the transformer is $V_{\phi 2} = V_{L2}/\sqrt{3}$.

The voltage ratio is then given by

$$V_{L1} = \frac{\sqrt{3}\ V_{\phi 1}}{\sqrt{3}\ V_{\phi 2}} = a \tag{3.41}$$

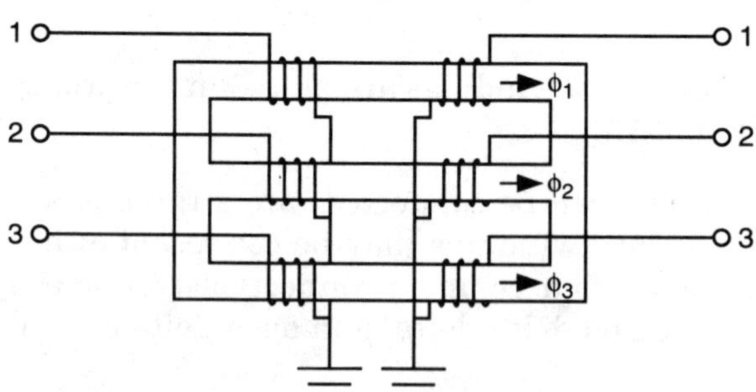

Figure 3.16 Single-core three-phase transformer.

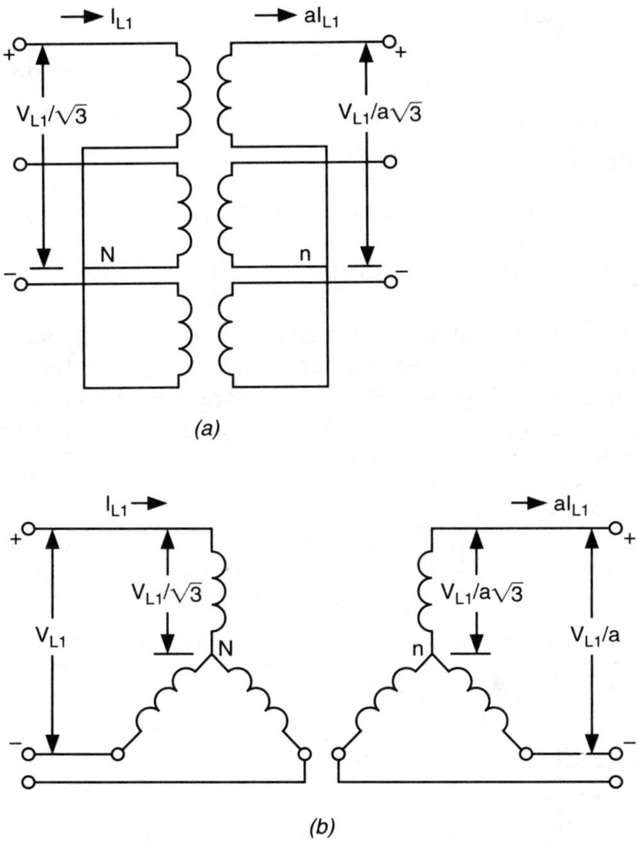

Figure 3.17 The wye-wye connection. (*a*) Physical arrangement of the three single-phase transformers. (*b*) Schematic diagram.

The Y-Y connection is seldom used because of the following drawbacks:

1. Third harmonic voltages
2. Unbalanced phase loads of the transformer

These two problems are solved by grounding the neutrals of the transformer and by adding a third or tertial winding connected in delta to the transformer bank. Under balanced three-phase conditions, the neutral conductor carries no fundamental component of the current because the fundamentals in the three phases are equal and 120° apart, and their sum is equal to zero. Thus,

$$I_N = I_a \angle 0^\circ + I_o \angle -120^\circ + I_c \angle 120^\circ = I_a(e^{j0^\circ} + e^{-j120^\circ} + e^{j120^\circ}) = 0 \tag{3.42}$$

where under balanced conditions

$$I_a = I_b = I_c \tag{3.43}$$

However, the third harmonics are displaced from each other by 3 × 120° = 360°. Therefore,

$$I_N = I_a(e^{j0^\circ} + e^{-j360^\circ} + e^{j360^\circ}) = 3I_a \tag{3.44}$$

This is true for multiples of third harmonics.

Wye-delta connections. The wye-delta connection is shown in Fig. 3.18. In this figure the primary line voltage of the transformer is related to the primary phase voltage by $V_{L1} = \sqrt{3}V_{\phi 1}$, while the secondary line voltage is equal to secondary phase voltage $V_{L2} = V_{\phi 2}$. The ratio of the line voltages to the phase voltages is given by

$$\frac{V_{L1}}{V_{L2}} = \frac{\sqrt{3}V_{\phi 1}}{V_{\phi 2}} = a\sqrt{3} \tag{3.45}$$

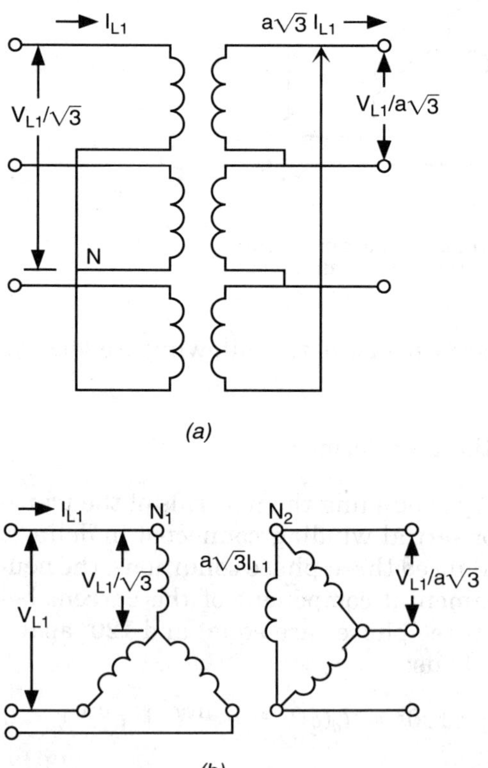

Figure 3.18 Wye-delta connection. (*a*) Physical arrangement of three single-phase transformers. (*b*) Schematic diagram.

The wye-delta connection does not contain third harmonics because the harmonics are consumed by the circulating current in the delta-connected side of the transformer.

Delta-wye connection. Figure 3.19 shows a delta-wye bank of transformers. Such connection is found in distribution networks where the delta side is connected in the high-voltage transmission network and the wye side is connected to provide low-voltage users. The single-phase loads are connected between line and neutral and three-phase motors are connected between lines.

In the delta-wye connection the primary line voltage is equal to the primary phase voltage $V_{L1} = V_{\phi 1}$, while the line voltage of the secondary is equal to the cube root times the phase voltage of the secondary side of the transformer ($V_{L2} = \sqrt{3}\, V_{\phi 2}$).

The ratio between the line-to-line voltages of the primary and sec-

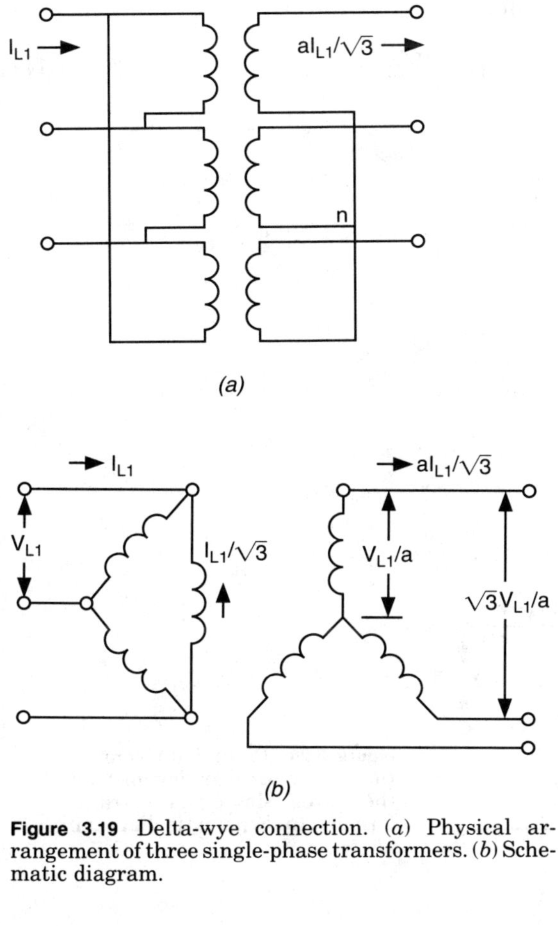

Figure 3.19 Delta-wye connection. (*a*) Physical arrangement of three single-phase transformers. (*b*) Schematic diagram.

ondary side of the transformer is given by

$$\frac{V_{L1}}{V_{L2}} = \frac{V_{\phi 1}}{\sqrt{3}\, V_{\phi 2}} = \frac{a}{\sqrt{3}} \tag{3.46}$$

Delta-delta connection. The delta-delta connection has the advantage that one transformer may be removed for repair or maintenance while the remaining two continue to function as a three bank with the rating reduced to 58 percent, at unity power factor of the original bank. This type of connection is known as open delta or Y connection. Under this condition the transformer impedances introduce a slight unbalance to the system. This connection is also useful when a growing load is anticipated, that is, a three-phase transformer bank is designed with the anticipated maximum load, but two units are installed initially. Figure 3.20 shows the delta-delta connection. In this connection the line voltage is equal to the phase voltage of the primary and secondary side of the transformer, therefore,

$$V_{L1} = V_{\phi 1} \qquad V_{L2} = V_{\phi 2} \tag{3.47}$$

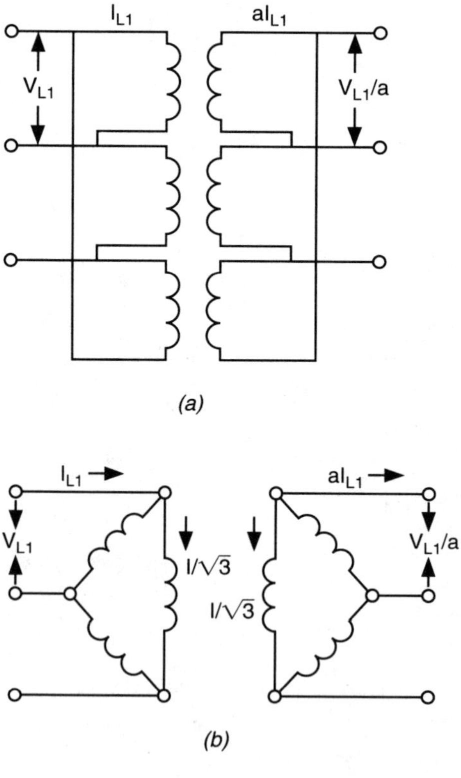

Figure 3.20 Delta-delta connection. (*a*) Physical arrangement of the three single-phase transformers. (*b*) Schematic diagram.

Therefore, the relationship between the primary and secondary line voltages is

$$\frac{V_{L1}}{V_{L2}} = \frac{V_{\phi 1}}{V_{\phi 2}} = a \tag{3.48}$$

The delta-delta connected transformer has no phase shift associated with it and has no problems with unbalanced loads or harmonics.

Example 3.5 Three single-phase transformers connected in delta-wye are each rated 100 kVA, 24,000/2400 V, and 60 Hz. The high-voltage windings are connected in delta and the low-voltage winding is connected in wye. Calculate the rated line and phase voltages and currents on both sides of the transformer.

solution The voltage and currents on the delta connection are

$$V_{\phi 1} = V_{\text{Line 1}} = 24{,}000 \text{ V}$$

$$I_{\phi 1} = \frac{100 \times 10^{-3}}{2400} = 4.17 \text{ A}$$

$$I_{\text{Line 1}} = \sqrt{3}\, I_{\phi 1} = \sqrt{3}(4.17) = 7.22 \text{ A}$$

The voltages and currents on the wye connection are

$$V_{\phi 2} = \frac{V_{\phi 1}}{a} = \frac{24{,}000}{(24{,}000/2400)} = 2400 \text{ V}$$

$$V_{L2} = \frac{\sqrt{3} V_{\phi 1}}{a} = 4156.9 \text{ V}$$

$$I_{\phi 1} = I_{L2} = \frac{100 \times 10^{3}}{2400} = 41.67 \text{ A}$$

Open delta or *V*-*V* connection. The delta-delta connection, also called V-V connection, is usually used in moderate-voltage systems, since the windings operate at full line-to-line voltage. In the case of a bank consisting of three single-phase transformer units, one of the advantages of the delta-delta connection is that if one transformer breaks down it can be removed from the circuit and the rest of the power is shared by the remaining two transformers. Figure 3.21 shows the open delta connection. This connection is also used initially in growing load centers, the full growth of which may take several years, at which time a third unit is added to the existing open delta connection to make it into delta-delta.

Disconnecting one transformer from Fig. 3.20*b* gives us Fig. 3.21, the secondary line voltages do not change, V_{ab} and V_{bc} are the same as before, and V_{ca} is determined by phasor addition.

Performing phasor addition on Fig. 3.22, we find that V_{ca} is the same whether connected in delta-delta or open delta. Since the three sec-

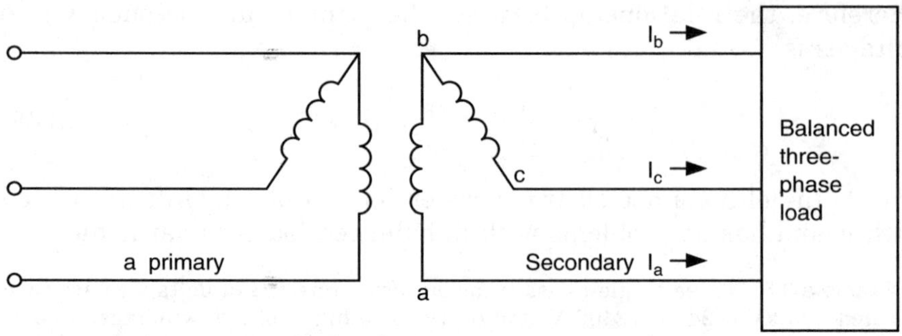

Figure 3.21 Open delta or V-V connection.

ondary line voltages are the same whether connected in delta-delta or open delta and the load impedance does not change, line currents I_a, I_b, and I_c must also be the same whether operating as delta-delta or open delta.

In Fig. 3.22, the currents in the two remaining transformers must increase to equal the line currents, that is, $I_{ab} \rightarrow I_a$ and $I_{bc} \rightarrow I_c$. In this

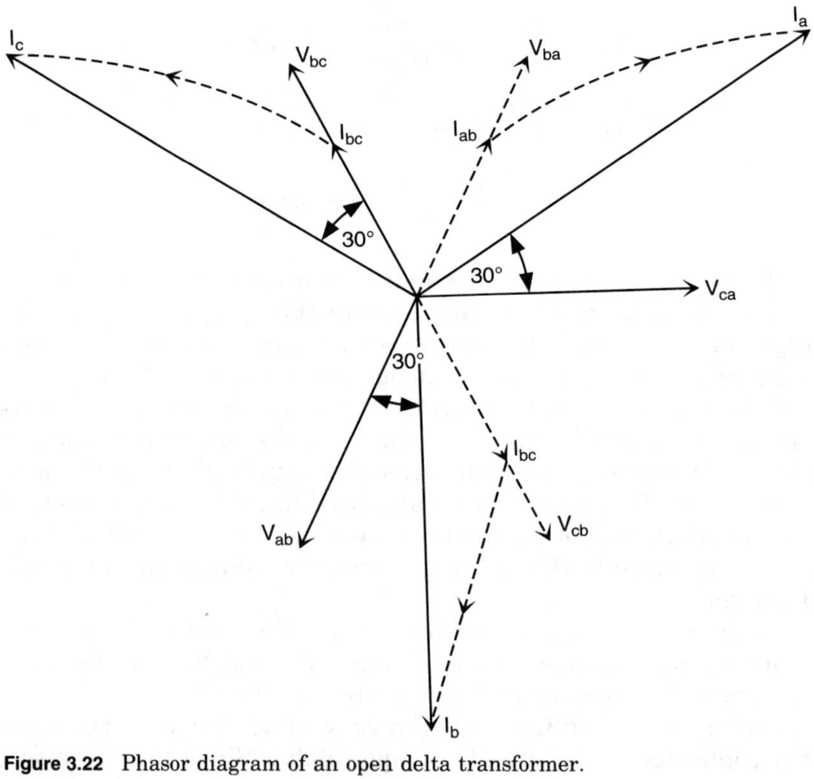

Figure 3.22 Phasor diagram of an open delta transformer.

figure it shows that phase current I_{bc} increases in magnitude and shifts its phase by 30° to coincide with that of line current I_c, and phase current I_{ab} increases in magnitude and shifts its phase by 30° to coincide with that of line current I_a. Therefore, if a delta-delta bank is operating at a rated load with one transformer removed, the current in the two remaining transformer coils will increase to 1.7321 times its normal rating. To prevent overheating when operating open-delta the transformer bank current and the bank apparent power must be rerated to reflect the lower kVA capacity.

$$I_{\text{open}\,\Delta} = \frac{I_{\Delta\text{-}\Delta\,\text{rated}}}{\sqrt{3}} = 0.5774 I_{\Delta\text{-}\Delta}$$

The transformer bank rating when connected open delta is

$$S_{V\text{-}V\,\text{rated}} = \frac{S_{\Delta\text{-}\Delta\,\text{rated}}}{\sqrt{3}} = 0.5774 S_{\Delta\text{-}\Delta\,\text{rated}}$$

3.4 Summary

In this chapter the fundamental principles of the power transformer have been introduced and discussed in detail. The analytical methods to calculate parameters of single-phase and three-phase transformers have been examined. The effect of power transformers will be encountered during discussion on load flow in electric power systems. Selective transformer problems and worked-out examples have been included.

Problems

1 A single-phase transformer rated 20 kVA, 2400/240 V, 60 Hz has an approximate equivalent circuit with the following parameters: $Y_{f1} = G_{c1} + jB_{m1} = 14.2 - j30.1$ S and $Z_{e1} = R_{e1} + jX_{e1} = 7.5 + j12.6\ \Omega$. Find the voltage regulation at a 0.8 power factor lagging.

2 The result of open-circuit and short-circuit tests on a 20-kVA, 4400/220-V, 60-Hz single-phase transformer are as follows:

Open-circuit test with the input to the low-voltage side: $V_{oc} = 220$ V, $I_{oc} = 25$ A, and $P_{oc} = 230$ V.

Short-circuit test with the input to the high-voltage side: $V_{sc} = 100$ V, $I_{sc} = 16$ A, and $P_{sc} = 1500$ W.

Find the equivalent circuit parameters R_{e1}, X_{e1}, G_{c1}, and B_{m1}.

3 A 200-kVA, 1100/2200-V, 60-Hz single-phase transformer has the following equivalent circuit parameters referred to the high-voltage side: $R_1 = 4.1$

Ω, $X_1 = 30$ Ω, $X_m = 60.2$ Ω, $R_2 = 6.1$ Ω, $X_2 = 30$ Ω, $R_c = 120$ Ω. The transformer is supplying a load circuit of 40 Ω impedance at 2200 V and a lagging loading power factor of 0.7. Calculate the voltage and power factor at high-voltage terminals of the transformer.

4 A 100-kVA, 1100/2200 V, 60-Hz single-phase power transformer gave the following test results:

Open-circuit test, low-voltage winding excited: $V_{oc} = 2200$ V, $I_{oc} = 1.28$ A, $P_{oc} = 950$ W.

Short-circuit test high-voltage winding excited: $V_{sc} = 550$ V, $I_{sc} = 8.91$ A, $P_{sc} = 1200$ W.

a. Find the equivalent circuit of the transformer referred to the low-voltage side.
b. Find the voltage regulation of the transformer when supplying rated load at 2200 V and a lagging power factor of 0.8.

5 A 10-kVA single-phase transformer has a primary voltage of 440 V and a rated secondary voltage of 120 V (Fig. 3.23). Determine
a. The turns ratio
b. Rated secondary current
c. Transformer core losses
d. Transformer efficiency at half load

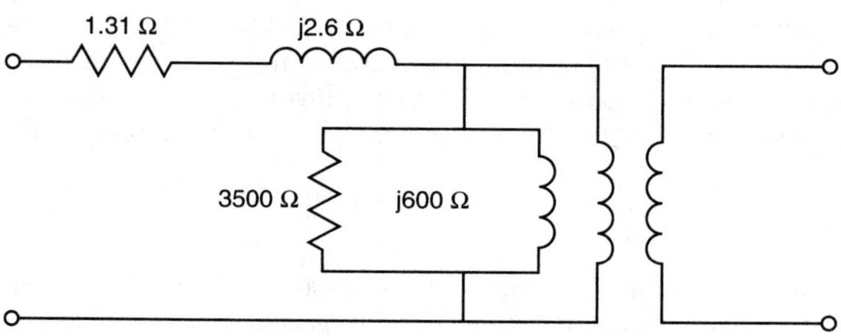

Figure 3.23 Transformer equivalent circuit referred to the primary winding for Prob. 5.

6 A 500-kVA, 2400/120-V, 60-Hz single-phase transformer has no-load loss of 150 W and a short-circuit loss of 7000 W. Determine the efficiency of the transformer under the following conditions of load:
a. Rated current at 0.8 power factor lagging
b. 300 kW at 0.8 power factor leading
c. 100 kW at 0.8 power factor lagging

7 A balanced three-phase load of 330 kVA, 440 V, is connected to a 2400/440-V, three-phase system by a delta-wye bank of transformers. Determine the current, voltage, and kilovoltampere rating of the winding for each transformer.

8 Three single-phase transformers are connected to make a three-phase delta-wye bank. The line voltages of the bank are 2400 V and 240 V on the primary and secondary, respectively. The load on the transformer is 20 kW, 0.8 power factor lagging wye-connected. Calculate the primary and secondary line and phase currents.

9 A 150-kVA, 2200/440-V wye-wye three-phase 60-Hz transformer bank has an iron loss of 1400 W and a copper loss of 250 W at full load. Find the efficiency of the transformer for 65 percent full load at 0.8 power factor.

10 A three-phase 80-kVA, 220-V, 0.8 lagging power factor lead is to be supplied by single-phase transformers having a turns ratio of 2. Find the currents in the windings of each transformer and the corresponding power factor if the transformers are connected in

a. Open delta

b. Delta-delta

Bibliography

Bergeseth, F. R., and S. S. Venkata, *Introduction to Electric Energy Devices,* Prentice-Hall, Englewood Cliffs, N.J., 1987.

Lindsay, J. F., and M. H. Rashid, *Electromechanics and Electrical Machinery,* Prentice-Hall, Englewood Cliffs, N.J., 1986.

Mablekos, Van E., *Electric Machine Theory for Power Engineers,* Harper and Row, New York, 1980.

McLaren, P. G., *Elementary Electric Power and Machines,* Ellis Horwood Limited Publishers Chichester halsted Press a division of John Wiley and Sons, New York, 1984.

Matsch, L. W., and J. D. Morgan, *Electromagnetic and Electromechanical Machines,* Harper and Row, New York, 1986.

Nasar, Syed A., *Electric Machines and Transformers,* Macmillan, New York, 1984.

Chapter

4

Synchronous Machines

4.1 Introduction

In electric power systems, the bulk of electric power is produced by three-phase synchronous generators. These machines are called synchronous machines because they operate at nearly constant speeds and constant frequencies under steady-state conditions. Like most rotating machines, synchronous machines can be operated both as motors and as generators. The operation of synchronous machines is based on Faraday's law of electromagnetic induction discussed in Chap. 2. The generation of emf is made possible by the relative motions of conductors and magnetic flux. The synchronous machine has two major parts, the magnetic field carrying a dc excited winding and the armature.

The armature winding of synchronous generators is on the stator and the field winding is on the rotor. The field is excited by a direct current conducted by means of brushes bearing on the slip rings.

4.2 Synchronous Generator Construction Features

Synchronous generators fall into two general classifications:

1. Cylindrical rotor synchronous machines
2. Salient-pole synchronous machines

The cylindrical rotor generators are driven by fast steam turbines and are known as turbogenerators. Figure 4.1 shows a two-pole cylindrical rotor synchronous machine. The salient-pole rotors are used for low-speed synchronous generators. These machines are driven by water

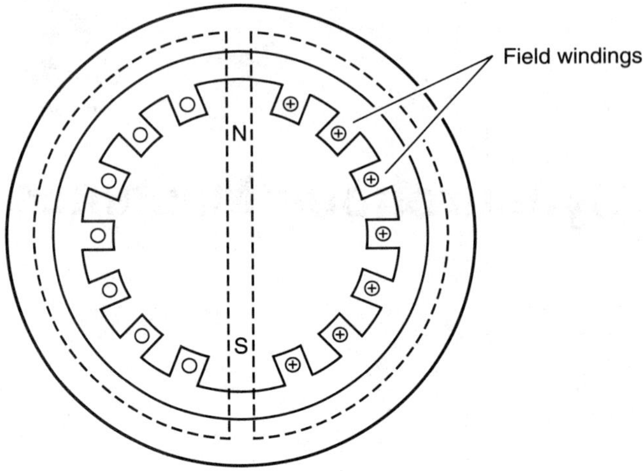

Figure 4.1 A two-pole cylindrical rotor synchronous machine.

turbines (water wheels). Figure 4.2 shows a 4-pole salient-pole synchronous machine.

In the introductory remark it was stated that the dc excited winding is placed on the rotating structure of the rotor which produces magnetic flux. It is customary to mount a dc exciter on the same shaft as that of the synchronous machines. The direct current is then fed to the field

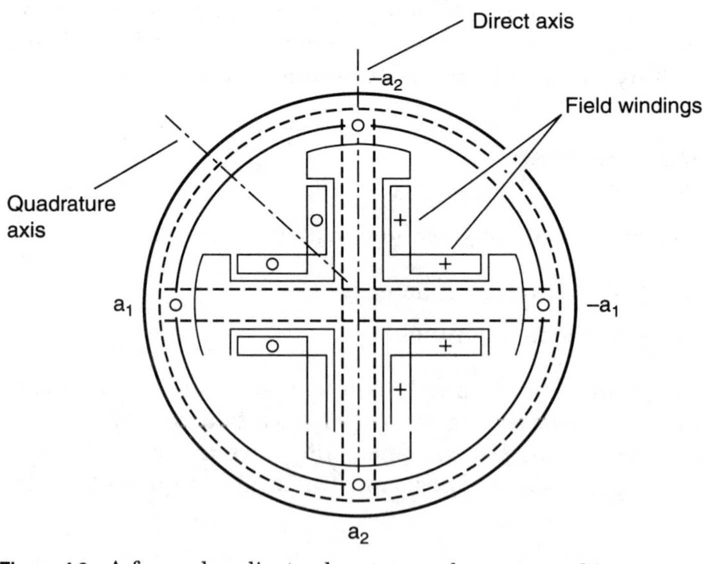

Figure 4.2 A four-pole salient-pole rotor synchronous machine.

winding of the synchronous machine. Large ac generators usually have exciters consisting of an ac source with solid-state rectifiers.

The relationship between the synchronous speed and frequency is given by

$$n_{sy} = \frac{120f}{P} \tag{4.1}$$

where f is the frequency of the induced voltage. The motor speed is directly proportional to the generator's frequency, and inversely proportional to the number of poles.

Therefore the speed of the synchronous machine depends on the number of poles. A two-pole synchronous machine, for instance, develops a speed of 3600 rpm at 60 Hz while a four-pole machine develops a speed of 18,000 rpm. This shows that the rotor structure depends on the speed rating of the machine.

For high-speed synchronous machines the cylindrical rotor construction is more appropriate because it can withstand the centrifugal force developed in it. The speed range of 1800 to 3600 rpm is common to steam turbines, gas turbines, and nuclear turbines.

For low-speed synchronous machines the salient-pole rotor is used. Such machines are driven by slow-speed prime movers such as hydroturbines or water wheels. The salient-pole synchronous machines have a large number of poles. In order to obtain a large number of poles a large diameter is commonly used. The speed range can reach as low as 100 rpm for a 72-pole machine.

When the synchronous machines have more than two poles as shown in Fig. 4.2 the field coils are connected in such a way that the poles are of alternate north and south polarity. There are two complete wavelengths or cycles in the flux distribution around the periphery. In Fig. 4.2 the armature winding consists of two coils, a_1-a' and a_2-a_2', connected in series by their end connections. The span of each coil is ½ wavelength of the flux, therefore the generated voltage goes through two complete cycles per revolution of the rotor. The frequency then is twice the speed in revolutions per second. It is convenient therefore to express angles in electrical degrees instead of expressing it in mechanical degrees. One pair of poles in a P-pole machine or cycle of flux distribution equals 360 electrical degrees or 2π electrical radians. Since there are $P/2$ complete wavelengths or cycles in one revolution, it follows that

$$\alpha_{el} = \frac{P}{2}\,\alpha_m \tag{4.2}$$

Figure 4.3 shows the relationship between mechanical degrees and electrical degrees of the space distribution of flux density in a four-pole synchronous generator.

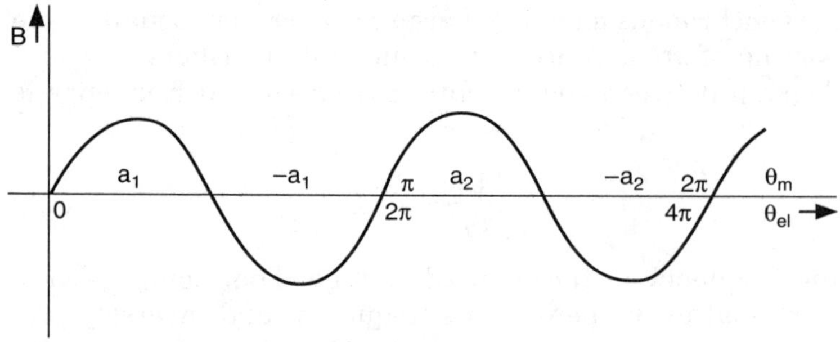

Figure 4.3 Relationship between mechanical and electrical degrees.

4.3 AC Three-Phase Armature Windings

The armature winding in ac machines is the source of the induced voltage. The armature windings are interconnected so that their electrical and magnetic effects are additive or cumulative. The coils may have full pitch or fractional pitch. A full pitch spans 180 electrical degrees while the fractional pitch spans less than 180 electrical degrees, but seldom less than 120 electrical degrees. In this section, only a full pitch machine will be discussed.

Since the main concern here is the energy conversion, all ac machines are built as three-phase machines. Three-phase winding results from the addition of another two sets of the armature coils displaced 120 and 240 electrical degrees from the first phase. This connection produces a system of three-phase voltages equal in magnitude but displaced from each other by 120 electrical degrees. Figure 4.4*a* and *b* shows the phasor diagrams of the positive sequence *abc* and the negative sequence *acb*, and Fig. 4*c* shows wye and delta connections of synchronous machines.

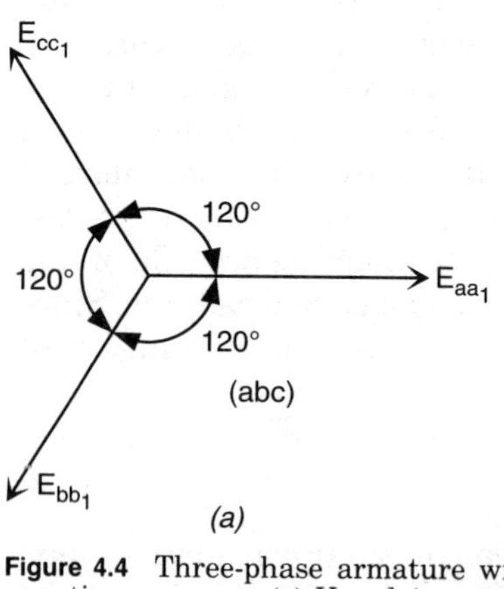

Figure 4.4 Three-phase armature winding development. (*a*) Positive sequence; (*b*) negative sequence; (*c*) Y and Δ connections.

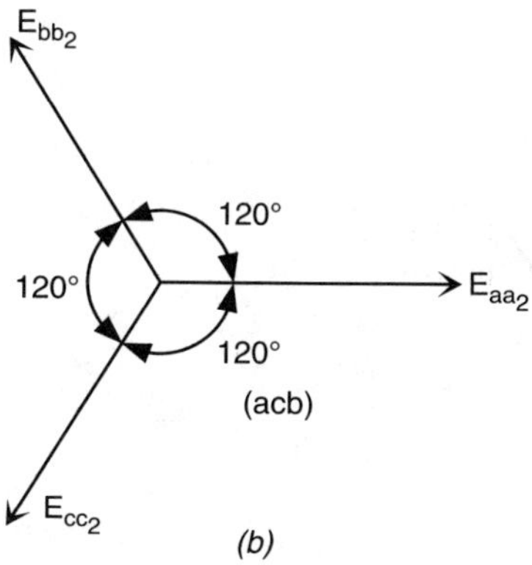

(b)

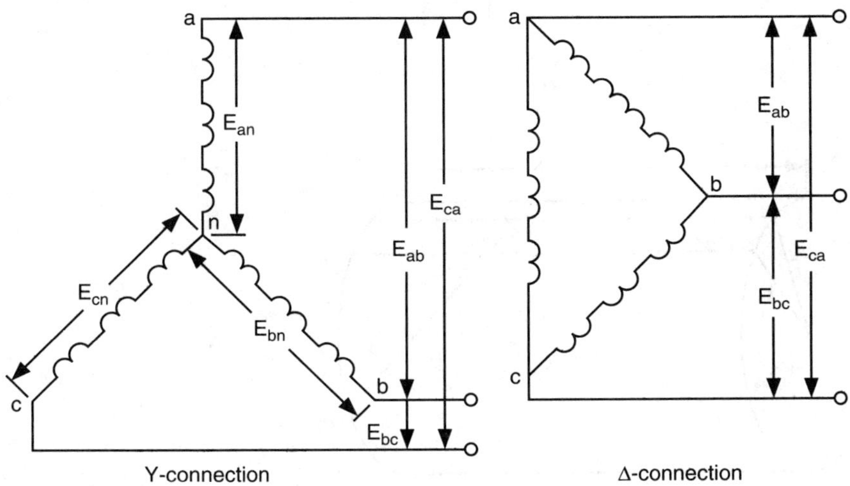

(c)

Figure 4.4 *(Continued)*

4.4 Induced Voltages in the Stator Windings of Synchronous Machines

The stator winding of the synchronous generator is the source of all the ac energy utilized by industry and by household appliances of modern society. The synchronous generator is universally used by all electric power industries for supplying three-phase, as well as single-phase, power to consumers. The synchronous motor, however, is used for power-factor correction. The main task here is to develop expressions for induced voltages on the periphery of the stator. In Fig. 4.5*a* the *N* and *S* poles and the rotor field plot *B* move to the right while

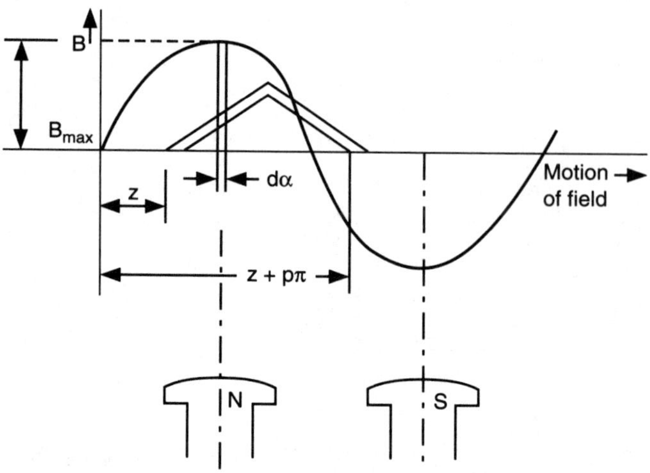

(a)

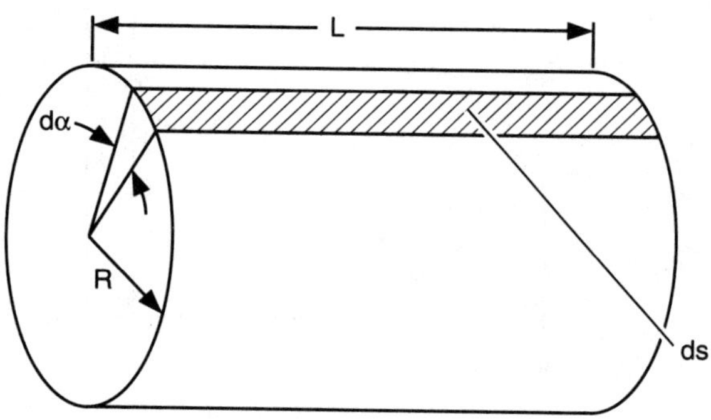

(b)

Figure 4.5 (*a*) A concentrated winding of a three-phase synchronous generator in two poles. (*b*) Stator structure.

stator windings are stationary. The relative angular position of the rotor with respect to the stator winding is Z. In Fig. 4.5*b*, L is the effective axial length of the stator, R is the inner radius of the stator, and α_m is the mechanical angle between 0 and 360 degrees. This mechanical angle is related to the electrical angle of the magnetic field density B by Eq. (4.2), that is,

$$\alpha_m = \frac{2}{P}\alpha_{el} \tag{4.3}$$

The flux density at a given point is given by

$$B = B_{\max} \sin \alpha_{\mathrm{el}} \tag{4.4}$$

The incremental area ds is given by

$$ds = RL\, d\alpha_m \tag{4.5}$$

Therefore, the flux $d\phi$ of the stator coil is given by

$$d\phi = B\, ds = RLB_{\max} (\sin \alpha_{\mathrm{el}})\, d\alpha_m \tag{4.6}$$

Substituting Eq. (4.3) into Eq. (4.6), we have

$$d\phi = RLB_{\max} (\sin \alpha_{\mathrm{el}})\left(\frac{2}{P}\right) d\alpha_{\mathrm{el}} = \frac{2RL}{P} B_{\max} \sin \alpha_{\mathrm{el}}\, d\alpha_{\mathrm{el}} \tag{4.7}$$

Integrating Eq. (4.7) from Z to $Z + p\pi$ we have

$$\phi = \int_Z^{Z+p\pi} \frac{2RL}{P} B_{\max}(\sin \alpha_{\mathrm{el}})\, d\alpha_{\mathrm{el}}$$

$$= \left(\frac{2RL}{P}\right) B_{\max}[\cos Z - \cos (Z + p\pi)] \tag{4.8}$$

Using the trigonometric identity, Eq. (4.8) becomes

$$\phi = \left(\frac{4RL}{P}\right) B_{\max}\left[\sin \left(Z + p\frac{\pi}{2}\right) \sin \left(p\frac{\pi}{2}\right)\right] \tag{4.9}$$

Let

$$\lambda = N\phi \tag{4.10}$$

Therefore substituting Eq. (4.9) into Eq. 4.10, we have

$$\lambda = \frac{4RLN}{P} B_{\max} \left[\sin \left(p\frac{\pi}{2}\right) \sin \left(Z + p\frac{\pi}{2}\right)\right] \tag{4.11}$$

where N is the number of stator coils and P is the pitch. Let

$$Z = -\omega t \tag{4.12}$$

Substituting Eq. (4.12) into Eq. (4.11) we get

$$\lambda = \left(\frac{4RLN}{P}\right) B_{\max} \left[\sin \left(\frac{p\pi}{2}\right) \sin \left(\omega t - \frac{p\pi}{2}\right)\right] \tag{4.13}$$

Using Faraday's law of electromagnetic induction we have

$$e(t) = -\frac{d\lambda}{dt} = \frac{4RLN\omega}{P} B_{\max} \left[\sin\left(\frac{p\pi}{2}\right) \cos\left(\omega t - \frac{p\pi}{2}\right) \right] \quad (4.14)$$

where sin $(p\pi/2)$ is the pitch factor. It is symbolized by K_p:

$$K_p = \sin\left(\frac{p\pi}{2}\right) \quad (4.15)$$

Substituting $4RLB_{\max}/P = \phi$ into Eq. (4.14) we get

$$e(t) = \omega N_\phi \phi K_p \cos\left(\omega t - \frac{p\pi}{2}\right) \quad (4.16)$$

Setting $\omega = 2pf$, the effective value of the coil voltage is

$$|E_1| = \frac{2\pi f N_\phi \phi K_p}{\sqrt{2}} = 4.44 f K_p N_\phi \phi \quad (4.17)$$

4.5 Voltage Induced in a Distributed Winding

A *distributed winding* is defined as the winding in which each phase occupies more than one slot per pole. The flux linkages of coils occupying different slots under a given pair of poles differ because the angle α must have different values for the different coils. The resultant flux linkage of a series-connected coil group may be found by phasor addition as long as the flux density is distributed sinusoidally.

The ratio of the magnitude of the vector sum of the voltage (or the phasor sum) to the arithmetical sum is called the breadth factor and the formula of the factor is given by

$$K_b = \frac{E_{\text{group}}}{nE_{\text{coil}}} \quad (4.18)$$

where E_{group} = phasor sum of the individual voltage induced in each coil, as shown in Fig. 4.6*c*

E_{coil} = coil voltage, as shown in Fig. 4.6*c*

n = number of coils per pole and phase

From Fig. 4.6*a* the phasors *AB*, *BC*, *CD*, and *DE* represent the coil voltage of the group or the resultant phasor voltage of the individual coil voltages. The angle occupied under one pole by one phase is rep-

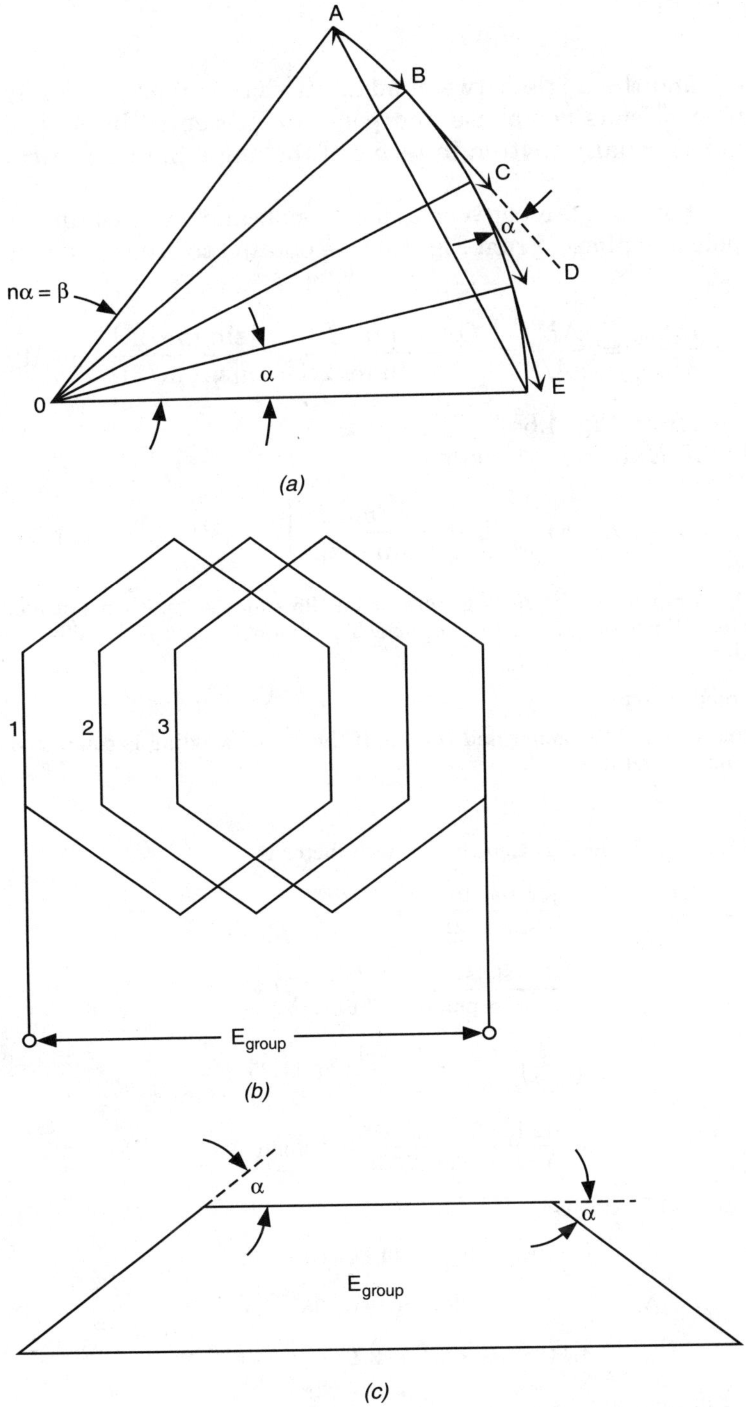

Figure 4.6 (*a*) Phasor diagram for obtaining breadth factor. (*b*) Group of three coils in a distributed winding. (*c*) Phasor diagram of induced voltage.

resented by β and the angle between adjacent slots is represented by α. The number of slots per phase and pole is represented by n. The breadth factor is usually written in terms of the angle β, in electrical degrees.

Figure 4.6*a* shows angle α between adjacent slots and n is the number of slots per pole and phase. From Fig. 4.6*a* we obtain the breadth factor as

$$K_b = \frac{E_{\text{group}}}{nE_{\text{coil}}} = \frac{AE}{nAB} = \frac{OA[\sin(\beta/2)]}{nOA[\sin(\alpha/2)]} = \frac{[\sin(n\alpha/2)]}{n[\sin(\alpha/2)]} \qquad (4.19)$$

where $\beta = n\alpha$, from Fig. 4.6*a*.

The total emf (E) in rms value is

$$|E| = |E_1|\,K_b = |E_1|\left[\frac{\sin(n\alpha/2)}{n\sin(\alpha/2)}\right] \qquad [\text{V}] \qquad (4.20)$$

Example 4.1 A 6-pole synchronous generator has 96 stator slots. Each slot has two conductors. Each slot spans 180 degrees. The magnetic flux is 3.1 Wb per pole. Calculate

1. The breadth factor
2. The rms value of the generated voltage if the stator winding is connected as a three-phase winding.

solution

1. From Eq. (4.18) and Fig. 4.5*a* the breadth factor is

$$K_b = \frac{\sin(n\alpha/2)}{n\sin(\alpha/2)}$$

$$n = \frac{\text{slots}}{\text{pole} \times \text{phase}} = \frac{96}{3(6)} = 5.33$$

$$\alpha_e = \left(\frac{P}{2}\right)\frac{\alpha_m}{\text{slots}} = \frac{6}{2}\left(\frac{360}{96}\right) = 11.25$$

$$K_b = \frac{\sin[5.33(11.25/2)]}{5.33\sin(11.25/2)} = 0.9565$$

2. From Eq. (4.17) we have

$$|E_1| = 4.44fK_pN_\phi\Phi = 4.44(60)(1)(N_\phi)(3.1)$$

$$N_\phi = nPN_c = 5.33(6)(2) = 63.96 \approx 64$$

$$|E_1| - 4.44 \times 60 \times 64 \times 3.1 = 52.34 \text{ kV}$$

The total emf in rms value is

$$|E| = |E_1|\,K_b = 52.34(0.9565) = 50.067 \text{ kV}$$

4.6 The Steady-State Equivalent Circuit for a Cylindrical-Rotor Synchronous Machine

The cylindrical-rotor synchronous machine may be regarded as a smooth air-gap machine. Figure 4.7 shows a schematic diagram of a two-pole three-phase cylindrical-rotor synchronous machine.

In Fig. 4.7 each pair of conductors, a-a', b-b', and c-c', produces sinusoidally distributed mmf around the air gap. When the machine is operating at synchronous speed, the position of the rotor winding at any instant is

$$\theta = \omega t + \theta_0 \text{ electrical radians} \tag{4.21}$$

Expressing the flux linkages with the armature phases a,b,c, and field winding f in terms of inductances and currents, the induced voltages can be found by using the electromagnetic law of induction. The equation for flux linkages, expressed in matrix form, becomes

$$\begin{bmatrix} \lambda_a \\ \lambda_b \\ \lambda_c \\ \lambda_f \end{bmatrix} = \begin{bmatrix} L_{aa} & L_{ab} & L_{ac} & L_{af} \\ L_{ba} & L_{bb} & L_{bc} & L_{bf} \\ L_{ca} & L_{cb} & L_{cc} & L_{cf} \\ L_{fa} & L_{fb} & L_{fc} & L_{ff} \end{bmatrix} \begin{bmatrix} i_a \\ i_b \\ i_c \\ i_f \end{bmatrix} \tag{4.22}$$

where like double subscripts denote self-inductances and unlike subscripts denote mutual inductances.

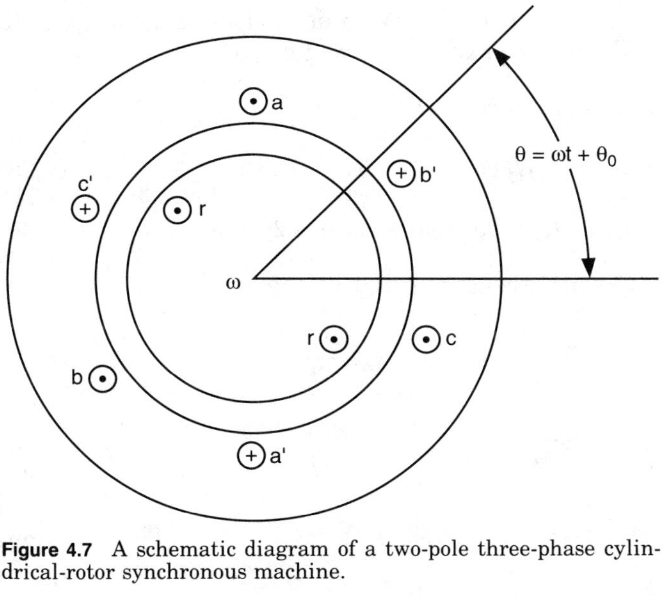

Figure 4.7 A schematic diagram of a two-pole three-phase cylindrical-rotor synchronous machine.

Investigating the inductances of the cylindrical stator we realize that the self-inductance of the field winding is independent of the rotor position 8, thus

$$L_{ff} = L_{ffo} + L_{fl} \tag{4.23}$$

where L_{ffo} is the space fundamental component and L_{fl} is the self-inductance due to field winding leakage flux.

The stator-to-rotor mutual inductance varies with the rotor position by an angle θ between the field winding and armature phase a. The mutual inductance between phase a and field winding f varies as cos θ. Therefore,

$$L_{af} = L_{af} \cos \theta \tag{4.24}$$

For phases b and c the mutual inductances are

$$L_{bf} = L_{af} \cos (\theta - 120°) \tag{4.25}$$

and

$$L_{cf} = L_{af} \cos (\theta + 120°) \tag{4.26}$$

where θ is the angle in electrical degrees.

With the rotor rotating at synchronous speed ω

$$\theta = \omega t + \theta_0 \tag{4.27}$$

where θ_0 is the rotor angle at $t = 0$. With dc excitation I_f in the field winding the flux linkages of phases a, b, and c are

$$l_{af} = L_{af} I_f \cos (\omega t + \theta_0) \tag{4.28}$$

$$l_{bf} = L_{af} I_f \cos (\omega t + \theta_0 - 120°) \tag{4.29}$$

and

$$l_{cf} = L_{af} I_f \cos (\omega t + \theta_0 + 120°) \tag{4.30}$$

The terminal voltages are also generated voltages:

$$V_{an} = e_{an} = -I_{af} \frac{dL_{af}}{dt} \tag{4.31}$$

$$V_{bn} = e_{bn} = -I_{af} \frac{dL_{bf}}{dt} \tag{4.32}$$

$$V_{cn} = e_{cn} = -I_{af} \frac{dL_{cf}}{dt} \tag{4.33}$$

$$V_f = r_f = I_f \tag{4.34}$$

Since the sinusoidal mmf and flux density distribution waves are assumed for the idealized polyphase machine, the mutual inductances are therefore given by Eqs. (4.24) to (4.26).

Substituting Eqs. (4.24) to (4.26) into (4.31) to (4.33) and performing the derivations, we have

$$e_{an} = I_{af}\omega L_{af} \sin(\omega t + \theta_0) \tag{4.35}$$

$$e_{bn} = I_{af}\omega L_{af} \sin(\omega t + \theta_0 - 120°) \tag{4.36}$$

$$e_{cn} = I_{af}\omega L_{af} \sin(\omega t + \theta_0 + 120°) \tag{4.37}$$

The effective value of E_{af} is

$$E_{af} = \frac{\omega L_f I_{af}}{\sqrt{2}} \tag{4.38}$$

Equation (4.17) gives the rms voltage generated in the armature phases by a rotating flux. It can be shown that Eqs. (4.38) and (4.17) are identical. In the steady-state sinusoids it is very convenient to use the phasor method. Equation (4.35), for instance, shows a voltage whose rms is given by Eq. (4.38).

The phase terminal voltage is the sum of the armature resistance voltage drop $R_a i_a$ and the induced voltages.

$$V_{ta} = R_a i_a + \frac{d\lambda_a}{dt} = R_a i_a + L_s \frac{di_a}{dt} + \frac{d\lambda_{af}}{dt} \tag{4.39}$$

Changing Eq. (4.39) into phasors:

$$V_{ta} = R_a I_a + jX_s I_a + E_{af} \tag{4.40}$$

where $X_s = \omega L_s$ is the synchronous reactance.

Equation (4.40) gives the cylindrical-rotor synchronous machine in motor operation as shown in Fig. 4.8. The terminal voltage in generator

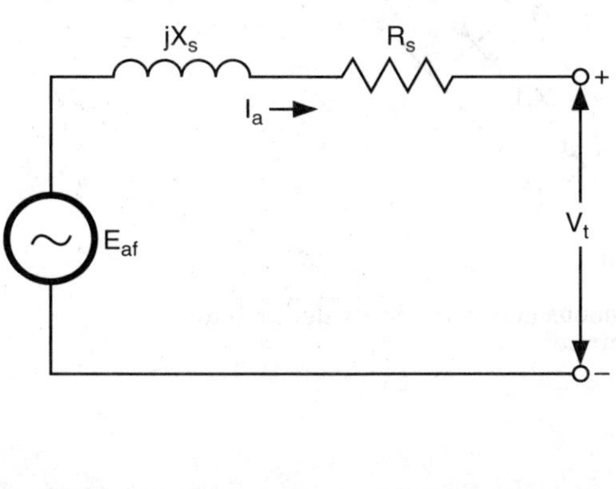

Figure 4.8 Cylindrical-rotor synchronous generator equivalent circuit.

operation is equal to

$$V_t = E_{af} - R_a I_a - jX_s I_a \tag{4.41}$$

An alternative form of synchronous machine equivalent circuit is found by splitting the synchronous reactance into magnetizing reactance X_a and the leakage reactance X_{al}. Therefore, the corresponding synchronous reactance is

$$X_s = X_{al} + X_a \tag{4.42}$$

From Fig. 4.9*a* the generated terminal voltage is

$$E_r = V_t + (R_a + jX_{al})I_a \tag{4.43}$$

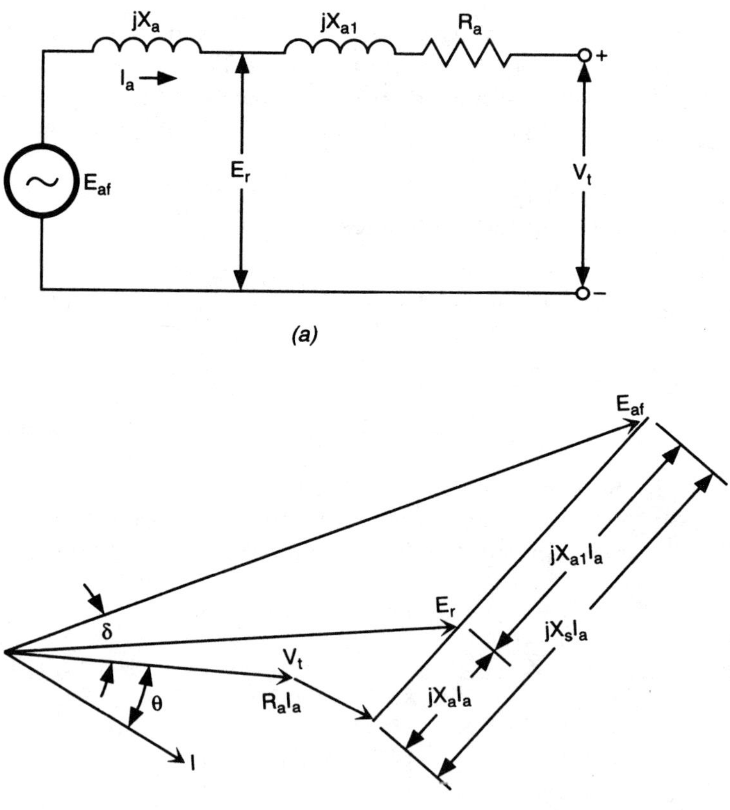

(b)

Figure 4.9 (*a*) Cylindrical-rotor synchronous generator equivalent circuit. (*b*) Phasor diagram of the equivalent circuit.

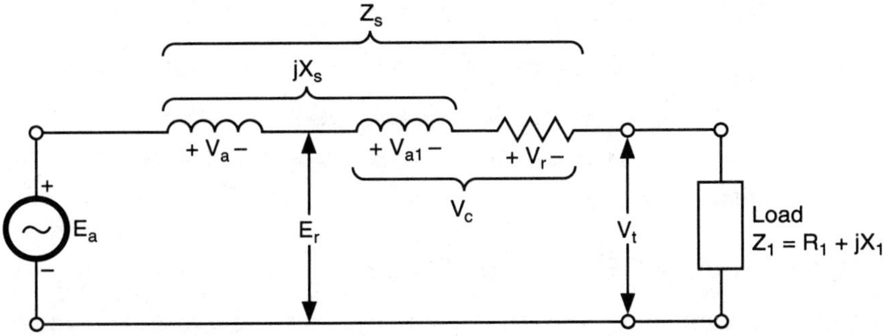

Figure 4.10 For Example 4.2.

The armature voltage or generated voltage is

$$E_{af} = V_t + R_a I_a + j(X_a + X_{al}) \tag{4.44}$$

Figure 4.9*b* shows the phasor diagram of a synchronous generator.

Example 4.2 A three-phase synchronous generator has the following per-phase parameters. The generated voltage $E_a = 1500e^{j30.60}$ V, the magnetizing reactance $X_a = 6.0\ \Omega$, the leakage reactance $X_{al} = 0.4\ \Omega$, and synchronous resistance $R_s = 0.15\ \Omega$. The load impedance $Z_L = 3.7422 + j1.8171 = 4.16e^{j25.90}$, losses neglected. Calculate

1. The synchronous reactance per phase X_s
2. The synchronous impedance Z_s
3. The stator coil impedance Z_c
4. V_a, V_{al}, V_R, V_c, V_s, V_t, and the air-gap voltage E_g

solution Using Fig. 4.10*a* we have

1. $X_s = j(X_a + X_{al}) = j(6.0 + 0.4) = j6.4\ \Omega$
2. $Z_s = R_s + jX_s = 0.15 + j6.4 = 6.4018\underline{/88.66^\circ}\ \Omega$
3. The coil impedance

$$Z_c = R_s + jX_{al} = 0.15 + j0.4 = 0.4272\underline{/69.64^\circ}\ \Omega$$

4. $$I_a = \frac{E_a/\sqrt{3}}{Z_s + Z_L} = \frac{1500/\sqrt{3}}{0.15 + j6.4 + 3.7422 + j1.8171} = 95.25\underline{/-64.65^\circ}$$

$$V_a = jX_a I_a = (6\underline{/90^\circ})(95.25\underline{/-64.65^\circ}) = 571.5\underline{/25.35^\circ}\ \text{V}$$

$$V_{al} = jX_{al} I_a = (0.4\underline{/90^\circ})(95.25\underline{/-64.65^\circ}) = 38.1\underline{/25.35^\circ}\ \text{V}$$

$$V_R = R_s I_a = (0.15)(95.25\underline{/-64.65^\circ}) = 14.29\underline{/-64.65^\circ}\ \text{V}$$

$$V_c = (R_s jX_{al})(I_a) = (0.15 + j0.4)(95.25\underline{/-64.65^\circ}) = 40.69\underline{/4.79^\circ}\ \text{V}$$

$$V_s = Z_s I_a = (6.401\angle 88.66^\circ)(95.25\angle -64.65^\circ) = 14.29\angle -64.65^\circ \text{ V}$$

$$V_t = Z_L I_a = (4.16\angle 25.9^\circ)(95.25\angle -64.65^\circ) = 396.24\angle 38.75^\circ \text{ V}$$

$$E_g = E_a - V_a = 866.03\angle 30.6^\circ - 571.5\angle 25.35^\circ = 396.24\angle 38.75^\circ$$

$$= 745.43 + j440.85 - 467.66 - j221.56 = 277.77 + j219.29$$

$$= 353.89\angle 38.28^\circ$$

4.7 Open-Circuit and Short-Circuit Tests

The magnetization curve and the air-gap line are the two synchronous machine characteristics, which determine the saturation effects and machine parameters. The following discussion will apply for both cylindrical-rotor and salient-pole synchronous machines.

4.7.1 Open-circuit characteristic

The open-circuit characteristic of a synchronous machine is obtained by driving the machine at its rated speed without load (armature terminal open). By varying the field current, the line-to-line voltages are measured and recorded. A voltage versus field current curve is plotted as shown in Fig. 4.11. If it were not for the magnetic saturation of the

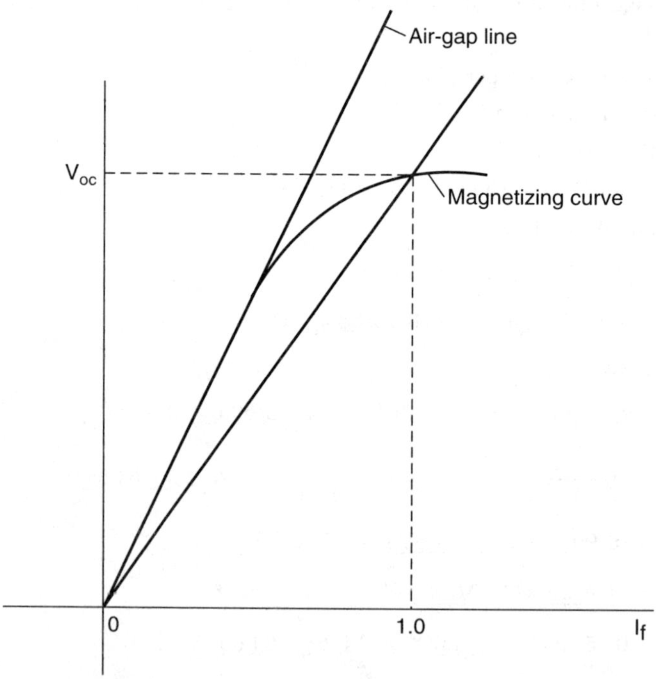

Figure 4.11 Open-circuit characteristics–volts versus field current.

iron the open-circuit characteristic would be linear as represented by the air-gap line shown in Fig. 4.11.

The no-load losses are obtained by measuring the mechanical power when the machine is running at synchronous speed. The no-load losses are the friction, windage, and core losses.

4.7.2 Short-circuit test

Figure 4.12 shows the connection for a short-circuit test. The three armature terminals of the synchronous generator are short-circuited through current-measuring instruments or current transformers with ammeters in their secondary windings. The machine is then driven at a rated speed and measurements of short-circuit currents for various values of field current are recorded. The short-circuit characteristic is linear because the iron is unsaturated up to the rated armature current. Figure 4.13 shows open-circuit and short-circuit characteristics.

In Fig. 4.13 the field current *Oa* produces the open-circuit line-to-line voltage *Oc* on the air-gap line, which would have been the open-circuit voltage if there were no saturation. The field current *Oa* pro-

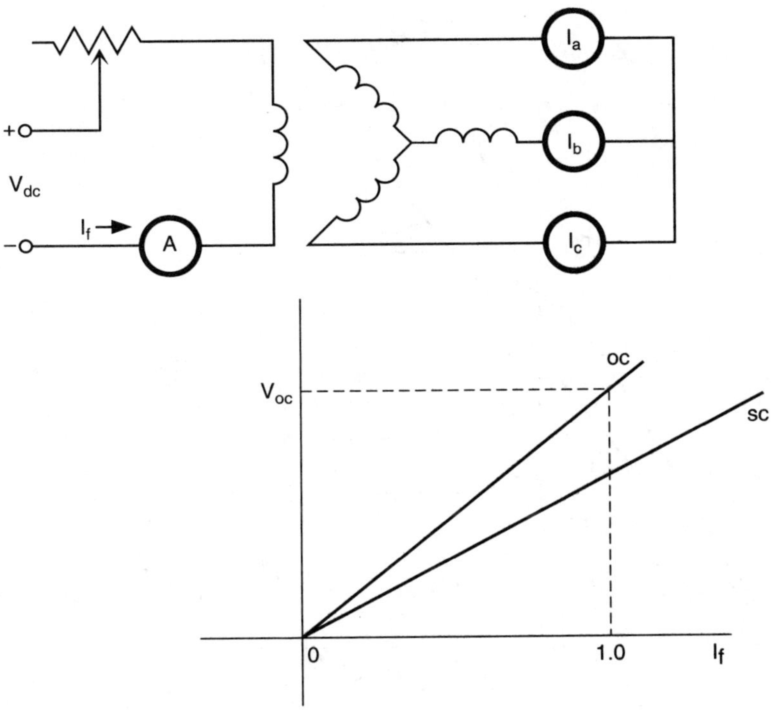

Figure 4.12 Connection for short-circuit test.

duces the armature current Od, and the unsaturated synchronous impedance is given by

$$Z_{us} = \frac{Oc}{\sqrt{3}\,O'd} = \frac{E_{af(\text{air gap})}}{\sqrt{3}\,I_{sc}} \quad \Omega/\text{phase} \tag{4.45}$$

where $E_{af\text{air gap}}/\sqrt{3}$ is the air-gap phase voltage and Z_{us} is the unsaturated impedance. Let Oc be the rated voltage, then Ob is the required no-load field current which produces the armature short-circuit current $O'e$.

The saturated synchronous impedance is found by using the field resistance line. This line represents the saturated machine and is different from the air-gap line which represents the unsaturated machine. Therefore, synchronous impedance is equal to

$$Z_s = \frac{Oc}{\sqrt{3}\,O'e} = \frac{V_t}{\sqrt{3}\,I_{sc}} \tag{4.46}$$

where $Oc = V_t$ is the rated voltage and $O'e = I_{sc}$ is the armature current read from the short-circuit characteristic.

The synchronous reactance is given by

$$X_s = \sqrt{Z_s^2 - R_a^2} = Z_s \tag{4.47}$$

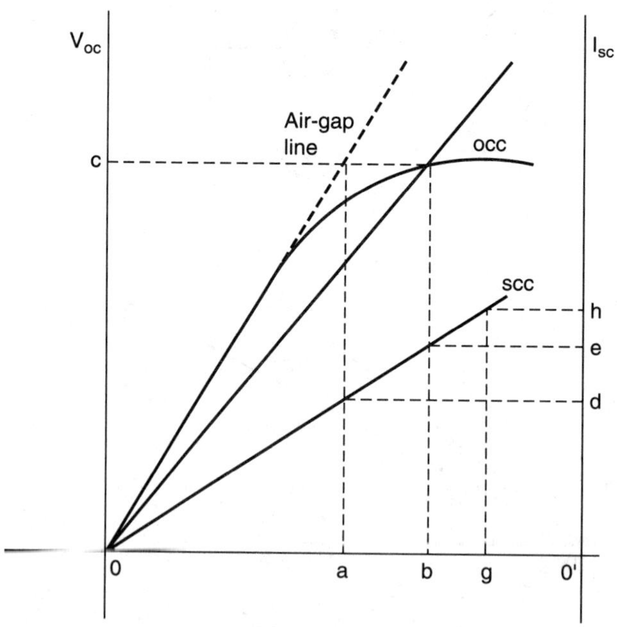

Figure 4.13 Open-circuit and short-circuit characteristics.

where R_a is the armature resistance, which is much smaller than the reactance $R_a << X_s$.

By using the field resistance line the field current can be found for a given terminal voltage, load current and the power factor. Therefore,

$$\mathbf{E}_{af} = \mathbf{V}_t + X_s\mathbf{I} \tag{4.48}$$

where $\mathbf{E}_{af}$, $\mathbf{V}_t$, and $\mathbf{I}$ are phasors.

The short-circuit ratio (SCR) is the ratio of the field current required for the rated voltage on open circuit to the field current required for armature current on short circuit. From Fig. 4.13 the short-circuit ratio (SCR) is

$$\text{SCR} = \frac{Ob}{Og} \tag{4.49}$$

4.7.3 Voltage regulation

If the field excitation of a generator is not adjusted with the changing load, the terminal voltage cannot remain constant. The voltage regulation is found by a change from full load (rated load) to no load (open circuit) for any given power factor. The *voltage regulation* is then defined as the difference between the full-load and no-load values of the terminal voltage.

$$V_R = \frac{V_{NL} - V_{FL}}{V_{FL}} \times 100\% \tag{4.50}$$

Example 4.3 A 100 MVA, 0.85 power factor, three-phase, Y-connected cylindrical rotor synchronous generator has a negligible resistance. The following data were obtained from the open-circuit and short-circuit tests:

Open-circuit test

Field current (A)	200	300	400	700	800	900
Line-to-line terminal voltage (kV)	3.8	5.8	9.8	9.8	11.3	12.6

Short-circuit test

Armature current (A)	4043	8086
Field current (A)	350	700

Calculate

1. The unsaturated synchronous impedance.
2. Saturated synchronous impedance at rated voltage.
3. Short-circuit ratio (SCR).

4. Estimate the field current and voltage regulation.

solution

1. The current required for a rated line-to-line voltage of 10 kV on the air-gap line is

$$\frac{10(200)}{3.800} = 526.315$$

This field current produces a short-circuit current armature line current of

$$\frac{10(200)}{3.8} = 526.315 \text{ A}$$

$$X_{s(ag)} = \frac{10 \times 10^3}{\sqrt{3}\ (6079.6)} = 0.9496\ \Omega/\text{phase}$$

2. Saturated synchronous reactance:

$$X_s = \frac{10(1000)}{8086\sqrt{3}} = 0.714\ \Omega$$

3. Short-circuit ratio:

$$\text{SCR} = 350/700 = 0.5$$

4. Applying Eq. (4.48) we obtain:

$$E_a = V_t + jX_sI = \frac{10{,}000}{700\sqrt{3}} + (0.714\angle 90°)(8086\angle -31.78°)$$

$$= 10{,}088.3\angle 29.10°$$

$$E_{\text{(line to line)}} = \sqrt{3}(10{,}088) = 17{,}473 \text{ V}$$

The field current required to produce this voltage is

$$\text{Voltage regulation} = \frac{10{,}086 - 5773.5}{5773.5} = 0.747$$

Example 4.4 A 200 kVA, 4400 V, three-phase, Y-connected synchronous generator has a synchronous resistance of 0.19 Ω and a synchronous reactance of $X_s = 30\ \Omega$ per phase. Calculate

1. The generated voltage per phase at a power factor 0.8 lagging and 0.8 leading
2. The corresponding voltage regulation

solution

$$V_t = \frac{4400}{\sqrt{3}} = 2540.34 \text{ V} \qquad I = \frac{200{,}000}{\sqrt{3}(4400)} = 26.2 \text{ A}$$

1. At power factor 0.8 lagging

$$E_g = V_t + I(R_s + jX_s) = 2540.34 + (26.2\angle -36.87^\circ)(1.9 + j30)$$
$$= 3051.8 + j598.86 = 3109.99\angle 11.1^\circ$$

At power factor 0.8 leading

$$E_g = V_t + I(R_s + jX_s) = 2108.64 + j658.7 = 2209.13\angle 17.34^\circ$$

2. The corresponding voltage regulation is

$$VR_{(0.8\,\text{lag.pf})} = \frac{V_{NL} - V_{FL}}{V_{FL}} \times 100\% = \frac{3109.99 - 2540.34}{2540.34} \times 100 = 22.4\%$$

$$VR_{(0.8\,\text{lead.pf})} = \frac{V_{NL} - V_{FL}}{V_{FL}} \times 100\% = \frac{2209.13 - 2540.34}{2540.34} \times 100 = 13\%$$

4.8 Power Angle Characteristics

The apparent power S delivered to the generator can be expressed as a function of the terminal voltage V_t, generated voltage E, synchronous reactance X_s, and the power angle or torque angle δ. If the power angle δ increases, the real power output increases reaching its maximum when $\delta = \pi/2$. This is the steady-state power limit. The maximum torque or pull-out torque of synchronous motors occurs at $\delta = \pi/2$.

An increase in mechanical power to the generator, or mechanical output of the synchronous motor after δ has reached 90%, produces a decrease in real electric power, and the generator accelerates while the motor decelerates resulting in loss of synchronism.

The complex power output of the generator in volt-amperes per phase is given by

$$S = P + jQ = VI^* \tag{4.51}$$

Using Fig. 4.14, the terminal voltage is given by

$$V = V\angle 0^\circ \tag{4.52}$$

The generated voltage is

$$E_{af} = E_{af}/\angle\delta \tag{4.53}$$

and the current is

$$IZ_s = E_{af} - V \tag{4.54}$$

Substituting Eqs. (4.52) and (4.53) into Eq. (4.54) and solving for the current I, the current equation is given by

$$I = \frac{E_{af} - V}{Z_s} = \frac{(E_{af}\cos\delta - V) + j(E_{af}\sin\delta)}{Z_s} \tag{4.55}$$

Taking the conjugate of current I and synchronous impedance we have

$$I^* = \frac{(E_{af}\cos\delta - V) - j(E_{af}\sin\delta)}{Z_s^*} \tag{4.56}$$

where $Z_s^* = R_a - jX_s$. Multiplying the numerator and denominator

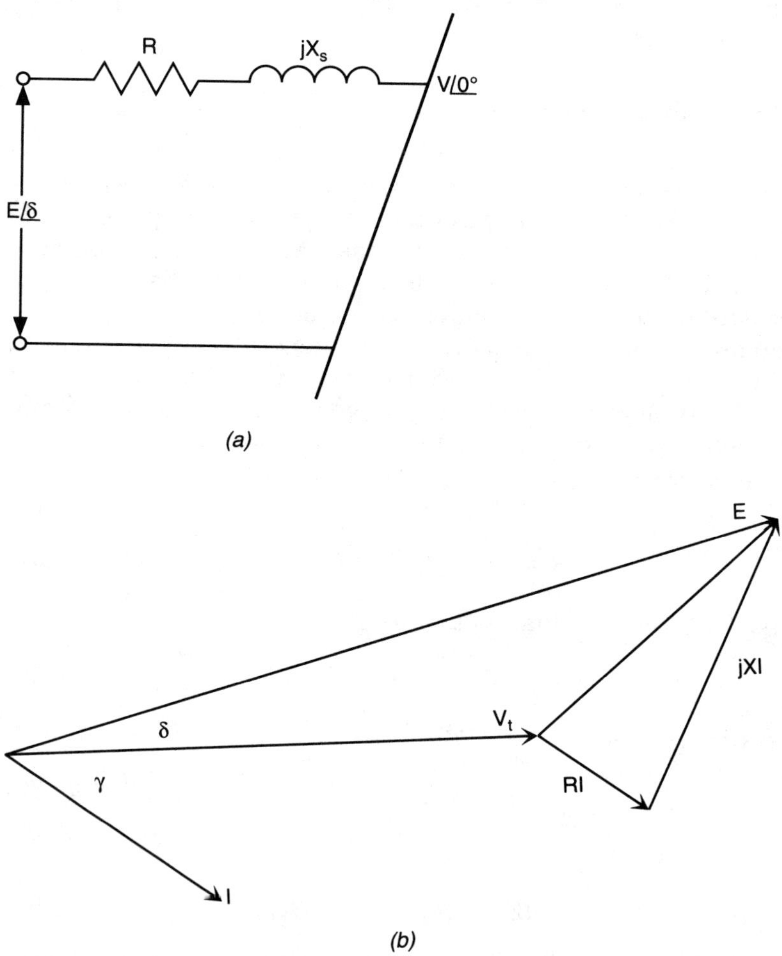

Figure 4.14 (*a*) Synchronous machine connected to an infinite bus. (*b*) The phasor diagram.

by Z_s, the conjugate current is given by

$$I^* = \frac{R_s(E_{af}\cos\delta - V) - X_s(E_{af}\sin\delta)}{(Z_s^*)(Z_s)} + \frac{jX_s(E_{af}\cos\delta - V) - R_s(E_{af}\sin\delta)}{(Z_s^*)(Z_s)} \tag{4.57}$$

In practical synchronous machines, the synchronous resistance is much smaller than the synchronous reactance ($R_s << X_s$); therefore substituting Eq. (4.57) into Eq. (4.51) we have

$$S = \frac{R_s(VE_{af}\cos\delta - V^2) + X_s(VE_{af}\sin\delta)}{R_s^2 + X_s^2} + \frac{jX_s(VE_{af}\cos\delta - V^2) - R_s(VE_{af}\sin\delta)}{R_s^2 + X_s^2} \tag{4.58}$$

where the real and reactive power is given by

$$P = \frac{R_s(VE_{af}\cos\delta - V^2) + X_s(VE_{af}\sin\delta)}{R_s^2 + X_s^2} \tag{4.59}$$

$$Q = \frac{S_s(VE_{af}\cos\delta - V^2) - R_s(VE_{af}\sin\delta)}{R_s^2 + X_s^2} \tag{4.60}$$

For $R_s << X_s$ Eqs. (4.59) and (4.54) become

$$P = \frac{VE_{af}\sin\delta}{X_s} \tag{4.61}$$

$$Q = \frac{VE_{af}\cos\delta - V^2}{X_s} \tag{4.62}$$

The maximum real power output per phase is reached when $\delta = \pi/2$. Therefore,

$$P_{\max} = \frac{VE_{af}}{X_s} \tag{4.63}$$

The power angle equation can be obtained by connecting the synchronous machine to an external large system. Figure 4.15 shows the synchronous machine connected to the Thevenin voltage V_{TH} and Thevenin reactance X_{TH}.

$$P = \frac{E_{af}V_{TH}\sin\delta}{X_s + X_{TH}} \tag{4.64}$$

Any further increase in the prime mover input to the generator causes the real power output to decrease, the excess power going into acceleration causing the generator to increase its speed and to pull out of synchronism. Hence the steady-state stability limit is reached when $\delta = \pi/2$. The transient stability limit is generally of greater interest in power system operation.

The power angle or torque angle characteristic is shown in Fig. 4.16. For normal steady operating conditions, the torque is well under 90°, for instance the torque angle for practical values is 31.7°. The maximum torque or pull-out torque per phase that a cylindrical-rotor synchronous motor can develop for a gradually applied load is

$$T_{\max} = \frac{P_{\max}}{\omega} = \frac{3VE_{af}}{(2\pi n_s/60)X_d} = \frac{90VE_{af}}{\pi n_s X_d} \tag{4.65}$$

Example 4.5 The per-phase generated volume of a six-pole, three-phase, Y-connected, synchronous generator is $E_a = 4400e^{j45°}$ V and its frequency is 60 Hz. The per-phase synchronous impedance of the winding is $Z_s = 3.5 + j14$ and the per-phase load impedance is $Z_L = 3.5 + j16.4$. Determine the speed and torque of the prime mover so that the generated voltage and frequency remain constant and all losses are 15 kW.

solution From Eq. (4.1) the speed of the prime mover is

$$n_s = \frac{120f}{P} \qquad f = \frac{n_s P}{12} \qquad n = 20f = 1200 \text{ rpm}$$

Speed in radians per second is

$$\omega_s = \frac{2\pi n}{60} = \frac{2\pi(1200)}{60} = 125.66 \text{ rad/s}$$

$$I_a = \frac{E_a}{Z_s + Z_L} = \frac{4.400\underline{/4.5°}}{(3.5 + j14) + (31.5 + j16.4)} = 94.911\underline{/4.0°}$$

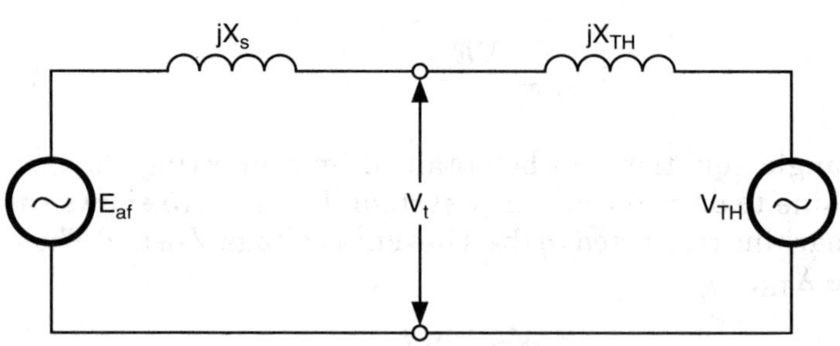

Figure 4.15 Synchronous machine connected to an external large system.

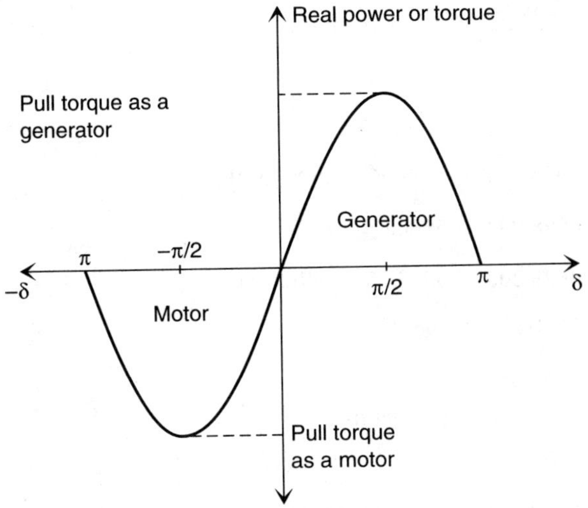

Figure 4.16 Steady-state power or torque angle characteristic of a cylindrical-rotor synchronous machine.

Using Eq. (4.51) the apparent power per phase is

$$S = E_a I_a^* = (4400\underline{/45°})(94.911\underline{/-4.023°}) = 417{,}608.4\underline{/40.98°}$$

$$= 315{,}268.68 + j273{,}865.73 \text{ VA}$$

Total power is

$$S_T = 35 = P_T + jQ_T = 945{,}806.04 + j821{,}597.19 \text{ VA}$$

$$P_{\text{in}} = P_T + P_{\text{losses}} = 945{,}806.04 + 15{,}000 = 960{,}806.04 \text{ W}$$

The torque required from the prime mover is

$$T = \frac{P_{\text{in}}}{\omega} = \frac{960{,}806.04}{125.66} = 7646.08 \text{ N} \cdot \text{m}$$

Example 4.6 A three-phase, 4-pole, Y-connected, 60-Hz, 12 kV generator is connected to an infinite bus. Neglecting the stator resistance, the per-phase synchronous reactance is 10 Ω. The power delivered to the system has a lagging power factor of 0.8. Determine

1. The generated voltage E_g
2. The power angle δ
3. The maximum power $P_{\max}$

solution

$$P\phi = \frac{10 \times 10^6}{3} = 3.33 \text{ MVA}$$

$$V_\phi = \frac{12 \times 10^3}{\sqrt{3}} = 6928.203 \text{ V}$$

$$I = \frac{10 \times 10^6}{3(6928.203)(0.85)} = 566.04 \text{ A}$$

1. $E_g = V_t + jIX_s = 6928.203 + (566.04\underline{/-36.87^\circ})(10\underline{/90^\circ})$

$$= 6928.203 + 5660.4\underline{/53.13^\circ}$$

$$= 9628.203 + 3396.25 + j4528.31$$

$$= 11{,}273.86\underline{/23.68^\circ}$$

2. Power angle

$$\delta = 23.68^\circ$$

3. Maximum power

$$P_{\max} = \frac{3|V_\phi|\ |E_\phi|}{X_s} = \frac{3(6928.203)(11{,}273.86)}{10} = 23.43 \text{ MW}$$

Example 4.7 Two identical 500 kVA, 4400 V, synchronous generators are connected in parallel to supply a load. The prime movers have different speed characteristics. When the field currents of the two generators are equal, one of the generators delivers 300 A at 0.9 power factor lagging, while the other one delivers 200 A at 0.73 power factor lagging, Calculate

1. The real and reactive power supplied by each generator to the load
2. The overall power factor

solution

1. $P_1 = \sqrt{3}\, V_L I_L \cos\phi = \sqrt{3}\,[(440)(300)(0.9)] = 205.767 \text{ kW}$

$P_2 = \sqrt{3}\, V_L I_L \cos\phi = \sqrt{3}\,[(440)(200)(0.73)] = 111.267 \text{ kW}$

$Q_1 = \sqrt{3}\, V_L I_L \sin\phi = \sqrt{3}\,[(440)(300)(0.436)] = 99.658 \text{ kVAR}$

$Q_2 = \sqrt{3}\, V_L I_L \sin\phi = \sqrt{3}\,[(440)(200)(0.683)] = 104.17 \text{ kVAR}$

2. The total load power:

$$P_T = P_1 + P_2 = 205.767 + 111.267 = 317.034 \text{ kW}$$

The total reactive power:

$$Q_T = Q_1 + Q_2 = 99.658 + 104.17 = 203.83 \text{ kVAR}$$

The overall power factor:

$$\text{Pf} = \cos\phi = \cos\left(\tan^{-1}\frac{Q_T}{P_T}\right) = \cos\left(\tan^{-1}\frac{203.83}{317.034}\right)$$

$$= 0.8412 \text{ lagging}$$

4.9 Salient-Pole Synchronous Machines

Synchronous machines with cylindrical rotors have a uniform air-gap length, as has been shown in Fig. 4.1. Salient-pole synchronous machines are the topic of this section. The synchronous machine has a nonuniform air-gap length as shown in Fig. 4.2.

In salient-pole machines the rotor has protrusions. The air-gap length is the smallest at the tip of the poles, and the largest at midway between the poles. The maximum flux is found at the smallest air gap and the minimum flux is found midway between the poles. The magnetic axis midway between the poles is called the direct axis. The direct axis synchronous reactance X_d is produced in the direct axis and the quadrature axis synchronous reactance X_a is produced in the quadrature axis.

4.9.1 The flux and mmf waves

Since the salient-pole synchronous machine has a nonuniform air-gap length, the sinusoidal mmf produces a distorted flux density in the direct axis, as shown in Fig. 4.17.

The analytical calculations of a salient-pole machine are quite complicated. In order to avoid these complications, we usually use the cylindrical-rotor theory with some modifications. In this case the armature current I_{ad} lags the generated emf E_g by 90°. The armature-reaction flux ϕ_{ar} is directly opposite to the field flux ϕ_f, as shown in Fig. 4.18.

Figure 4.18 also shows the phasor relation between generated emf terminal voltage and impedance as given by

$$E_g = V_t + (R_a + jX_d)\, I_{ad} \tag{4.66}$$

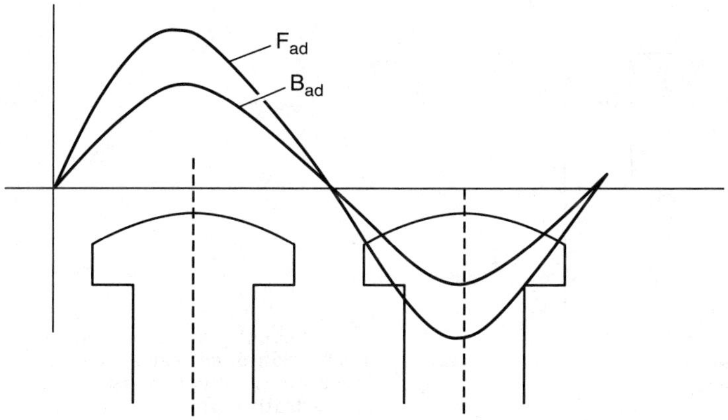

Figure 4.17 mmf and flux density waves in the direct axis.

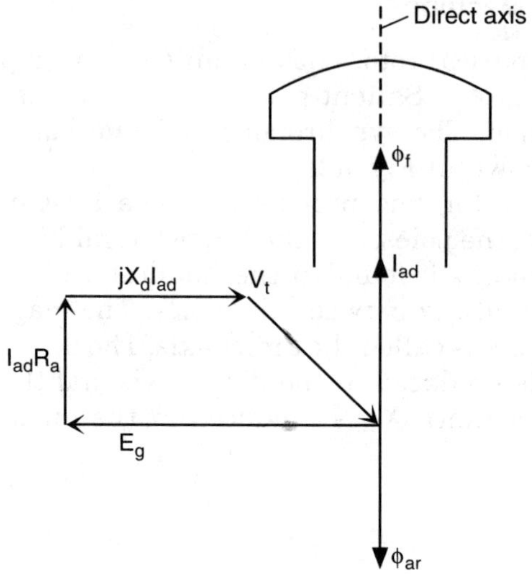

Figure 4.18 Direct-axis phasor diagram of a salient-pole synchronous machine.

In the quadrature axis the armature current is in phase with the generated voltage, and the axis of the armature reaction is opposite to an interpolar space as shown in Fig. 4.19*a*. The flux density wave in the quadrature axis has a different shape from that produced in the direct axis. Figure 4.19*b* shows the mmf and the flux density waves in the quadrature axis.

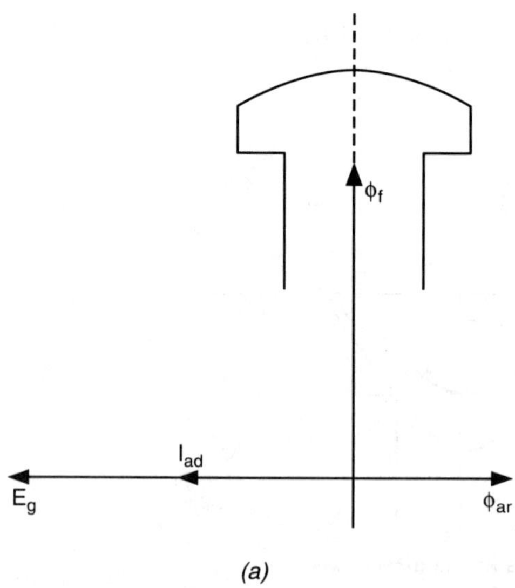

Figure 4.19 (*a*) Phasor diagram of a salient-pole synchronous machine in quadrature axis. (*b*) The mmf and flux density waves in quadrature axis.

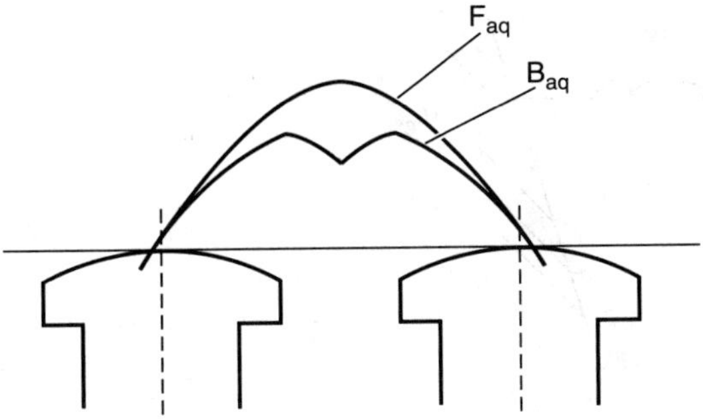

(b)

The effects of salient poles can be studied by resolving the armature current into two components. One component is in quadrature axis and the other in direct axis. Figure 4.20 shows the two axes and the resultant flux which is a vector sum of the armature reaction flux ϕ_{ar} and the field flux ϕ_f.

4.9.2 Power angle characteristics of salient-pole synchronous machines

Let the salient-pole synchronous machine be connected to an infinite bus, through an equivalent reactance X_{eq} and synchronous reactance X_s, as shown in Fig. 4.21. Figure 4.22 shows a phasor diagram of a salient-pole synchronous machine.

From Fig. 4.19*a* the armature mmf along the quadrature axis and quadrature axis synchronous reactance X_q and the armature current

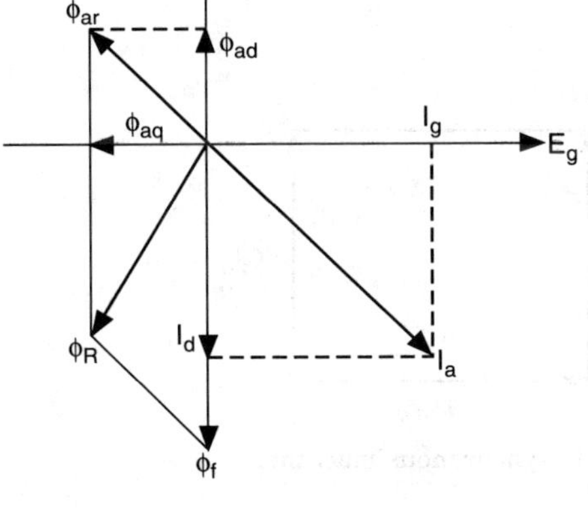

Figure 4.20 Salient-pole synchronous machine with angular relationship of current and generated voltage.

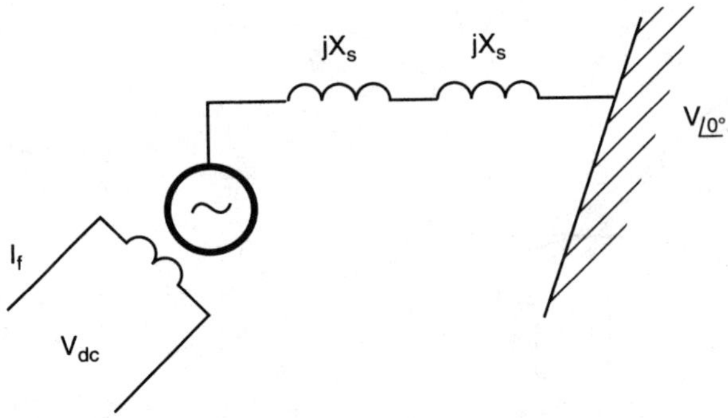

Figure 4.21 One-line representation of a salient-pole synchronous machine connected to an infinite bus.

I_q, the generated voltage is given by

$$E_q = V_t + (R_a + jX_q)I_q \tag{4.67}$$

In keeping with Eqs. (4.66) and (4.67) the generated voltage of the salient-pole generator is

$$E_q = V_t(R_a + jX_d)I_d + (R_a + jX_q)I_q \tag{4.68}$$

The components I_d and I_q are found from Fig. 4.23, thus

$$I_d = I \sin (\gamma + \delta) \tag{4.69}$$

$$I_q = I \cos (\gamma + \delta) \tag{4.70}$$

where $I_d + I_q = I$.

Figure 4.23 gives us the basis of determining the value of the power angle δ. Assuming the current components I_d and I_q to be unknown,

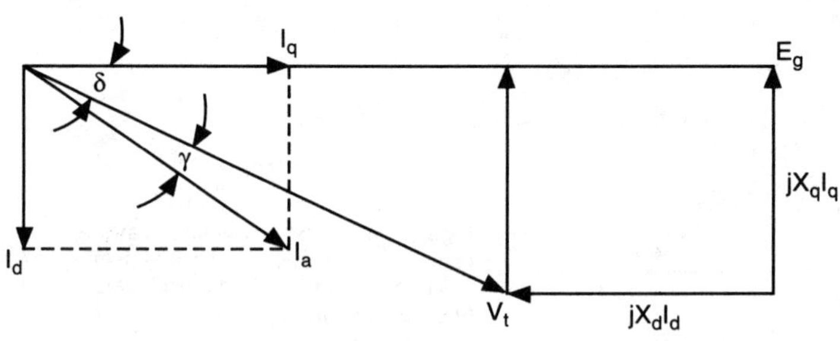

Figure 4.22 Phasor diagram of a salient-pole synchronous machine.

the voltage triangle *ABC* from Fig. 4.23 is similar to the current triangle *abc*. From Fig. 4.23, where *OC* is the distance from *O* to *C* along E_g,

$$OC = V_t + \mathbf{I}(R_a + jX_q) \tag{4.71}$$

$$= V_t + \mathbf{I}(R_a + jX_q)(\cos\gamma - j\sin\gamma)$$

$$= V_t + \mathbf{I}R_a\cos\gamma - jIR_a\sin\gamma + jIX_q\cos\gamma$$

$$+ IX_q\sin\gamma$$

$$= V_t + \mathbf{I}(R_a\cos\gamma + X_q\cos\gamma)$$

$$+ jI(X_q\cos\gamma - R_a\sin\gamma)$$

$$\tan\delta = \frac{I_m\{OC\}}{R_c\{OC\}} = \frac{IX_q\cos\gamma - IR_a\sin\gamma}{V_t + I(R_a\cos\gamma + X_q\sin\gamma)} \tag{4.72}$$

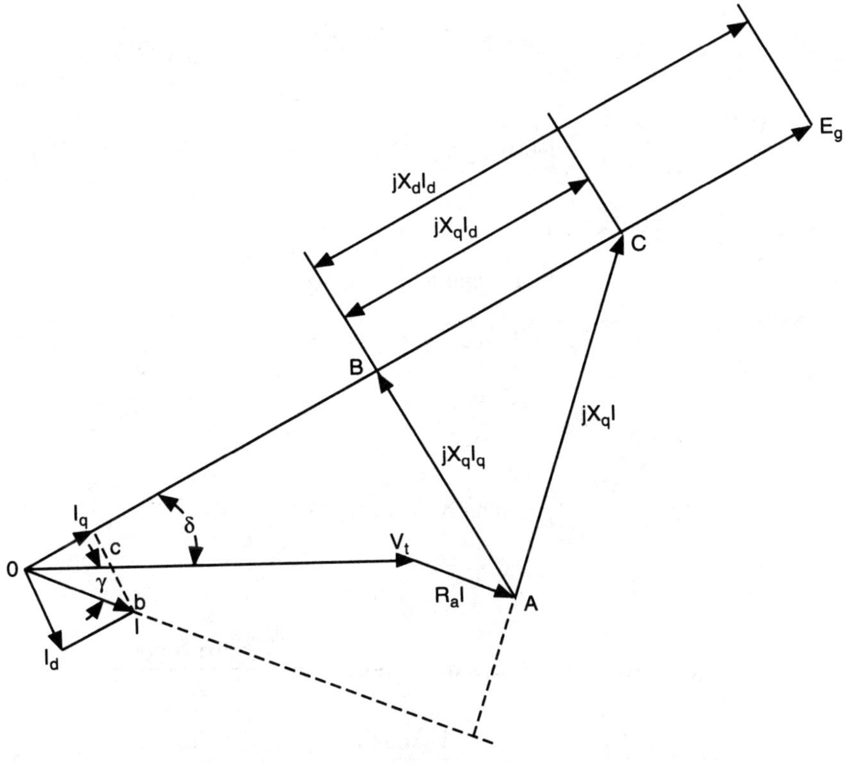

Figure 4.23 Phasor diagram for determining power angle.

Neglecting the armature resistance from Eq. (4.72) we get

$$\tan \delta = \frac{IX_q \cos \gamma}{V_t + IX_q \sin \gamma} \tag{4.73}$$

Power is obtained by finding the apparent power in the direct and quadrature axes:

$$S_d = P + jQ = V_t \mathbf{I}_d$$

$$\mathbf{I}_d = I_d (\sin \delta - j \cos \delta) \tag{4.74}$$

$$\mathbf{I}_d = I_d (\sin \delta + j \cos \delta)$$

From Fig. 4.23

$$E_g = V_t \cos \delta + X_d I_d \tag{4.75}$$

$$I_d = \frac{E_{af} - V_t \cos \delta}{X_d} \tag{4.76}$$

$$S_d = P + jQ = V_t I_d^* = V_t \left(\frac{E_g - V_t \cos \delta}{X_d}\right)(\sin \delta + j \cos \delta)$$

$$= \left(\frac{V_t E_q - V_t^2 \cos \delta}{X_d}\right)(\sin \delta + j \cos \delta) \tag{4.77}$$

For the quadrature axis:

$$\mathbf{I}_q = I_q (\cos \gamma + j \sin \delta)$$

$$I_q^* = I_q (\cos \gamma - j \sin \delta) \tag{4.78}$$

$$I_q = \frac{V_t \sin \delta}{X_q}$$

$$\therefore S_q = V_t I_q^* = \frac{V_t^2 \sin \delta}{X_q} (\cos \delta - j \sin \delta) \tag{4.79}$$

$$S = S_d + S_q$$

$$= \frac{V_t E_q}{X_d} \sin \delta - \frac{V_t^2}{X_d} \cos \delta - j \sin \delta + \frac{V_t^2 \sin \delta \cos \delta}{X_d}$$

$$+ j\left(\frac{V_t E_q}{X_d} - \cos \delta - \frac{V_t^2 \cos \delta}{X_d} - \frac{V^2 \sin^2 \delta}{X_q}\right) \tag{4.80}$$

Using trigonometric identities we get the final result:

$$S = \frac{V_t E_q}{X_d} \sin \delta - \frac{V_t^2 \cos \delta}{2X_d X_q}(X_d - X_q) \sin 2\delta + j\,\frac{V_t E_q}{X_d} \cos \delta$$

$$+ \left(\frac{V_t^2}{2X_d X_q}\right)[(X_d + X_q) - (X_d - X_q) \cos 2\delta] \tag{4.81}$$

Therefore the real and reactive powers are given as

$$P = \frac{V_t E_g}{X_d} \sin \delta - \frac{V_t^2}{2X_d X_q}(X_d - X_q) \sin 2\delta \tag{4.82}$$

and

$$Q = \frac{V_t^2}{2X_d X_q} \cos \delta - \frac{V_t^2}{2X_d X_q}[(X_d + X_q) - (X_d - X_q) \cos 2\delta] \tag{4.83}$$

In Eq. (4.82) above the first term is the power due to excitation and the second part is power due saliency. When the machine reactances $X_d = X_q = X_s$, the second term is zero and we obtain the power expression for a cylindrical-rotor machine. Maximum power at a fixed excitation can be found by differentiating Eq. (4.82) with respect to power angle δ. Then

$$\frac{dP}{d\delta} = \frac{E_g V \cos \delta}{X_d} + \frac{V^2 (X_d - X_q) \cos 2\delta}{X_d X_q} = 0 \tag{4.84}$$

That is,

$$\frac{V (X_d - X_q^2 \cos^2 \delta)}{X_q} + \frac{E_q \cos \delta - V (X_d - X_q)}{X_d X_q} = 0 \tag{4.85}$$

Example 4.8 A 70-kVA, 400-V, three-phase, Y-connected salient-pole synchronous generator runs at 70 percent full load at 0.92 leading power factor. The per-phase direct and quadrature axis reactances are 1.6 Ω and 0.82 Ω, respectively, with the synchronous armature resistance neglected. Calculate

1. The power angle
2. The developed power

solution

1. $I_a = \dfrac{(70 \times 10^3)(0.7)}{400\sqrt{3}} = 70.725 \text{ A}$

$\theta = \cos^{-1} 0.92 = 23.074°$

From Eq. (4.73)

$$\tan \delta = \frac{I_a X_q \cos \phi}{V_t - I_a X_q \sin \phi} = 0.2985$$

Hence, from Eq. (4.75)

$$E_q = V_t \cos \delta + I_d X_d = V_t \cos \delta + X_d I_a \sin (\delta + \theta) = 293.56 \text{ V}$$

2. Developed power from Eq. (4.73) is

$$P_d = \frac{V_t E_q \sin \delta}{X_d} + \frac{V_t^2 (X_d - X_q) \sin 2\delta}{2 X_d X_q} = 20.806$$

4.10 The Phasor Diagram Analysis of the Synchronous Generator

A balanced three-phase, P-pole synchronous generator, its phasor representation is shown in Fig. 4.21. The equivalent circuit diagrams of the synchronous machine are represented by Fig. 4.9*a* and 4.9*c*. The load in Fig. 4.9*c* is given by the expression $Z_L = R_L + jX_L$ or sometimes is defined by volt-ampere rating and its power factor, $\cos \phi_L$. From Fig. 4.9*c* the generated current I_a is given as

$$I_a = \frac{E_a}{R_s + j(X_a + X_{al}) + (R_L + jX_L)} = \frac{E_a}{Z_s + Z_L} \tag{4.86}$$

Once the generated current is known, other voltage drops, such as

Magnetizing voltage, V_a

Leakage voltage, V_{al}

Resistance voltage, V_R

Coil voltage, V_c

Synchronous voltage, V_s

Load voltage or terminal voltage, V_t

are defined as indicated in Fig. 4.24. Using Fig. 4.9*c* the phasor representation of the voltages are

$$V_a = jX_a I_a \tag{4.87}$$

$$V_{al} = jX_{al} I_a \tag{4.88}$$

$$V = R_s I_a \tag{4.89}$$

$$V_c = (R_s + jX_a I_a) \tag{4.90}$$

$$V_s = [R_s + j(X_a + X_{al})]\, I_a = (R_s + jX_s)\, I_a \tag{4.91}$$

$$V_t = Z_L I_a = (R_L + jX_L)\, I_a \tag{4.92}$$

and

$$E_g = E_a - V_a \tag{4.93}$$

Phase power calculations in a synchronous machine can be made with the help of Eq. (4.51). The powers are

1. Power generated by the generator: S_a
2. Power in the stator winding: S_s
3. Power demanded or load power: S_L

Expressing the above powers in phasors we get

$$S_a = P_a + jQ_a = E_a I_a^* = S_a \cos \Phi + jS_{al} \sin \Phi \tag{4.94}$$

$$S_s = P_s + jQ_s = Z_s I_a^2 = (R_s + jX_s)\, I_a^2 = I_a^2\, [R_s + (X_a + X_{al})] \tag{4.95}$$

and

$$S_L = P_L + jQ_L = V_L I_a^* = I_a^2 Z_L$$

$$= S_L \cos \Phi_L + jS_L \sin \Phi_L \tag{4.96}$$

When phase powers S_a, S_s, and S are calculated then the total powers are

$$S_{aT} = P_{aT} + jQ_{aT} = 3S_a = 3P_a + j3Q_a \tag{4.97}$$

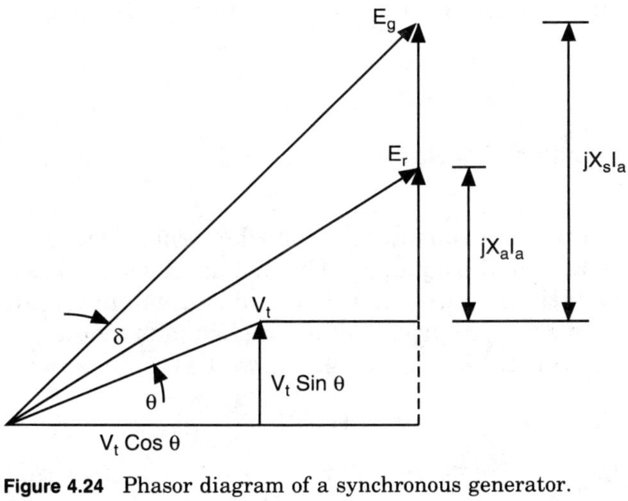

Figure 4.24 Phasor diagram of a synchronous generator.

$$S_{sT} = P_{sT} + jQ_{sT} = 3S_s = 3P_s + j3Q_s \tag{4.98}$$

and

$$S_{LT} = P_{LT} + jQ_{LT} = 3S_L = 3P + j3Q \tag{4.99}$$

The total mechanical power provided by the prime mover is given by

$$P_a = P_{\text{in}} - (P_w + P_{fr} + P_s) \tag{4.100}$$

where P_{in} = input power
P_w = windage power losses
P_{fr} = bearing-friction power losses
P_c = core power losses

Power is related to torque by the relationship:

$$P_{\text{in}} = T\omega_s \tag{4.101}$$

Efficiency:

$$\eta = \frac{P_{\text{out}}}{P_{\text{in}}} \times 100 \tag{4.102}$$

Voltage regulation:

$$V = \frac{|V_{NL}| - |V_{FL}|}{|V_{FL}|} \times 100 \tag{4.103}$$

where V_{NL} is the voltage drop across the output terminals of the generator, when the load terminals are open-circuited, thus giving $V_{NL} = E_g$, and V_{FL} is the voltage drop across the output terminals of the generator, when the full load Z_L is converted to it, thus giving $V_{FL} = Z_L I_a$.

4.11 Performance of Cylindrical-Rotor Synchronous Motors

A synchronous motor delivers a constant power when energized from a source of constant voltage and frequency. The armature current is a function of the field excitation. Figure 4.25 shows an equivalent circuit of one armature phase of the cylindrical-rotor synchronous motor.

Applying Kirchhoff's voltage law to Fig. 4.25*a* we have

$$V_t = E_{af} + I_a[R_a + j(X_{ar} + X_l)]$$

$$= E_{af} + I_a(R_a + jX_s] = E_{af} + I_a Z_s \tag{4.104}$$

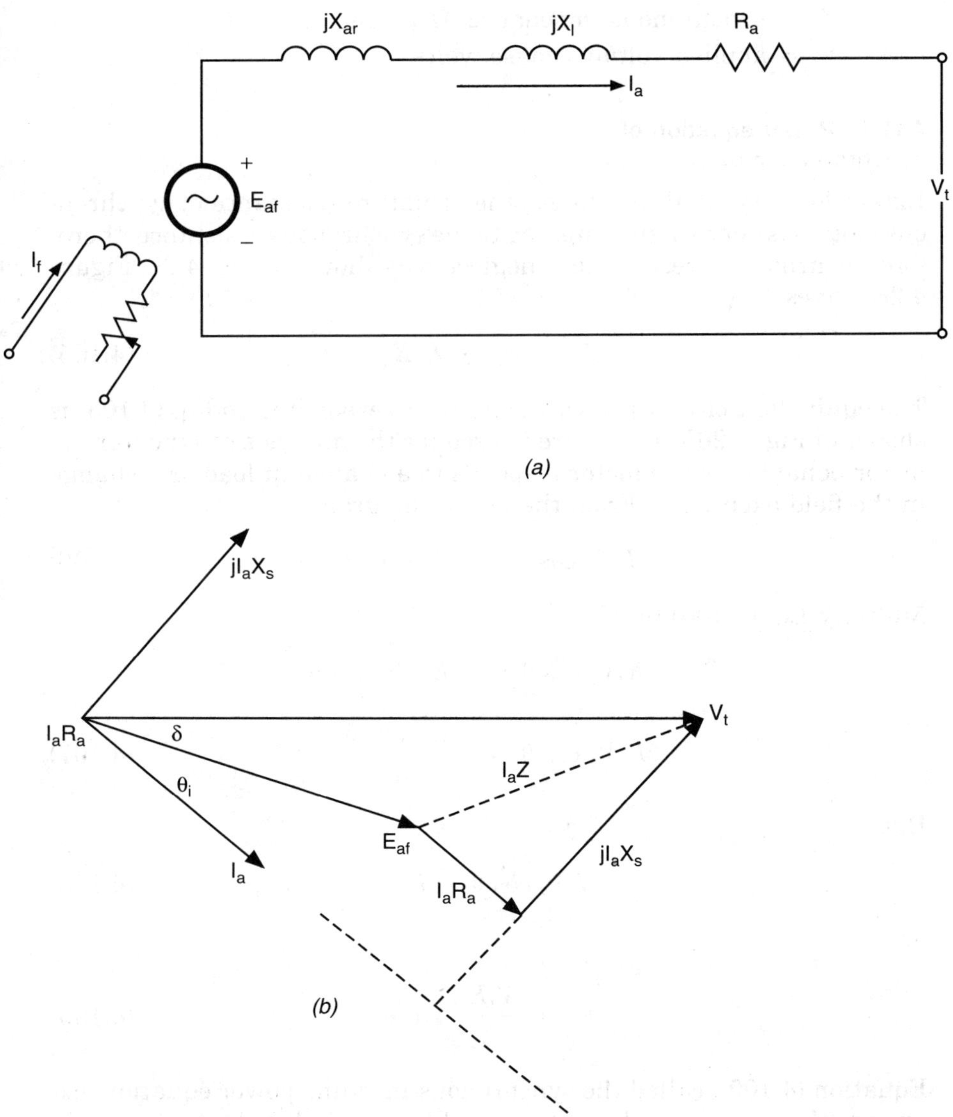

Figure 4.25 (*a*) Equivalent circuit of a cylindrical-rotor synchronous motor. (*b*) Phasor diagram of the equivalent circuit.

where $X_s = X_{ar} + X_l$

R_a = armature resistance, Ω/phase

X_l = leakage reactance, Ω/phase

X_{ar} = armature reaction reactance, Ω/phase

X_s = synchronous resistance, Ω/phase

Z_s = synchronous impedance, Ω/phase
V_t = applied voltage/phase, volts

4.11.1 Power equation of synchronous motors

Except for very small machines, the armature resistance of synchronous motors is very small compared to the synchronous resistance, therefore the armature resistance is neglected as shown in Fig. 4.26. Figure 4.26*a* gives

$$V_t = E_{af} + I_a X_s \tag{4.105}$$

The equivalent circuit phasor diagram corresponding to Eq. (4.105) is shown in Fig. 4.26*b*. The figure is used for the analysis of synchronous motor behavior as the motor responds to a change in load or a change in the field excitation. From the phasor diagram,

$$I_a X_s \cos\theta = -E_{af} \sin\delta \tag{4.106}$$

Multiply Eq. (4.106) by V_E

$$I_a X_s V_t \cos\theta = -E_{af} V_t \sin\delta$$

$$I_a V_t \cos\theta = -\frac{E_{af} \sin\delta}{X_s} \tag{4.107}$$

But

$$I_a V_t \cos\theta = P_{l\phi} \tag{4.108}$$

or

$$P_{l\phi} = \frac{V_t E_{af}}{X_s} \sin\theta \tag{4.109}$$

Equation (4.109), called the synchronous-machine power equation, expresses the power per phase developed by a cylindrical-rotor motor in terms of its excitation voltage and power angle.

4.11.2 Effect of changes in shaft load on armature current, power angle, and power factor

Figure 4.27 illustrates the effect that changes in shaft load have on armature current, power angle, and power factor. The stator voltage, frequency, and field excitation are assumed constant. From Fig. 4.27,

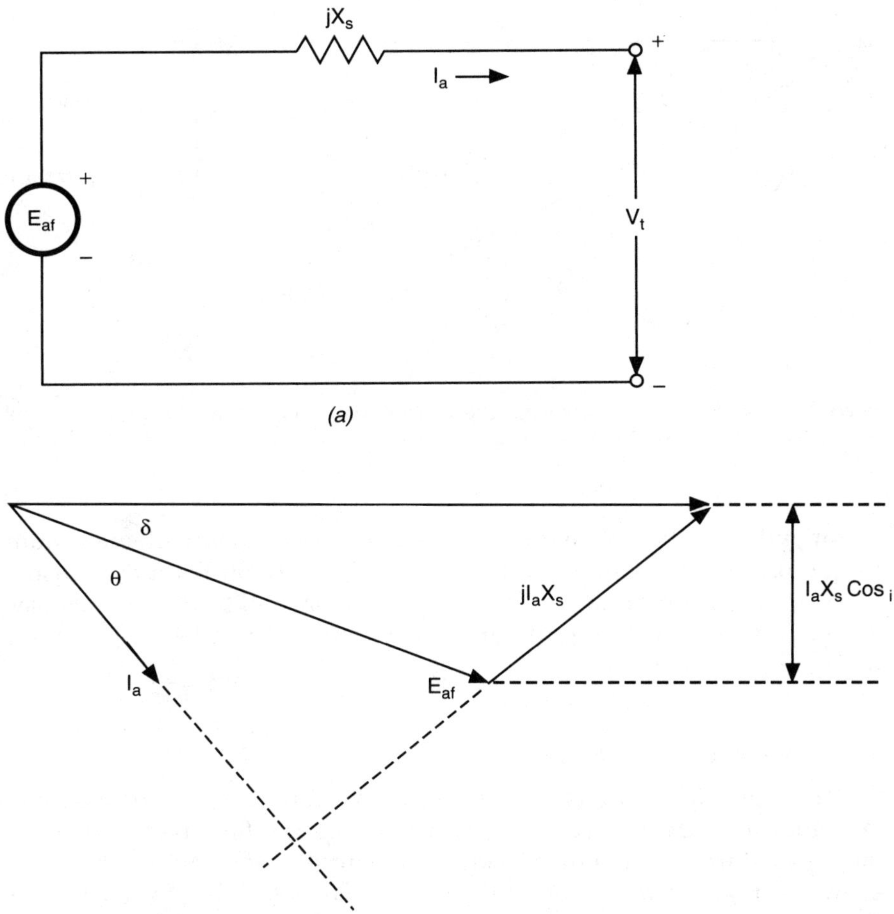

(b)

Figure 4.26 (*a*) Equivalent circuit and (*b*) phasor diagram of synchronous motor.

the heavy lines indicate the initial load conditions, and the light lines indicate the new steady-state conditions that correspond to doubling the shaft load. Doubling the shaft load doubles both Eqs. (4.108) and (4.109). Furthermore, as shown in Fig. 4.27, if the excitation is not changed, increasing the shaft load causes the load of the E_{af} phasor to describe a circular arc. It should be noted that an increase in shaft load is accompanied by a decrease in θ resulting in an increase in power factor. Further increasing the load a point is reached at which δ will fail to cause a corresponding increase in motor torque, and the

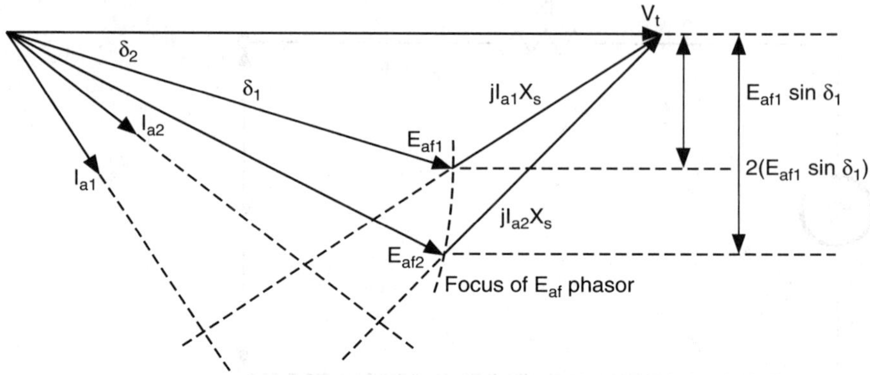

Figure 4.27 Phasor diagram showing the effect of changes in shaft load on I_a, δ, and power factor of a synchronous motor.

motor will finally pull out of synchronism. The point of maximum torque occurs at a power angle of approximately 90 for a cylindrical machine as shown in Eq. (4.63). The critical value of torque that causes a synchronous motor to pull out of synchronism is called the pull-out torque.

4.11.3 V-curves

Curves of armature currents of synchronous motor versus field currents or armature currents versus excitation voltage, for given values of mechanical powers, result in a family of curves known as V-curves, as shown in Fig. 4.28*a*. The solid lines represent a family of V-curves for an unsaturated synchronous motor. The dashed lines are loads for constant power factor and are known as compounding curves. In order to generate reactive power the synchronous motor's excitation is adjusted. Any overexcited synchronous machine whether it operates as a motor or as a generator produces reactive power. In large power supply systems, condensers are used to generate reactive power. The operation power factor of a synchronous motor is controlled by varying the field current I_f thusly:

1. When I_f is adjusted so that the generated voltage is smaller than terminal voltage, the machine is said to be underexcited.
2. When I_f is increased, the synchronous machine becomes overexcited.

Figure 4.28*b* shows the two cases represented by a phasor diagram.

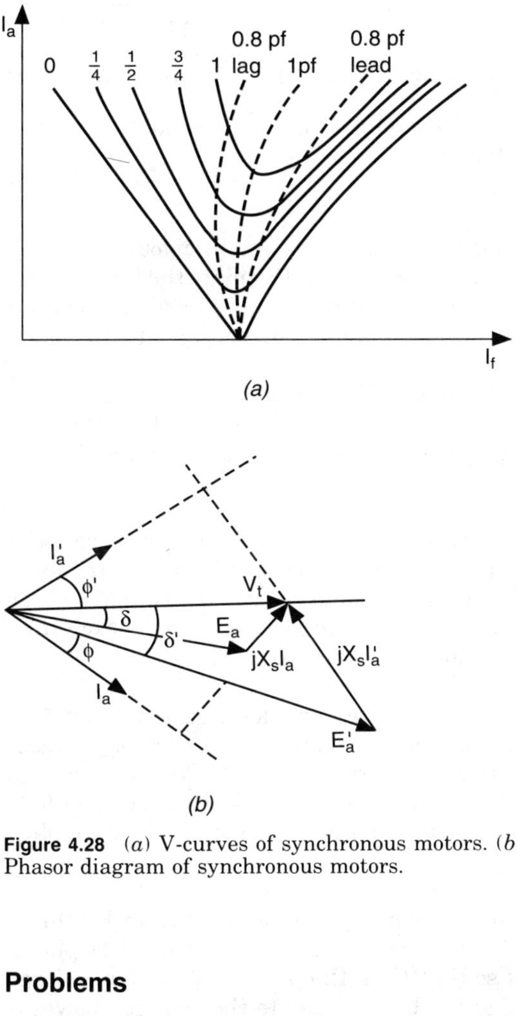

Figure 4.28 (*a*) V-curves of synchronous motors. (*b*) Phasor diagram of synchronous motors.

Problems

1 A 22-kV, three-phase, Y-connected turbogenerator with a synchronous reactance of 0.0112 Ω/phase is delivering 250 MW to the grid at 22 kV and unity power factor.

With the turbine power kept constant the machine excitation voltage is increased by 15 percent. Assuming a linear model for the machine determine the current and power factor.

At a new excitation voltage level the turbine power is increased until the machine delivers 300 MW. Calculate the new power factor.

2 A three-phase 10-MVA, 11 kV, Y-connected synchronous machine gave the following results on test:

Open-circuit test voltage (kV)	4	6	8	10	11	12	13
Field current (A)	60	90	126	171	300	302	324
Short-circuit test	Line current of 566 A at a field current of 200 A						

Assuming R the stator winding resistance to be small, calculate the reactance X from the air-gap line and short-circuit test. Estimate the field current if the machine operates as a generator supplying a 10-MVA load of 0.8 power factor lagging and 11 kV. What is the open-circuit line voltage at this field?

3 A three-phase, 60-Hz, 4-pole, Y-connected synchronous generator has 800 conductors distributed in 50 slots. The coil spans 180 degrees. Calculate the fundamental frequency line voltage on no-load when the fundamental flux per pole is 0.01 Wb.

4 A three-phase, Y-connected, salient-pole synchronous generator rated at 20 kVA, 220 V has a reactance of $X_d = 3\ \Omega$/phase and $X_q = 1.2\ \Omega$/phase. When the machine is delivering full-load current at the rated terminal voltage of 220 V and a power factor of 0.8 lagging, calculate the power angle, excitation voltage, and maximum power that can be generated.

5 A synchronous generator has a three-phase, 1000-kVA unit rated at 13.8 kV. The synchronous impedance is $9.45\underline{/89.2°}$ per unit. The generator produces its rated voltage of 13.8 kV with the field current of 30 A. Suppose we wish to supply power of 1000 kW to the bus with a voltage of 13.8 kV. The field current is 40 A. Find the power angle δ and the reactive voltamperes supplied to the bus.

6 A three-phase lossless synchronous generator is connected to a 15-kV three-phase infinite bus. The machine has a synchronous reactance of 11 Ω/phase. The machine's field current is set so that E_g is 120 percent of the bus voltage V_t. The machine delivers 12 MW to the bus. Calculate the complex power S and the current magnitude I.

7 A 20-kVA, 220-kV, 60-Hz, Y-connected, three-phase salient-pole synchronous generator, operating under balanced steady-state conditions at rated voltage, is connected to an infinite bus. The reactances per unit are given by $X_d = 1.25$ per unit and $X_q = 0.75$ per unit.

a If the field circuit is open, find the maximum real power per unit.

b Calculate the reactive power in units, current in per-unit power factor.

8 Two identical three-phase, 11-kV, Y-connected synchronous generators are operating in parallel and sharing equally the load of 12 kW at 0.8 power factor lagging. The synchronous impedance of each machine is 6.52 Ω/phase, and the armature resistance is negligible. If one of the machines has a field excitation adjusted such that it delivers 120 A lagging current, compute the armature current, power factor, excitation voltage, and power of each machine.

9 A 1200-kVA, 4600-V (line), three-phase, Y-connected synchronous generator has a synchronous resistance of $R_s = 3\ \Omega$/phase and a synchronous reactance of $X_s = 30\ \Omega$/phase. Calculate:

a The full-load generated voltage E_g per phase at power factors lagging: 0.4, and 0.8; and power factor leading: 0.4, and 0.8

b The power angle, δ

c The voltage regulation

Bibliography

Brown, D., and E. P. Hamilton III: *Electromechanical Energy Conversion,* Macmillan, New York, 1984.

Chapman, S. J.: *Electric Machinery Fundamentals,* McGraw-Hill, New York, 1985.

Elgerd, O. I.: *Basic Electric Power Engineering,* Addison-Wesley, Reading, Mass., 1977.

Hunt, W. Jr., and R. Stein: *Electric Power Systems—Transformers and Rotating Machine,* Van Nostrand-Reinhold, New York, 1979.

Kingsley, C. Jr., and A. E. Fitzgerald: *Electric Machinery,* McGraw-Hill, New York, 1980.

Mablekos, Van E.: *Electric Machine Theory for Power Engineers,* Harper and Row, New York, 1980.

McLean, P. G.: *Electromagnetic and Electromechanical Machines,* Ellis Horwood, Chichester, U.K., 1984.

Matsh, L. W., and J. D. Mogan: *Electromagnetic and Electromechanical Machines,* Harper and Row, New York, 1986.

Nasar, S. A., and L. E. Unnewehr: *Electromechanics and Electric Machines,* John Wiley, New York, 1983.

Sarma, M. S.: *Electric Machines,* Wm. C. Brown, Dubuque, Ia., 1985.

Chapter

5

DC Machines

5.1 Introduction

Dc machines are of two types, namely, the dc generators which convert mechanical energy to dc electric energy, and the dc motors that convert electric energy to mechanical energy. This chapter will be subdivided into two parts: The first part will deal with principles of the dc machines and the second part will deal with the performance and operation of dc machines.

5.2 Principles of DC Machines

To demonstrate the characteristics of a normal rotating dc generator or motor, it is useful to discuss a model of a linear dc machine. The dc linear machine exhibits the same behavior as real generators and motors. The linear dc motor shown in Fig. 5.1 consists of a battery, a switch, conducting rails, and the conducting rod across the rail. The flux density B is directed into the page.

The behavior of dc machines can be determined by applying four basic equations. These are

1. The electromagnetic force equation on the conducting bar is given by

$$\mathbf{F} = i(\mathbf{L} \times \mathbf{B}) \qquad \text{N} \tag{5.1}$$

where $\mathbf{F}$ = force on the bar
i = current flowing in the bar
$\mathbf{L}$ = effective length of the bar
$\mathbf{B}$ = magnetized flux density

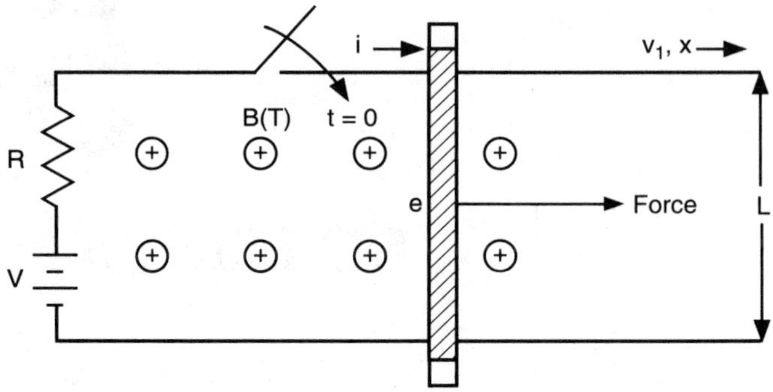

Figure 5.1 Representation of a linear dc motor.

2. The induced emf in the bar moving in the magnetic field is

$$e_{\text{ind}} = (\mathbf{V} \times \mathbf{B})\mathbf{L} \qquad \text{V} \tag{5.2}$$

where e_{ind} = induced emf at the conductor
$\mathbf{V}$ = velocity of the conductor
$\mathbf{L}$ = effective length of the conductor
$\mathbf{B}$ = magnetic flux density

3. Kirchhoff's voltage law (KVL) for the linear dc machine circuit, as shown in Fig. 5.1, is

$$V = e_{\text{ind}} + Ri \tag{5.3}$$

4. Newton's law of motion for the conductor bar across the rails is given by

$$F = m\frac{dv}{dt} + bv \tag{5.4}$$

where m = mass of the bar
$\frac{dv}{dt}$ = acceleration
bv = force of inertia which the bar is trying to overcome

5.2.1 Force and torque in a conventional DC machine

The linear motor discussed in Sec. 5.2 served as an introduction to the study of rotating machines. To demonstrate the fundamental aspects of a conventional dc machine, Fig. 5.2 will be used. In Fig. 5.2 the field circuit is in the rotor. The field produced in the stator is a stationary

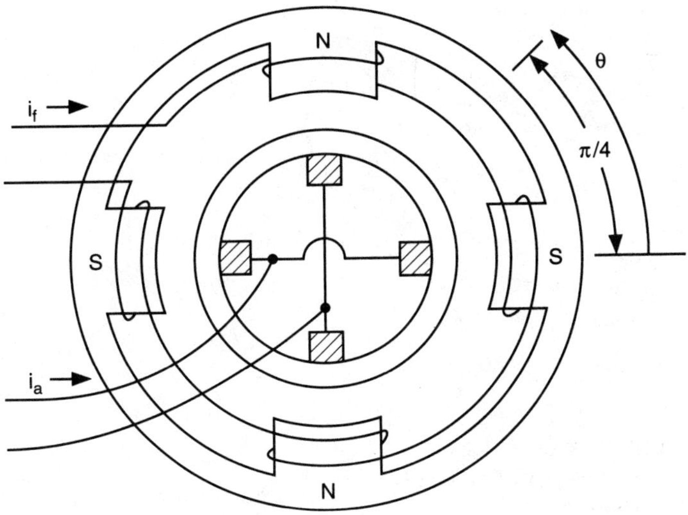

Figure 5.2 A four-pole dc machine.

magnetic field whose radial component of flux density, at radius R, is a periodic function of the angle with half-wave symmetry as shown in Fig. 5.3. The armature winding of a machine is the winding that has the major voltage induced in it. In the case of a dc machine, the armature winding is located on the rotor.

The torque in any dc machine depends on the flux in the machine, the armature current, and the design constants. The approach to determine the torque will be by the use of Figs. 5.3 and 5.4. If Z coils are distributed on the circumference of the rotor, the circumference defined by $R\ d\theta$ is $(Z \cdot R\ d\theta)/2\pi R$. The torque developed on the $Z\ d\theta/2\pi$ coils due to a sinusoidally distributed field is given by

$$dT = RB_\theta Li_c \frac{Z\ d\theta}{2\pi} \qquad \text{N} \cdot \text{m} \tag{5.5}$$

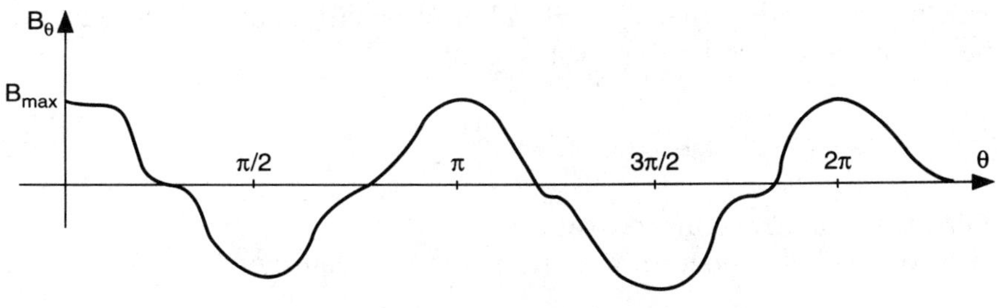

Figure 5.3 Air-gap magnetic flux density in a four-pole dc machine.

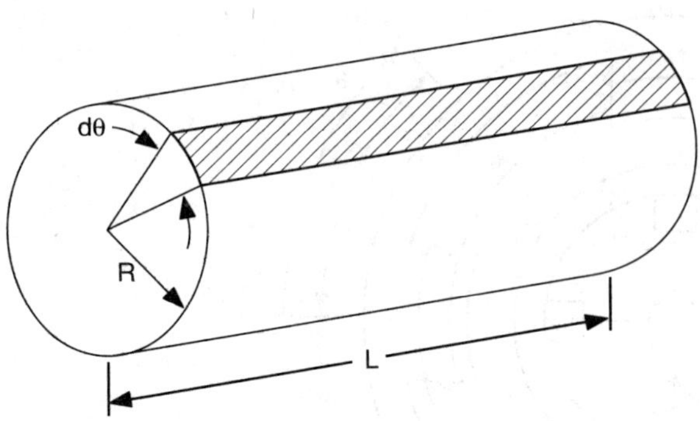

Figure 5.4 Developed torque on a rotor.

where B_θ = periodic flux density
L = effective length of the rotor
i_c = coil current

Integrating from $\theta = -\pi/4$ to $\theta = \pi/4$ for a four-pole dc machine or in general for a P-pole machine, the limits of integration are $\theta = -\pi/P$ to $\theta = \pi/P$. Therefore torque per pole is given by

$$T_p = \int_{-\Pi/p}^{\Pi/p} RB_\theta Li_c \frac{Z}{2\pi}\, d\theta = \frac{Z}{2\pi} i_c \int_{-\Pi/p}^{\Pi/p} B_\theta LR\, d\theta \qquad \text{N} \cdot \text{m/pole} \tag{5.6}$$

The total flux per pole is

$$\phi = \int_{-\Pi/p}^{\Pi/p} B_\theta LR\, d\theta \tag{5.7}$$

Substituting Eq. (5.7) into Eq. (5.6) the total torque is

$$T = PT_p = \frac{ZP}{2\pi} i_c\phi \qquad \text{N} \cdot \text{m} \tag{5.8}$$

where i_c is the current of one conductor. The actual winding of a dc armature generally involves a number of parallel paths denoted by a. Therefore the current in one conductor is

$$i_c = \frac{i_a}{a} \tag{5.9}$$

where i_a is the armature current.

The torque developed in a dc machine is given by

$$T = \frac{PZ}{2\pi a} \phi i_a \qquad \text{N} \cdot \text{m} \tag{5.10}$$

$$T = K_T \phi i_a \qquad \text{N} \cdot \text{m} \tag{5.11}$$

where $K_T = PZ/2\pi a$. To demonstrate the characteristics of a normal rotating dc motor it is useful to discuss a model of a linear motor, as shown in Fig. 5.1.

If the bar placed on the rails can slide and assuming that the magnetic field is uniform and constant, the electromechanical force on the bar is given by

$$F = BLi \qquad \text{N} \tag{5.12}$$

This force is acting in the direction of x. Writing the equation of motion for the bar we have

$$F = m\frac{dv}{dt} + bv \qquad \text{N} \tag{5.13}$$

where bv is the force of inertia which the bar is trying to overcome.

Equation (5.12) can be used to develop a familiar result for motional emf induced if a conductor moves through a magnetic field. If a positive charge q at any point on the conductor experiences a force, the work done in moving the charge q from one end to another is given by

$$W = qvBL \qquad \text{J} \tag{5.14}$$

Therefore the emf on work per unit charge is given by

$$e = \frac{w}{q} = BLv \qquad \text{V} \tag{5.15}$$

Applying Ohm's law on Fig. 5.1 we get

$$i = \frac{v - e}{R} = \frac{V}{R} - \frac{BLv}{R} \qquad \text{A} \tag{5.16}$$

Equating Eqs. (5.12) and (5.13) we have

$$\frac{dv}{dt} + \frac{B^2L^2V}{mR} - \frac{BLV}{mR} + \frac{bv}{m} = 0 \tag{5.17}$$

5.2.2 Power balance equations and efficiency in DC machines

The work done on a body in rotational motion is given by

$$dW_w = T_m d\theta \tag{5.18}$$

Dividing Eq. (5.18) by dt we have

$$P_m = \frac{dW_m}{dt} = T_m \frac{d\theta}{dt} = T_m\omega \tag{5.19}$$

Multiplying Eq. (5.16) by i we get

$$Vi = Ri^2 + ei \tag{5.20}$$

Comparing Eqs. (5.19) and (5.20) the power balance equation is

$$P_m = T_m\omega = ei = Vi - Ri^2 \tag{5.21}$$

where $Vi = P_{\text{in}}$ = input power to the rotor winding
$Ri^2 = P_R$ = power dissipated in the rotor resistance
P_m = developed electromagnetic power

The motor power provides the useful power to shaft, P_{sh}, and other losses such as bearing and friction P_{fr}, windage resistance P_{WR}, and core losses in rotor iron P_c.

Therefore the output power or the shaft power is

$$P_{\text{sh}} = P_{\text{in}} - P_{WR} - P_{\text{fr}} - P_c = P_{\text{in}} - P_{\text{loss}} = P_{\text{out}} \tag{5.22}$$

where P_{out} is the useful or output power, and $P_{\text{loss}} = P_{WR} + P_{\text{fr}} + P_c$.

Shaft power is usually measured in horsepower (hp). The performance of a motor is measured by the index of performance known as the percentage efficiency, which is defined as

$$\eta = \frac{P_{\text{out}}}{P_{\text{in}}} \times 100\% = \frac{P_{\text{sh}}}{P_{\text{in}}} \times 100\% \tag{5.23}$$

Example 5.1 Calculate the torque delivered by a 4-pole, lap-wound dc machine having 20 active conductors, if the armature current is $i_a = 20$ A and the flux per pole is equal to 0.02 Wb.

solution From Eq. (5.10)

$$K_T = \frac{PZ}{2\pi a} = \frac{4(20)}{2\pi(4)} = 3.1831$$

Therefore,

$$T = K_T\phi i_a = 3.1831(0.2)(20) = 12.7324 \text{ N} \cdot \text{m}$$

Example 5.2 A 40-hp, 400-V dc series motor operating at 550 rpm was under a test and the following results were recorded: armature current $I_a = 70$ A, armature resistance $R_a = 0.25\ \Omega$, field resistance $R_f = 0.2\ \Omega$, core resistance

$R_c = 0.1\ \Omega$, core losses $P_c = 1200$ W, windage, resistance losses $P_{WR} = 130$ W, bearing friction losses $P_{fr} = 50$ W. Calculate

1. The electromagnetic power P_m
2. The electromagnetic torque T_m
3. The shaft power P_{sh}
4. The motor efficiency η

solution From Eq. (5.20)

1. $P_m = T_m\omega_m = P_{in} - P_{loss}$

$$P_{in} = 400(70) = 28{,}000 \text{ W}$$

$$P_{R(total)} = R_a I_a^2 + R_f I_f^2 + R_c I_c^2 = 0.25(70)^2 + 0.2(70)^2 + 0.1(70)^2$$

$$= 2695 \text{ W}$$

$$P_{brush} = 2I_a = 2(70) = 140 \text{ W}$$

$$P_{total\ losses} = P_{r(total)} + P_{brush} = 2695 + 140 = 2835.0 \text{ W}$$

$$P_m = P_{in} + P_{total\ losses} = 28{,}000 - 2835 = 25{,}165 \text{ W}$$

2. $T_m\omega_m = P_m$

$$T_m = \frac{P_m}{\omega_m} = \frac{25{,}165}{2\pi n/60} = \frac{25{,}165(60)}{2\pi n} = 436.924 \text{ N} \cdot \text{m}$$

3. From Eq. (5.21) we have

$$P_{sh} = P_{in} - P_{losses}$$

$$P_{losses} = P_{WR} + P_{fr} + P_c = 130 + 50 + 1200 = 1380 \text{ W}$$

4. From Eq. (5.22)

$$\eta = \frac{P_{sh}}{P_{in}} \times 100\% = \frac{26{,}620(100)}{28{,}000} = 95.07\%$$

5.3 Induced Electromotive Force (EMF) in DC Machines

When the rotor of a dc machine rotates, its conductors cut the magnetic flux lines, thus resulting in the generation of emf in each conductor. The magnitude of the emf is found from Eq. (5.15). If the rotor speed is kept constant (n = const) then the tangential conductor speed v is also constant and the instantaneous value of the emf will be proportional to the flux density, as given by Eq. (5.15).

As the conductor travels the total distance of one pole pair, the generated conductor emf will complete a full ac cycle of the same waveshape as that of the flux density. The waveshapes of the emf are not sinusoidal, but because of commutation, the total emf between the brushes is constant, as shown in Fig. 5.5. In Fig. 5.5 the constant emf waveform still contains some ripples. The total dc emf which can be found within Z/P conductors per pole will be the sum of the instantaneous emfs existing in the conductor in question.

$$\sum_{\upsilon=1}^{Z/P} e_\upsilon = \sum_{\upsilon=1}^{Z/P} \upsilon L B_\upsilon = \upsilon L \sum_{\upsilon=1}^{Z/P} B_\upsilon \quad \text{V} \tag{5.24}$$

For the machine with P poles and parallel current paths, the total emf between the brushes is

$$e = \frac{P}{a} \sum_{\upsilon=1}^{Z/P} e_\upsilon = \frac{P\upsilon L}{a} \sum_{\upsilon=1}^{Z/P} B_\upsilon \quad \text{V} \tag{5.25}$$

But the total pole flux f is given by

$$\phi = \frac{\pi L D}{Z} \sum_{\upsilon=1}^{Z/P} B_\upsilon \tag{5.26}$$

Equation (5.26) can be derived by integrating Eq. (5.7) with respect to the flux density while keeping the speed of the rotor constant. From Eq. (5.26) we get

$$\sum_{\upsilon=1}^{Z/P} B_\upsilon = \frac{\phi Z}{L \pi D} \tag{5.27}$$

Substituting Eq. (5.26) into Eq. (5.14) we get

$$e = \frac{P Z \upsilon \phi}{\pi D a} \quad \text{V} \tag{5.28}$$

The tangential conductor speed υ is related to the angular motor speed

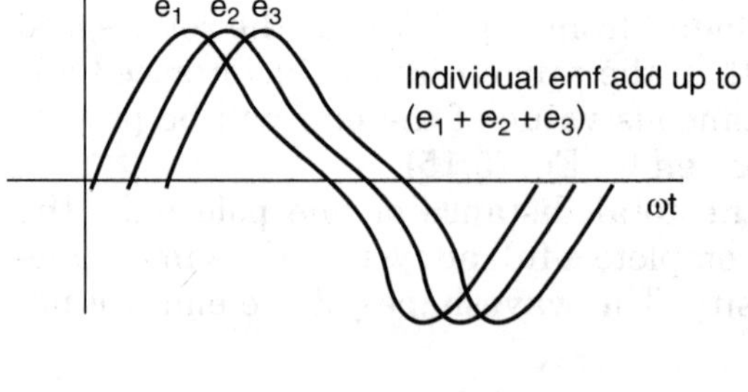

Figure 5.5 Voltage waveforms of a dc machine undergoing commutation.

by the following formula:

$$v = \frac{D\omega_m}{2} \qquad \text{m/s} \tag{5.29}$$

Therefore the total emf of a dc machine is given by

$$e = \frac{PZ\phi\omega}{2\pi a} \tag{5.30}$$

Let

$$K_T = \frac{PZ}{2\pi a}$$

Therefore,

$$e = K_T\phi\omega_m \qquad \text{V} \tag{5.31}$$

The emf found from Eq. (5.31) is shown graphically in Fig. 5.5. Equations (5.11) and (5.31) are the equations of the dc machines.

Example 5.3 A 4-pole dc machine with a lap winding has 16 conductors. The motor runs at 2500 rpm. The flux per pole is 0.2 Wb and the armature current is $i_a = 30$ A. Find the source voltage V_t required to supply the 30-A armature current. The armature resistance is $R_a = 5.501\ \Omega$.

solution From Eq. (5.31)

$$e = \frac{PZ\phi\omega_m}{2\pi a}$$

$$K_T = \frac{PZ}{2\pi a} = \frac{2(16)}{4\pi} = 2.546$$

$$\omega_m = \frac{2\pi n}{60} = \frac{2\pi(2500)}{60} = 261.8 \text{ rad/s}$$

$$e = K_T\phi\omega_m = 2.546(0.2)(261.8) = 133.3 \text{ V}$$

The supply voltage is

$$V_t = e + R_a i_a = 133.3(0.501)(30) = 148.33 \text{ V}$$

5.4 DC Armature Windings

In a two-pole dc machine (Fig. 5.6), the armature conductors are connected 180 electrical degrees apart. Such connection ensures that the direction of the induced voltage in conductor a and the induced voltage in conductor b have the opposite direction of the two conductors from one turn whose voltage is twice that of the individual conductor.

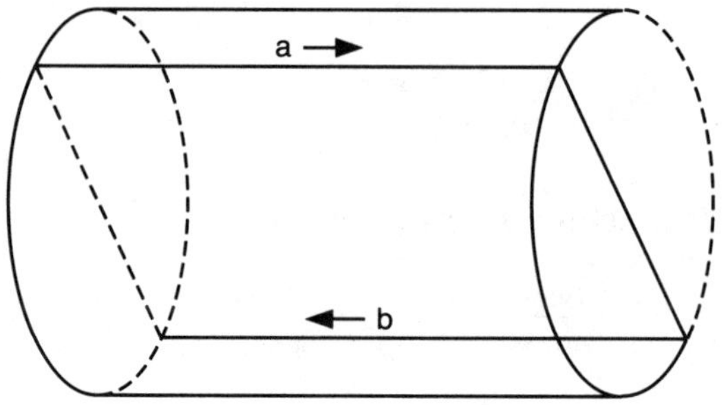

Figure 5.6 The armature conductors (a and b).

The windings of dc machines are continuous, without beginning or end. If the armature rotates for one slot pitch (a slot pitch is the total width of one slot and one tooth), the armature must be seen exactly as before the motion, even though each individual conductor has changed its position. In Fig. 5.6 let coil side a be the forward and coil side b the return coil side. Therefore, it follows that one-half of all coil sides must be forward coil sides, the other half return coil sides. These two kinds of coil sides appear alternating all the way around the periphery of the rotor. For a dc machine with more than two poles, each coil of armature winding consists of coil sides located at a distance of 180 electric degrees. The dc machine armature windings can be connected into two ways, namely:

1. The lap winding, as shown in Fig. 5.7
2. The wave winding, as shown in Fig. 5.8

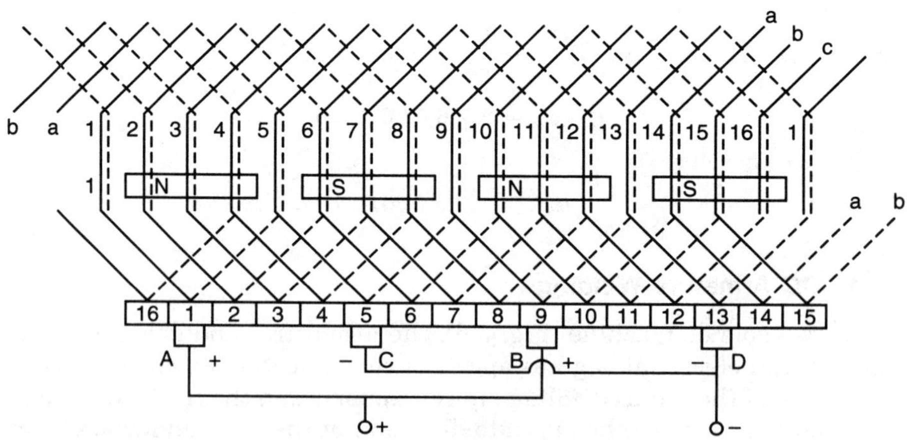

Figure 5.7 Lap winding of dc machine with commutator segments.

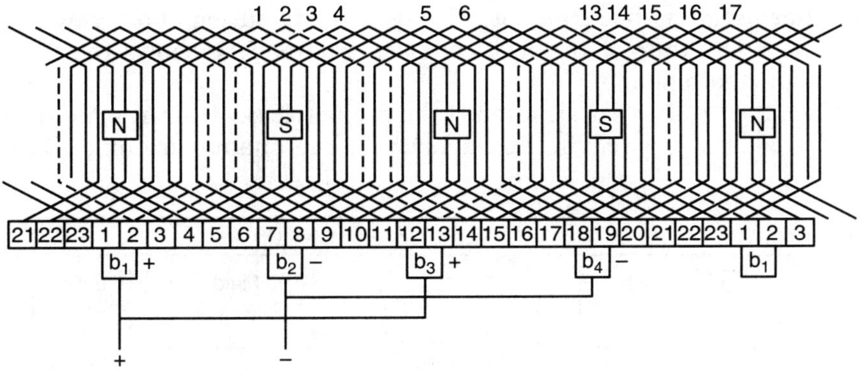

Figure 5.8 Wave winding of dc machine with commutator segments.

The lap winding has as many parallel current paths as there are poles, that is, $a = P$. In tracing the sequence of conductors through the armature the other coils are reached.

In the wave winding the tracing progresses in the same direction around the circumference going through one coil for each of the pairs of poles before returning to a coil near the first one.

The difference between lap and wave windings is very essential to note. A dc machine with P poles wound in lap winding has as many parallel current paths as there are number of poles, while the dc machine wound with wave winding has two parallel current paths regardless of the number of poles. Therefore,

1. For lap winding $a = P$
2. For wave winding $a = 2$

In other words, if the total number of conductors in the armature winding is Z/P conductors in series in each path then for the machine with a wave winding, the number of conductors will be equal to $Z/2$.

The choice of wave or lap windings depends on the current and voltage magnitudes. The lap winding is chosen when higher current magnitudes are required; whereas, the wave winding is chosen when higher voltage magnitudes are required.

5.5 Armature Reactions in DC Machines

When the dc machine is under load conditions, there is magnetomotive force (mmf) in the armature current. This mmf is determined by the position of brushes. When these brushes are set on a geometric neutral, the armature mmf is directed along the quadrature axis. The armature mmf has definite effects on both the space distribution of air-gap flux and the magnitude of the flux per pole. Figure 5.9*c* shows the mmf wave (sawtooth shape).

When the brushes are situated at the geometric neutral the armature mmf produces appreciable flux density at the quadrature axis. Both

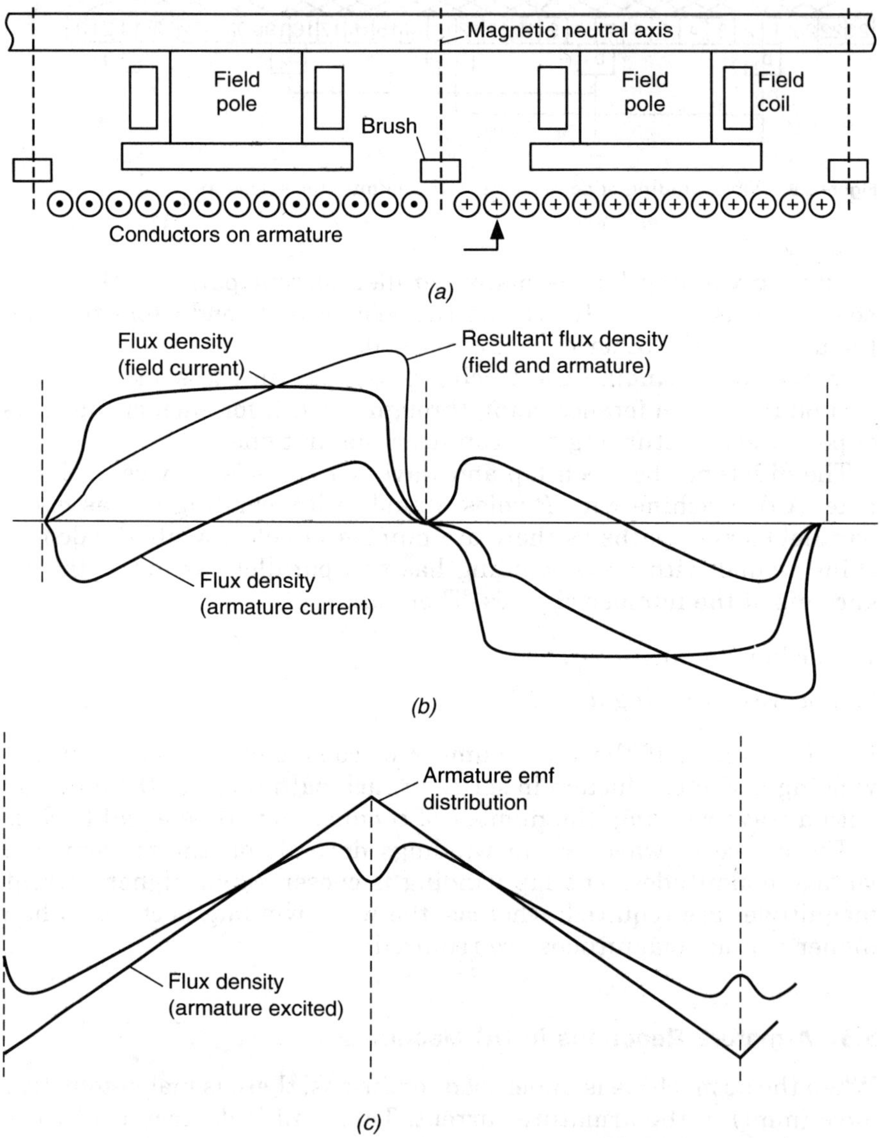

Figure 5.9 MMF and flux density distribution with the brushes on the magnetic neutral axis.

field and armature currents are present when the dc machine is under load conditions. The sum of the field and armature mmf's produce a flux density wave shown in Fig. 5.9*b*. The armature mmf produces armature reaction, while the major component of the armature mmf is in the quadrature axis. The armature reaction in the usual dc machine also has a demagnetizing effect on the field.

Commercially two methods are employed to eliminate the undesirable effects of the armature mmf during commutation. These methods are

1. Shifting of the brushes toward the electrical neutral zone
2. Placing of the interpoles in the geometric neutral axis

The interpoles have windings connected in series opposition with the armature. The mmf of the commutating poles is thus always proportional to the armature current.

Another effect which influences commutation is the reactance voltage produced by the reversal of flux linkages with the short-circuited armature coils, as the current reverses during commutation. These current reversals cause sparking between the commutator bars. To alleviate this problem, the brushes are shifted past the geometric neutral by an angle slightly greater than that which would set the brushes on electrical neutral. This method is good for either motors or generators.

5.5.1 Commutation in DC machines

The commutator in a dc machine is a mechanical rectifier, which is a cylinder formed of copper segments insulated from each other by mica and mounted on but insulated from the rotor shaft. The commutator changes or commutates the alternating voltages induced in the terminals. Let us discuss Fig. 5.10*a*,*b*, and *c* while we examine what is going on in the armature coil undergoing commutation.

In Fig. 5.10*a* the current flows from right to left and in Fig. 5.10*c* the current flows from left to right. At this instant, the process of communication for nearly sudden reversal of current direction is taking place. For coil 2 the process begins at the time when the brush first touches the edge of the commutator bar *b* and it ends when the brush loses contact with the commutator bar *a*. Disregarding the width of the thin insulating strip between bars the time interval available for commutation is

$$\Delta_t = \frac{d_B}{\omega} \tag{5.32}$$

where d_B is the width of the brush and ω is the velocity.

From Fig. 5.10*b* the current in coil 2 is zero. This is a linear commutation and it is represented by curve 1 in Fig. 5.11. The linear commutation is the desirable commutation because current distribution on the contact surface is uniform. Due to leakage inductance there is a delay in the change of the current, therefore the graph of current i versus time is shown as curve 2 in Fig. 5.11. During this interval (Δt) the current in coil 2 decays to zero. From Fig. 5.10*c* the current can flow only in commutator bar a to the brush by sparking. In all dc machines, interpoles are placed in the neutral zone between main poles. The interpoles induce voltage that controls the rate of change of current. The windings of the interpoles are connected in series with the armature winding, so that their mmf is always proportional to the armature current. If the interpoles have the right number of turns,

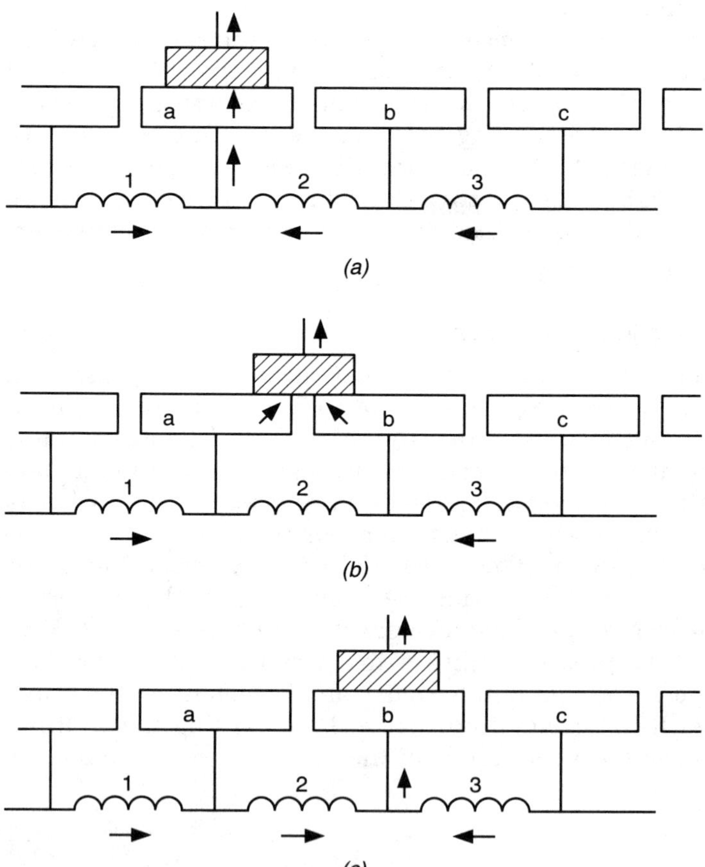

Figure 5.10 A dc machine undergoing commutation.

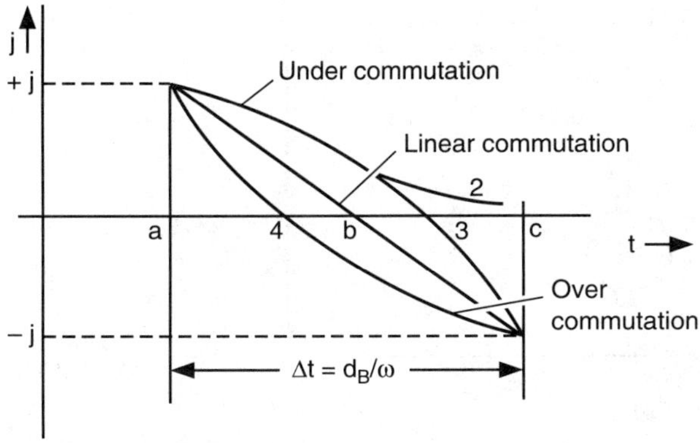

Figure 5.11 Current versus time relationship of rotor coils undergoing commutation.

the voltage induced in the coils undergoing commutation is just sufficient to make

$$\frac{dL_2}{dt} = \pm\frac{2i}{\Delta t}$$

and the desired linear commutation is obtained. Curve 3 in Fig. 5.11 is an approximate curve for eliminating sparking at the brushes. This curve is known as undercommutation. When compensating windings are placed in the pole faces of the main poles and are designed to compensate for the distortion effects of the armature reaction under the main pole, the commutation in this case is known as overcommutation as shown in Fig. 5.11.

5.6 Field Excitation of DC Machines

The performance characteristics of a dc machine depend largely on the type of connections between the armature and field windings. Such connections are classified as shunt, series, and compound dc machines. Figure 5.12 shows the schematic diagrams of different connections of dc machines.

Before discussing the different types of connections, a summary of the most important equations of dc machines will be reviewed.

$$e = K_T\phi\omega_m \qquad \text{V} \tag{5.33}$$

$$T = K_T\phi i_a \qquad \text{N}\cdot\text{m} \tag{5.34}$$

$$V = e \pm R_a i_a \qquad \text{V} \tag{5.35}$$

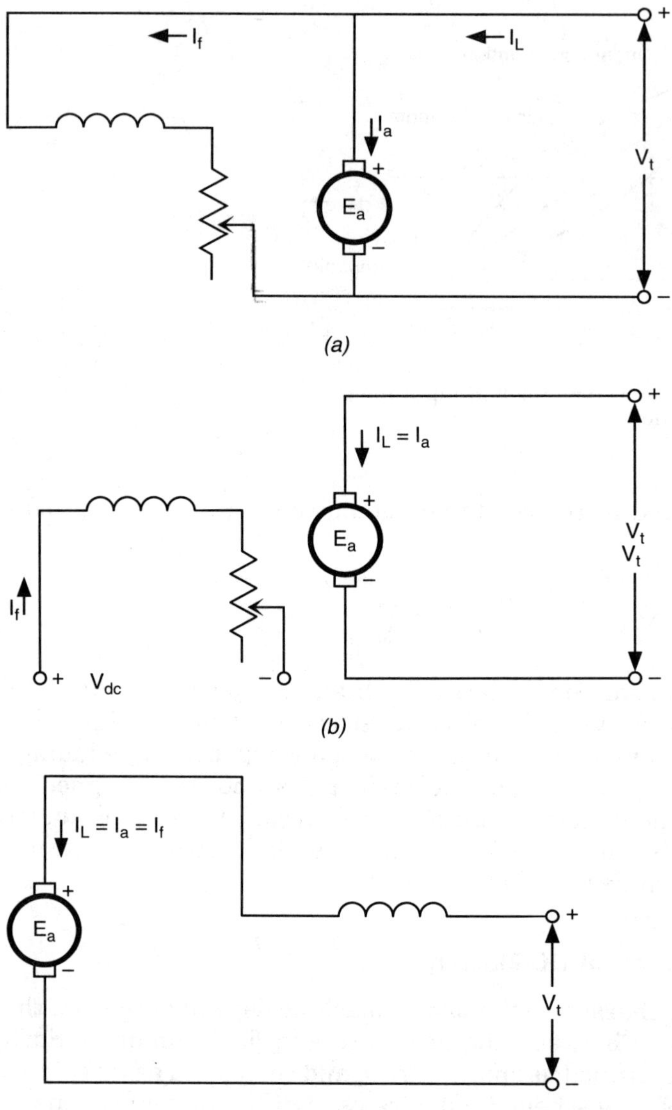

Figure 5.12 Schematic diagrams of connections for dc machines: (*a*) self-excited shunt, (*b*) separately excited shunt, (*c*) series, (*d*) long shunt, (*e*) short shunt.

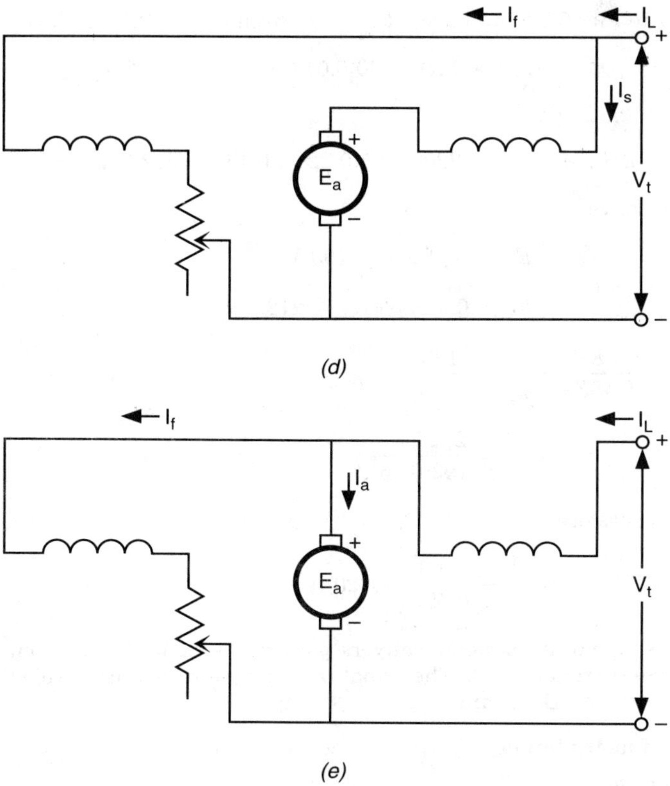

Figure 5.12 *(Continued)*

where R_a is the armature resistance and ± signs show the dc machine operating as a motor (positive) or generator (negative).

Equation (5.35) relates the armature terminal voltage to the armature induced voltage. The speed relationship can be found by substituting Eq. (5.34) into Eq. (5.35) and then solving for speed. We have

$$V = K_T\phi\omega_m \pm R_a i_a$$

$$\omega_m = \frac{V \mp R_a i_a}{K_T\phi} \qquad \text{rad/s} \tag{5.36}$$

Example 5.4 A dc series motor has a line current of 90 A from a 220-V grid. The motor rotates at 900 rpm. The series resistance is 0.12 Ω and the armature resistance is 0.18 Ω. The magnetic flux corresponding to 25 A is 45 percent of the flux corresponding to 90 A. Determine the speed of the motor at a line current of 25 A at 220 V.

solution Referring to Fig. 5.12*c* we have, at $I_a = 90$ A and $V_t = 220$ V,

$$E_1 = V_t - I_a(R_a + R_{ser}) = 220 - 90(0.012 + 0.18) = 193 \text{ V}$$

At $I_a = 25$ A and $V_t = 220$ V,

$$E_2 = V_t - I_a(R_a + R_{ser}) = 220 - 25(0.12 + 0.18) = 212.5 \text{ V}$$

From Eq. (5.30) we have

$$E_1 = K_T\phi\omega_{m1} = 193 \text{ V}$$

$$E_2 = 0.45K_T\phi\omega_{m2} = 212.5 \text{ V}$$

$$\frac{K_T\phi_{m1}}{0.45K_T\phi\omega_{m2}} = \frac{193}{212.5}$$

$$\omega_{m2} = \frac{212.5\omega_{m1}}{193(0.45)}$$

But $\omega = 2\pi n/60$. Therefore,

$$n = \frac{212.5(900)}{193(0.45)} = 220 \text{ rpm}$$

Example 5.5 A 200-V shunt dc motor delivers power of 40 kW at 1800 rpm, while drawing a line current of 50 A. The armature and field resistances are 0.6 Ω and 125 Ω, respectively. Determine

1. The developed motor torque
2. The motor efficiency

solution From Fig. 5.12*a*

$$I_f = \frac{V_t}{R_f} = \frac{208}{125} = 1.664 \text{ A}$$

$$I_a = I_L - I_f = 50 - 1.664 = 48.34 \text{ A}$$

$$E = V_t - I_aR_a = 208 - 48.34(0.06) = 205.9 \text{ V}$$

Using Eq. (5.21)

1. $$T_m = \frac{EI_a}{\omega_m} = \frac{205(48.32)(60)}{2\pi(1800)} = 52.55 \text{ N} \cdot \text{m}$$

2. $$P_m = 40{,}000 + V_fI_f + I_a^2R_a$$

$$= 40{,}000 + 208(1.664) + (48.32)^2(0.06) = 40.486 \text{ kW}$$

$$\eta = \frac{40 \times 100\%}{40.486} = 98.8\%$$

Example 5.6 A self-excited shunt generator, 15 kW, 220 V, delivers a rated load. The armature voltage drop is 5 percent of the rated voltage and the shunt field current equals 5 percent of the rated load current. Calculate the resistance of the field circuit and armature circuit.

solution Referring to Fig. 5.12*a*

$$V_t = E - I_a R_a \qquad E = V_t + I_a R_a$$

$$I_L = \frac{15 \times 10^3}{220} = 68.18 \text{ A}$$

$$I_f = 0.05 I_L = 0.05(68.18) = 3.409 \text{ A}$$

From Eq. (5.40)

$$I_a = I_L + I_f = 68.18 + 3.409 = 71.59 \text{ A}$$

The armature voltage is

$$I_a R_a = 0.05 V_t = 0.05(220) = 11 \text{ V}$$

$$R_a = \frac{11}{71.59} = 0.1537\ \Omega$$

The field voltage is

$$I_f R_f = V_t$$

$$R_f = \frac{V_t}{I_f} = \frac{220}{3.409} = 64.53\ \Omega$$

5.7 DC Generator Performance in Steady State

The steady-state performance of a dc generator is described by its load characteristics. During steady-state conditions the currents are constant or they change very slowly. The voltage drop due to conductive effects is therefore negligible.

The voltage at the armature terminals of a dc generator is given by

$$V_t = E - R_a I_a \tag{5.37}$$

where E is generated emf given in Eq. (5.33).

Assuming that the magnetic circuit is linear, then the magnetic flux is directly proportional to the field current $\phi = KI_f$. Therefore, Eq. (5.36) becomes

$$V_t = KK_T \omega_m I_f - R_a I_a = K_g I_F \omega_m - R_a I_a \tag{5.38}$$

where $K_g = KK_T$ is a new constant.

5.7.1 The self-excited shunt generator

In the self-excited shunt generator, the terminal voltage applied on the field circuit in Fig. 5.12*a* is

$$V_t = (R_f + R_c)I_f \tag{5.39}$$

where R_f is the field resistance and R_c is the resistance of the control rheostat.

If the field resistance is constant, the field current decreases with decreases in terminal voltage. The load characteristic for a self-excited shunt generator when the speed and field resistance are constant is shown in Fig. 5.13. In self-excited generators the armature current is equal to load current plus the field currents:

$$I_a = I_L + I_f \tag{5.40}$$

where I_L and I_f are load and field currents, respectively.

Note: In Fig. 5.12 all current directions are for motor operation; for generator operation the opposite direction should be taken.

If constant voltage is required, the self-excited generator is employed. The voltage drop from no-load to full load is very small. The self-excited generator is simpler and more economical than the separately excited generator, because it requires no external source for its field current.

5.7.2 Separately excited dc generator

In the separately excited dc generator the field and rotor circuit is not electrically connected. The equation for the rotor circuit for steady-

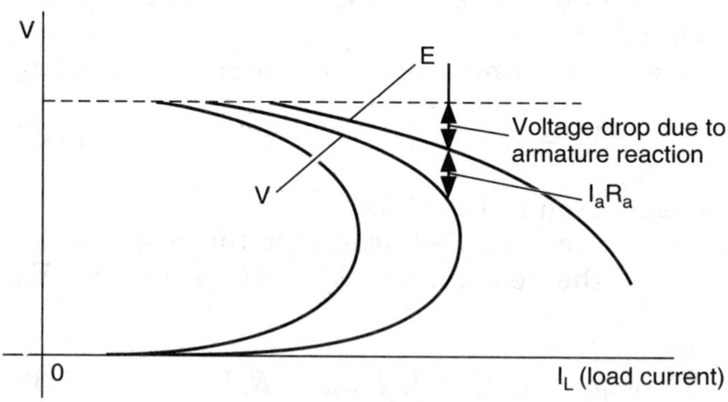

Figure 5.13 Voltage-current characteristics of self-excited dc shunt generators.

state operation becomes

$$V_t = E - R_a I_a \tag{5.41}$$

Using Eq. (5.33) and $f = KI_f$ enables us to write the generated voltage as

$$E = KK_T I_f \omega_m \tag{5.42}$$

The external characteristics of the separately excited generator plotted in Fig. 5.14 is made possible by holding the field current I_f and the angular velocity ω_m constant. At the same time armature reaction is ignored. Separately excited generators are used on loads that require a wide variation in the output voltage, such as motors that must operate through large speed ranges.

5.7.3 Compound dc generator

The performance of a dc generator is better described by its external characteristics. Figure 5.15 shows the external characteristics of different types of dc generators. A comparison of these curves shows that the terminal voltage of a shunt generator decreases faster than it does for a separately excited generator.

It is always desirable to operate a generator with nearly constant voltage at its terminals. The best way to achieve this is to combine the shunt and series winding to form a dc generator known as a compound generator, as shown in Fig. 5.12*d* and 5.12*e*. The external characteristics of a compound generator may possess a shape which is almost flat with the load current.

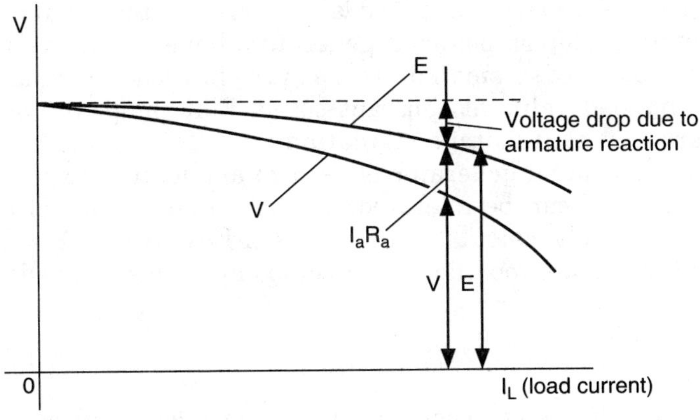

Figure 5.14 Voltage versus current characteristics of a separately excited dc generator.

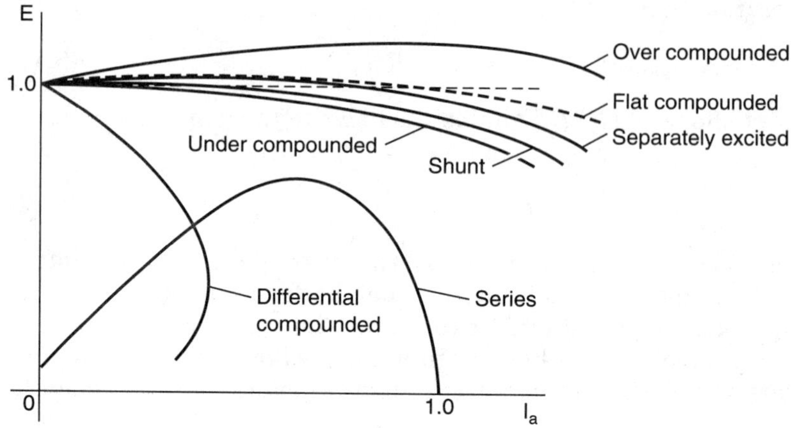

Figure 5.15 Comparison of external characteristics of various types of dc generators.

If the mmf's of the series and shunt windings are aiding, the dc generator is known as additive. If, on the other hand, the mmf's in the two windings are opposite one another, then the dc generator is subtractive. These windings give rise to two types of dc generators, namely, the cumulative compound and the differential compound generators.

Depending on the relative strengths of the mmf's in the two windings, the cumulative compound generator may be either overcompounded, flat compounded, or undercompounded. When the generator is overcompounded, the series field is strong enough to give a rising terminal voltage characteristic in the normal load range at constant speed. The field counteracts the effect of a decrease in the prime mover speed with increasing load. Overcompounding may also be used to compensate for the line drop when the load is at a considerable distance from the generator. Undercompounded generators have a relatively weak series field and there is a small decrease in the full-load terminal voltage from its no-load value at constant speed. Flat compounding maintains a nearly constant voltage regulation.

The differential compound generator is used in applications where a wide variation in load can be tolerated. The terminal voltage of a compound generator can be calculated by means of Eqs. (5.32), (5.39), and (5.40). Equation (4.50), modified for a compound generator, will be equal to

$$V_t = E - I_a(R_a + R_{\text{series}}) \tag{5.43}$$

Example 5.7 A 20-kW, 28-V, short-shunt compound generator has an armature circuit resistance including brushes of 0.055 Ω, series field resistance of 0.003 Ω, and shunt field current of 3 A. Calculate

1. The induced armature voltage at rated load and rated terminal voltage
2. The shunt field resistance

solution From Fig. 5.12*e*

$$I_L = \frac{20{,}000}{208} = 96.15 \text{ A}$$

$$I_a = I_L + I_f = 96.15 + 3 = 99.15 \text{ A}$$

1. $E = V_t + I_a R_a + I_L R_{\text{ser}} = 208 + 99.15(0.055) + 96.15(0.03)$

$= 216.334$ V

2. $I_f R_f = E - I_a R_a = 216.334 - 99.5(0.055) = 210.86$ A

$$R_f = \frac{210.86}{3} = 70.287\ \Omega$$

Example 5.8 A 7-kW, 208-V, 1150-rpm shunt dc generator has an armature resistance of 0.16 Ω. The armature current is 5 A when the generator is operated as a motor on no-load at rated speed and at rated terminal voltage. The no-load saturation curve of the generator driven at 1150 rpm is given by the following data:

Voltage, V	5	35	75	125	136	140
I_f, A	0	0.5	1.0	2	2.5	3

Calculate the full-load efficiency of the generator at rated speed and at rated terminal voltage.

solution From Fig. 5.12*a*,

1. When the machine is running as a motor:

$$V_t = E + I_a R_a$$

$$E = V_t - I_a R_a = 208 - 5(0.16) = 207.2 \text{ V}$$

Rotational losses

$$P_m = E I_a = 207.2(5) = 1036.0 \text{ W}$$

2. When the machine is operated as a generator:

$$I_a = I_L + I_f = 33.65 + I_f$$

$$E = V_t + I_a R_a = 208 + 0.16(33.65 + I_f) = 213.38 + 0.16 I_f$$

The above equation is a line drawn parallel with the abscissa and intersecting the magnetization curve. From the values of voltage, the field current is 2.22 A. Substituting $I_f = 2.22$ A in the equations obtained earlier, we have

$$I_a = I_L + I_f = 33.65 + 2.22 = 35.87 \text{ A}$$

$$E = V_t + I_a R_a = 208 + 0.16(35.87) = 213.74 \text{ V}$$

$$\text{Efficiency } (\eta) = \frac{P_{\text{out}}}{P_{\text{out}} + P_{\text{losses}}} \times 100\% = \frac{7000 \times 100\%}{7000 + 1036 + 667.25} = 80.42\%$$

5.7.4 Voltage buildup in self-excited dc generators

A dc generator is capable of generating its own field excitation at normal speed. Voltage buildup requires that some residual magnetism exist in the magnetic circuit of the generator in order that the process can start. This process can be explained clearly by using Fig. 5.16. In Fig. 5.16 the magnetizing curve has been plotted together with the field resistance line. The field resistance line is the plot of field voltage versus field current for the resistance of the field circuit of the self-excited generator.

The buildup process is as follows: When a shunt generator is driven with its field circuit open, a small residual voltage is generated in the armature of the dc generator if there is a residual flux. In Fig. 5.16, a residual voltage E_r is shown. When the shunt field is connected back

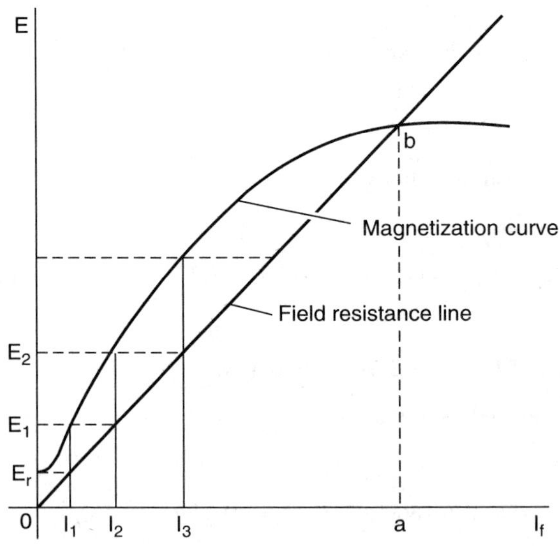

Figure 5.16 The voltage buildup in a self-excited shunt generator at constant speed.

to the armature of the dc generator, with the correct polarity, field current is initiated causing the flux to increase with the resultant increase in voltage to E_1. The increase of the armature voltage eventually increases the field current to I_2 which in turn builds up the armature voltage to E_2. The process continues until the current reaches Oa and the terminal voltage reaches the intersection of the magnetizing curve and field resistance line. At this point the terminal voltage ab equals the voltage drop in the field circuit. The shunt generator will then operate in stable condition supplying the load current.

5.8 Torque-Speed Characteristics of Various Types of Motors Under Steady-State Operation

Various ways of dc armature and field winding connections were shown in Fig. 5.12. Depending on the field excitation and rotor windings, the dc motor assumes different names.

5.8.1 Series motors

The field winding of a series motor is connected in series with the armature winding. In this case the field and armature currents are identical ($I_a = I_f$) as shown in Fig. 5.12c. The basic dc machine equations, given in Eqs. (5.32) and (5.33), will still apply but at this time, magnetic flux depends on load current. Assuming that the series motor operates on the linear portion of the magnetizing curve, then the flux will be equal to

$$\phi = K_1 I_f = K_1 I_a \tag{5.44}$$

where $I_f = I_a$ and K_1 is a constant.

The emf and torque equations become

$$E = K_T \phi \omega_m = \frac{K_T(2\pi n \phi)}{60} = \frac{K_T K_1(2\pi n)}{60} I_a = K_E n I_a \tag{5.45}$$

where $K_E = K_T K_1(2\pi)/60$ is the new constant and

$$T = K_T \phi I_a = K_T K_1 I_a^2 = K_m I_a^2 \tag{5.46}$$

where $K_m = K_T K_1$ is a new constant. Equation (5.46) indicates that the torque developed by a series motor is proportional to the square of the armature current.

In order to determine speed-torque characteristics, we note from Fig. 5.12*c* that

$$V_t = E + I_a(R_a + R_{ser}) = K_E n I_a + I_a(R_a + R_{ser})$$
$$= I_a(K_E n + R_T) \quad (5.47)$$

where $R_T = R_1 + R_{ser}$ is the total resistance of the armature circuit. Expressing Eq. (5.47) in terms of armature current I_a, and then substituting into Eq. (5.46), the equation for torque is given by

$$T = K_m \frac{V_t^2}{(K_e n + R_T)^2} \quad \text{N} \cdot \text{m} \quad (5.48)$$

For values of n, such that $K_E n \gg R_T$, the torque equation (5.48) will be inversely proportional to the speed, that is,

$$T = K_m \frac{V_t^2}{(K_e n)^2} \quad (5.49)$$

A plot of Eq. (5.49) is shown in Fig. 5.17.

In commercial aspects, the motors are the principal electric traction drives, and they are widely used in all types of electric vehicles, electric trains, industrial overhead cranes, etc.

5.8.2 Shunt motors

The shunt motor schematic diagram is shown in Fig. 5.12*a*. In this figure the motor line current supplies both the armature and field

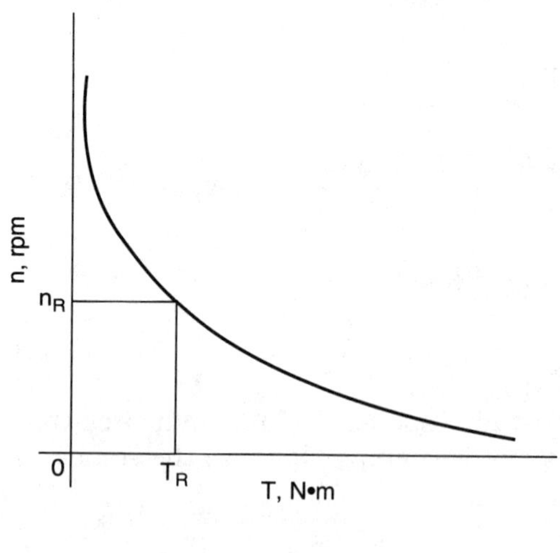

Figure 5.17 Speed versus torque characteristic of a series motor.

current. The line current is given by

$$I_L = I_a + I_f \tag{5.50}$$

Once again the basic dc machine equations (5.33) and (5.34) apply to the shunt motor. The basic equations are

$$E = K_T \phi \omega_m = \frac{K_T \phi (2\pi n)}{60} = K n \phi \tag{5.51}$$

where $\omega_m = 2\pi n/60$
n = speed, in rpm
K = new constant

and

$$T = K_T \phi I_a \tag{5.52}$$

From Fig. 5.12*a*, an expression for the motor terminal voltage is given by

$$V_t = E + I_a R_a = K n \phi + I_a R_a \tag{5.53}$$

Expressed in terms of speed, we have

$$n = \frac{V_t - I_a R_a}{K\phi} \tag{5.54}$$

An expression for speed as a function of developed torque is obtained by solving the torque equation in terms of the armature current and then substituting it into Eq. (5.54) to yield

$$n = \frac{V_t}{K\phi} - \frac{R_a T}{K\phi^2 K_T} = \frac{V_t}{K\phi} - \frac{R_a T}{K K_T \phi^2} = n_{n1} - mT \tag{5.55}$$

Equation (5.55) indicates a straight-line relation between speed and developed torque, as shown in Fig. 5.18.

The *speed regulation* (SR) of a dc motor is defined as the percent change in speed from no-load to rated load with respect to rated speed when operating at rated voltage and rated temperature from a constant voltage load.

$$\text{SR} = \frac{\omega_{\text{NL}} - \omega_{\text{FL}}}{\omega_{\text{FL}}} \times 100\%$$

$$(5.56)$$

or

$$\text{SR} = \frac{n_{\text{NL}} - n_{\text{FL}}}{n_{\text{FL}}} \times 100\%$$

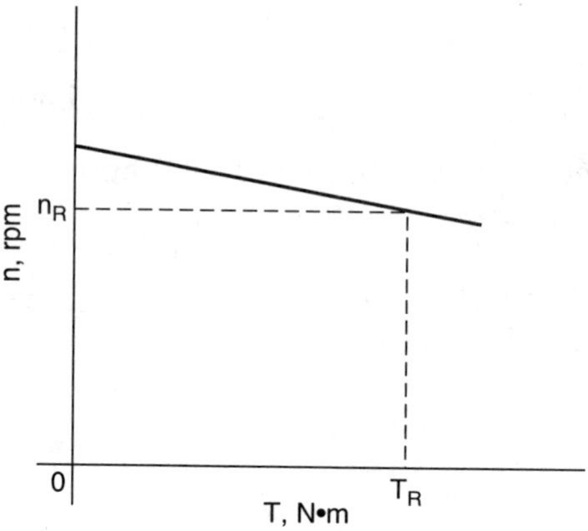

Figure 5.18 Speed versus torque characteristic of a shunt dc motor.

where n_{NL} = no-load speed, rpm
n_{FL} = full-load speed, rpm
SR = speed regulation

Speed regulation is an estimate of the shape of a motor's torque speed characteristic. When we have positive speed regulation this means that a motor's speed drops with the increasing load, and negative speed regulation means that the motor's speed increases with the increasing load.

Example 5.9 A 240-V, 1200-rpm, 4-hp dc motor operates at rated conditions and has a speed regulation of 21.4 percent. Calculate the no-load speed.

solution From Eq. (5.56)

$$SR = \frac{n_{NL} - n_{FL}}{n_{FL}} \times 100$$

$$N_{NL} = \frac{SR(n_{FL}) + n_{FL}(100)}{100} = \frac{21.4(1200) + 1200(100)}{100} = 1456.8 \text{ rpm}$$

5.8.3 Control of terminal voltage

The terminal voltage of separately excited dc generators can be controlled by changing the internal voltage E_a of the machine. Using Kirchhoff's voltage law, $V_t = E_a - I_a R_a$, it can be shown that, if E_a is increased, the terminal voltage V_t increases, and if E_a decreases, V_t will also decrease. From Eq. (5.30) we see that there are two possible ways to control the voltage of the generator:

1. *Change the speed of rotation* From Eq. (5.30) $E_a = K_d \omega_m$. If ω increases the generated voltage V_t will also increase. Since E_a in-

creases, V_t will also increase as shown by applying Kirchhoff's voltage law to Fig. 5.12*b*: $V_b = E_a - I_a R_a$.

2. *Change the field current* If R_f is decreased, then the field current increases, $I_f = V_f/R_f$, and the magnitude flux f in the machine increases. As the magnetic flux increases, the terminal voltage increases.

In most applications the speed of the prime mover is quite limited, so the terminal voltage is commonly controlled by changing the field current. Figure 5.19*a* shows a separately excited generator driving a resistance load and Fig. 5.19*b* shows the effect of a decrease in field resistance on the terminal voltage of the dc generator when operating under a load.

5.8.4 Graphical analysis of a separately excited dc generator

At this juncture it is proper to discuss briefly the graphical method of a separately excited dc generator. Since the internal generated voltage of a generator is a nonlinear function of its magnetomotive force, it is very difficult to predict analytically the value of E_a to be expected from a given field current. The output voltage can be calculated by means of the magnetization curve of a generator. If the machine has armature reaction its flux will be reduced with the increase in load, causing the decrease in generated voltage. The only method to determine the output voltage accurately is to use graphical analysis.

The total magnetomotive force (mmf) in a separately excited generator is

$$F_{\text{net}} = N_f I_f - F_{\text{AR}} \tag{5.57}$$

where F_{AR} is the mmf due to armature reaction. Since magnetization curves are plots of E_a versus I_f it is the same output voltage of the machine. Thus

$$I'_F = I_f - \frac{F_{\text{AR}}}{N_f} \tag{5.58}$$

Since each magnetization curve is drawn for a particular speed, usually at rated speed of the machine, it is possible to find the effects of a given field current even if the generator is turning at another speed than the rated speed.

The induced voltage in a dc generator is given by Eq. (5.30).

$$E_a = K_T \phi \omega_m = \frac{60K}{2\pi} \phi n = K'_T \phi n \tag{5.59}$$

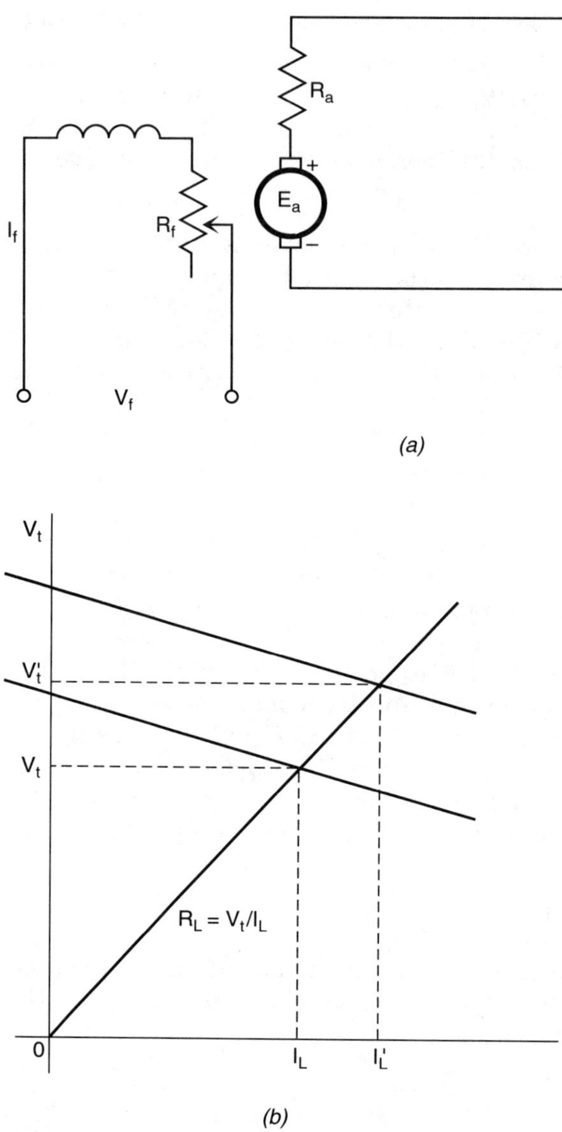

Figure 5.19 (*a*) A separately excited dc generator with a resistive load. (*b*) The effect of decreasing R_f on the output voltage of the generator.

where $K'_T = 60K/2\pi$ and n is the speed, in rpm.

For a given effective field current, the flux on the machine is fixed so the generated voltage in a machine is related to the speed by

$$\frac{E_a}{E_{a0}} = \frac{n}{n_0} \tag{5.60}$$

where E_{a0} and n_0 represent the reference values of voltage and speed. If the reference condition is taken to be the speed of the generator when the magnetization curve was made, and the other condition is taken to be the actual speed of the generator, then it is possible to find the actual voltage produced by the machine even if it is not rotating at the rated speed.

Example 5.10 A separately excited dc generator is rated at 20 kW, 230 V, 100 A, and 1800 rpm. The generator is shown in Fig. 5.20 and its magnetization curve is shown in Fig. 5.21. The machine has the following characteristics:

$$R_a = 0.5\ \Omega \qquad V_f = 430\ \text{V}$$

$$R_f = 20\ \Omega \qquad N_f = 100\ \text{turns/pole} \qquad R_{\text{adj}} = 80\ \Omega$$

Answer the following questions:

1. If the variable resistor is adjusted to 60 Ω and the generator prime mover is at 1600 rpm, find the generator's no-load terminal voltage.
2. What would its voltage be if a 2-Ω load was connected to its terminals.
3. State what adjustment could be made to the generator to restore its terminal voltage found in part 1.
4. Calculate the field current which would be required to restore the terminal voltage to its no-load value? Find the R_{adj} resistor.

solution

1. Total field resistance

$$R_f + R_{\text{adj}} = 20 + 60 = 80\ \Omega$$

Field current

$$I_f = \frac{V_f}{R_f + R_{\text{adj}}} = \frac{430}{80} = 5.137 \simeq 5.4\ \text{A}$$

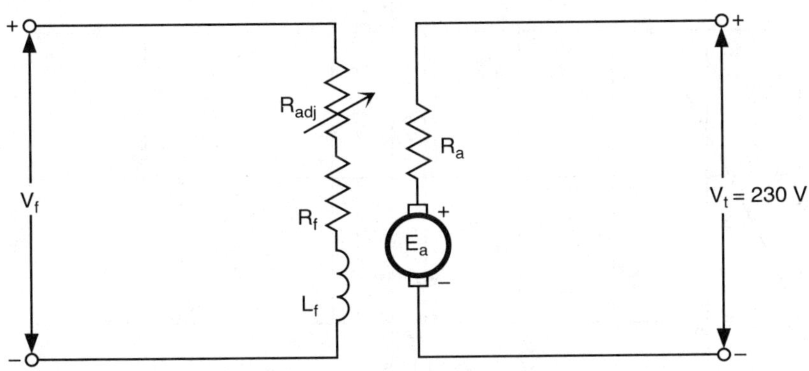

Figure 5.20 The separately excited dc generator for Example 5.20.

From Fig. 5.21 for $I_f = 5.4$ A $E_a = 430$ N at 1800 rpm. The generator actual speed is $n_m = 1800$, as the internal voltage E_a will be

$$\frac{E_a}{E_{a0}} = \frac{n}{n_0} = \frac{1600}{1800} = \frac{E_a}{430}$$

$$E_{a0} = \frac{430(1600)}{1800} = 382.22 \text{ V}$$

2. If a 2-Ω load was connected to the generator terminals, the total current flow should be

$$I_a = I_L = \frac{E_q}{R_a + R_{\text{Load}}} = \frac{382.22}{0.05 + 2} = 186.45 \text{ A}$$

$$V_t = I_L = R_a R_{\text{Load}} = 186.45(2) = 372.89 \text{ V}$$

3. Since the generator terminal voltage has fallen, to restore the original value the generator voltage must be increased, which means E_a must in-

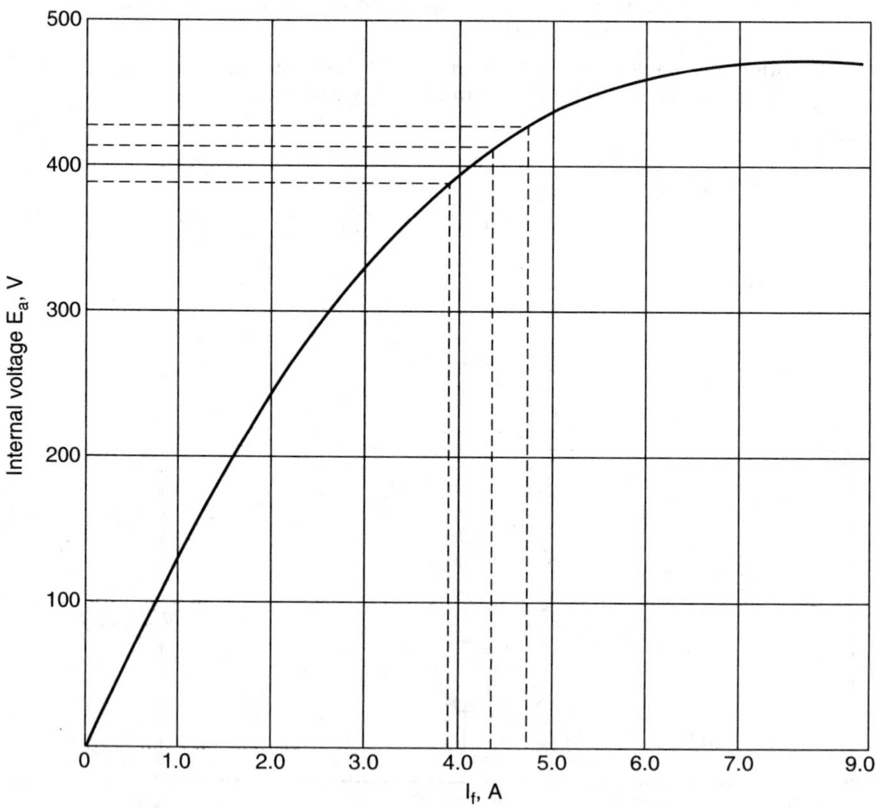

Figure 5.21 The magnetization curve for Example 5.10.

crease and R_{adj} must be decreased in order to increase the field current of the generator.

4. For the terminal voltage to return to 382.22 V the live current must rise to

$$I_L = \frac{382.22}{20} = 191.11 \text{ A}$$

Therefore,

$$E_a = V_t + I_a R_a = 382.22 + 191.11(0.05) = 391.8 \text{ V}$$

for $E_a = 391.8$ V at $n_m = 1600$ rpm. The equivalent voltage at 1800 rpm would be

$$\frac{E_a}{E_{n0}} = \frac{n}{n_0} = \frac{1600}{1800} = \frac{391.8}{E_{a0}}$$

$$E_{a0} = \frac{1800(3918)}{1600} = 440.75$$

From the magnetization curve for $E_a = 440.8$ V, $I_f = 6.0$ A. The total field circuit resistance is

$$R_f + R_{adj} = \frac{V_f}{I_f}$$

$$20 + R_{adj} = 430/6$$

$$R_{adj} = 430/6 - 20 = 51.67 = 52\ \Omega$$

5.8.5 Control of dc motors

Equation (5.36)

$$\omega_m = \frac{V_t - R_a I_a}{K_T \phi} \tag{5.36}$$

shows that the speed of a dc motor can be varied by control of the field flux, the resistance of the armature, and the armature applied voltage. The three most common motor speed-control methods are shunt field rheostart control, armature circuit resistance control, and armature terminal voltage control. The base speed of the machine is defined as the speed with rated armature voltage and normal armature resistance and field flux. Speed control above the base value can be obtained by varying the field flux. By inserting a series resistance in the shunt field circuit of a dc motor (or a compound motor), speed control over a wide range above the base speed can be achieved. In the case of a series motor, speed control above the base value can be achieved by placing a diverter resistance in parallel with the series winding so that the field current is less than armature current. The use of a shunt diverter in series dc motors is shown in Fig. 5.22. The line current is equal to the armature current, but the field current is less. The developed torque is given by Eq. (5.11)

$$T = K_T \phi I_a \tag{5.11}$$

Assuming that the dc series motor operates on the linear portion of the magnetizing curve, the magnetic flux will be equal to

$$\phi = K_I I_f \tag{5.61}$$

Therefore, Eq. (5.61) becomes

$$T = K_I K_T I_f I_a \tag{5.62}$$

From Fig. 5.22

$$I_f = \frac{R_d}{R_d + R_s} I_a \tag{5.63}$$

Therefore,

$$T = K_I K_T \left(\frac{R_d}{R_d + R_s}\right) I_a^2 \tag{5.64}$$

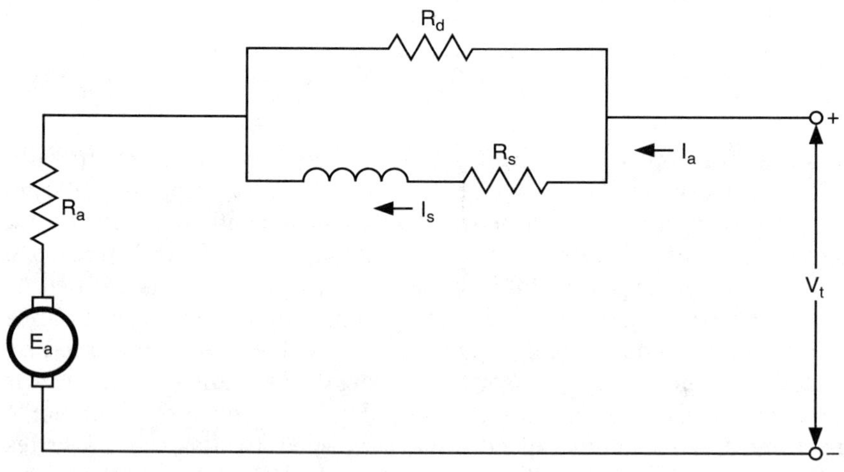

Figure 5.22 A diverter speed control in a dc series motor. The developed torque is given by Eq. (5.11).

From Fig. 5.22, using Kirchhoff's voltage law, we have

$$V_t = E_a + I_a\left(R_a + \frac{R_d}{R_d + R_s}\right) \tag{5.65}$$

But
$$E_a = K_I K_T I_a \left(\frac{R_d}{R_d + R_s}\right)\omega_m$$

Therefore,

$$V_t = K_I K_T I_a \left(\frac{R_d}{R_d + R_s}\right)\omega_m + I_a\left(R_a + \frac{R_d}{R_d + R_s}\right)$$

$$= K_I K_T I_a \beta\omega_m + I_a(R_a + \beta) \tag{5.66}$$

where $\beta = R_d/(R_d + R_s)$.

$$I_a = \frac{V_t}{K_I K_T \beta\omega_m + (R_a + \beta)} \tag{5.67}$$

Substituting Eq. (5.67) into Eq. (5.64):

$$T = K_I K_T \left(\frac{R_d}{R_d + R_s}\right)\left[\frac{V_t}{K_I K_T \beta\omega_m + (R_a + \beta)}\right]^2 \tag{5.68}$$

Assuming that R_a and R_s are negligible we can write

$$T = \frac{V_t^2}{\beta K_I K_T \omega_m^2} \quad \text{N} \cdot \text{m} \tag{5.69}$$

Since β is less than that one, we realize that for the same speed, the torque increases by the use of the diverter.

Example 5.11 A series dc motor is rated 120 V and has armature resistance of 0.05 Ω and series resistance of 0.1 Ω. The constant $K_I K_T$ is found to be 38.2 × 10^{-3} V/A · rad · s. The diverter resistance is 0.9 Ω for a speed of 55 rad/s. Find the resulting torque.

solution Using Eq. (5.68) we have

$$T = K_I K_T \left(\frac{R_d}{R_d + R_s}\right)\frac{V_t^2}{(K_I K_T[R_d/(R_d + R_s)]\omega_m + \{R_a + [R_d/(R_d + R_s)]\})^2}$$

$$\beta = \frac{R_d}{R_d + R_s} = \frac{0.9}{0.9 + 0.1} = 0.9$$

$$\therefore T = \frac{(38.2 \times 10^{-3})(0.9)(120)^2}{[(38.2 \times 10^{-3})(0.9)(55) + (0.05 + 0.9)]^2} = 61.3418 N_m$$

Note: This torque is higher than that without a diverter.

Problems

1 A 140-kW, 220-V shunt generator, as shown in Fig. 5.12*a*, has an armature resistance of 0.03 Ω, and a field resistance of 40 Ω. The generator operates at 1800 rpm. Determine

a The full-load current I_L

b The field and armature currents

c The developed electromagnetic torque T_m

2 A 20-kW, 120-V series generator, as shown in Fig. 5.12*c*, has an armature and field resistance of 0.1 Ω and 0.05 Ω, respectively. The generator operates at 1800 rpm. Determine

a The load current I_L

b The field and armature currents

c The generated emf E

d The loading resistance for rated conditions R_L

e The developed electromagnetic torque

3 The dc series motor has a rating of 80 hp, 1200 V. When tested at 2000 rpm, it exhibits the following results: armature current $R_f = 0.4$ Ω, core resistance $R_c = 0.02$ Ω, core loss $P_c = 1300$ W, friction loss $P_{\text{fr}} = 70$ W, and windage loss $P_{\text{WR}} = 150$ W. Determine

a The electromagnetic power P_m

b The electromagnetic torque T_m

c The shaft power P_{sh}

d The shaft torque T_{sh}

e The efficiency η

4 A 4-pole wave-wound dc machine has 100 conductors on the rotor. The flux per pole produced by the field current is $\phi = 0.03$ Wb. The induced emf of the machine is 124 V. Calculate the speed of the machine.

5 A 240-V shunt-connected dc motor takes a total current of 2.5 A on no-load, and 45 A on full load. The armature circuit resistance is 0.3 Ω including brushes, and the field resistance is 200 Ω. If the flux per pole is reduced to 3 percent, compared to its no-load value, calculate the percentage change in speed if the flux per pole remains constant.

6 Determine the resistance which could be placed in series with the armature of the machine discussed in Prob. 5, if the full-load speed is to be reduced by 50 percent with the total developed torque remaining constant. The field circuit resistance is 200 Ω.

7 An open-circuit test conducted on a shunt generator, yielded the following results:

Field current, A	1	3	4.5	7	9	12	15
Armature voltage, V	20	210	335	420	480	520	580

The generator is then shunt connected and runs at rated speed. The armature circuit resistance and the field resistance are 0.13 Ω and 43 Ω, respectively. Find the output voltage and current when the armature current is 350 A (neglecting armature reaction).

8 The test results taken on a dc shunt machine at 2500 rpm are recorded as follows:

Field current, A	0	1	2	3	4	5
Armature voltage, V	2	60	115	160	176	188

If the machine is operated as a self-excited shunt generator, calculate the shunt field resistance necessary to obtain a no-load voltage of 120 at a speed of 2500 rpm.

9 A 400-kW, 10-pole dc generator has an open-circuit armature terminal potential difference of 200 V when driven at 500 rpm. There are 500 conductors on the armature, which is lap wound. Calculate the flux per pole.

10 A shunt dc motor, 120 V, 50 A, has the following parameters: $R_a = 0.12$ Ω, $R_f = 100$ Ω, $R_{\text{adj}} = 110$ Ω. The motor runs at 1200 rpm. Determine

a The no-load speed of the motor when $R_{\text{adj}} = 110$ Ω

b The full-load speed

c The speed regulation

11 A shunt dc generator, 240 V, 50 A, has an armature resistance of 0.12 Ω, $R_f = 100$ Ω, and an adjustable resistance of 130 Ω. The generator rotates at 1200 rpm. The prime mover torque of 9.1 N · m is applied to the shaft of the generator at no load. A 90-A load is then connected to the terminals of the generator. Calculate the following:

a Terminal voltage V_t

b Generated voltage E_a

c Mechanical power P_m

d Copper losses

e Mechanical and core losses

f Torque applied to the generator by the prime mover

Bibliography

Brown, D., and E. P. Hamilton III: *Electromechanical Energy Conversion,* Macmillan, New York, 1984.

Elgerd, O. I.: *Basic Electric Power Engineering,* Addison-Wesley, Reading, Mass., 1977.

Hunt, W. Jr., and R. Stein: *Electric Power System Components—Transformers and Rotating Machines,* Van Nostrand-Reinhold, New York, 1979.

Kingsley, C. Jr., and A. E. Fitzgerald: *Electric Machinery,* McGraw-Hill, New York, 1983.

Mablekos, Van E.: *Electric Machine Theory for Power Engineers,* Harper and Row, New York, 1980.
McLean, P. G.: *Electromagnetic and Electromechanical Machines,* Ellis Horwood, Chichester, U.K., 1984.
Matsh, L. W.: *Electromagnetic and Electro-mechanical Machines,* A Dun-Donnelley Publisher, New York, 1977.
Nasar, S. A., and L. E. Unnewehr: *Electromechanics and Electric Machines,* John Wiley and Sons, New York, 1983.
Sarma, M. S.: *Electric Machines,* Wm. C. Brown, Dubuque, Ia., 1985.
Slemon, G. R., and A. Straughen: *Electric Machines,* Addison Wesley, Reading, Mass., 1980.

Chapter

6

Induction Machines

6.1 Introduction

Induction motors are the most widely used motors in industries. The induction motor is one in which the alternating current is supplied to the stator directly and the currents in the rotor windings are induced by the stator currents. In this chapter the induction machine will be analyzed in sufficient detail in order to develop equivalent circuit models for the prediction of their steady-state performance. The induction motor is equivalent to a static transformer with a secondary side capable of rotating with respect to the primary. The induction motor runs below synchronous speed and is known as an asynchronous machine. Its speed decreases with increasing load torque. There are two types of induction motors, according to their rotor construction. The first type is the cylindrical rotor induction motor which carries either conducting bars short-circuited at both ends or a polyphase winding with terminals brought out to slip rings for external connections. The former type is known as a squirrel cage rotor and the latter is known as a wound rotor. These two types are shown in Fig. 6.1.

6.2 Induction Machine Windings

The principles underlying the winding of ac machines are the same as those applied to dc machines. As in dc machines the pitch of each coil must be approximately the same as the pole pitch, and the coils must be connected together so that the emf induced in them acts in the same direction. The ac windings are open-circuit windings. There is a continuous path through the conductors of each phase, with the primary ending of the phase winding free. Below is a brief description of the main types of induction motor windings.

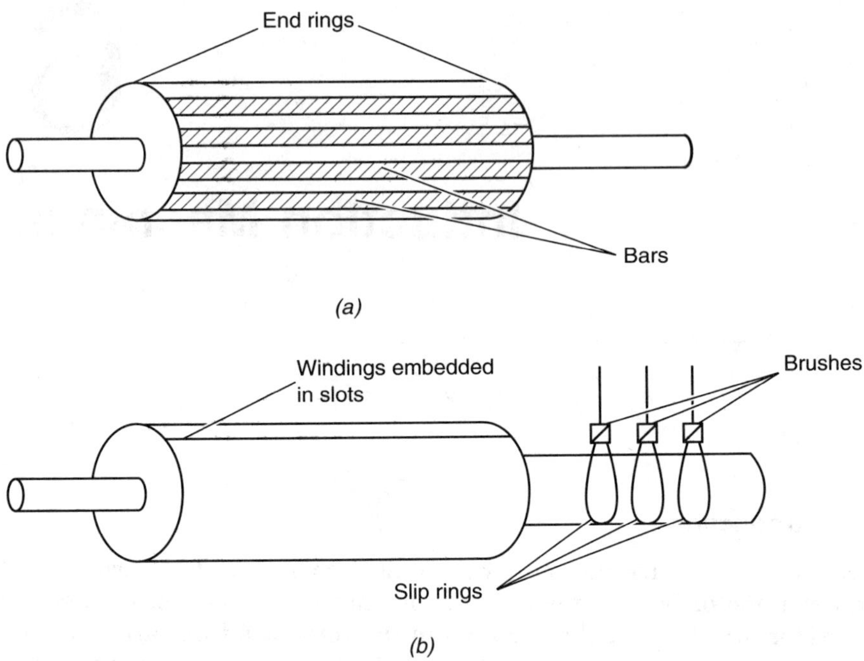

Figure 6.1 (*a*) Squirrel cage rotor. (*b*) Wound rotor.

6.2.1 Half-coil and whole-coil windings

Figure 6.2 shows the developed six-pole induction machine half-coil winding diagram. It shows that the windings are concentrated because all conductors on one side of each pole are grouped together in one slot. The diagram is said to be half-coil because there are half as many coils as there are poles. Figure 6.3 shows the whole-coil single phase concentrated winding. In this diagram the number of coils equals the number of poles. Since there are two coil sides in each slot this type of winding arrangement is known as a double-layer winding. The end connections of a whole-coil winding are uniformly distributed. The concentrated winding produces a rectangular-space mmf wave in the air gap. The rectangular waveform is usually approximated to a sinusoidal distribution by distributing the windings in a number of adjacent slots. The study of the waveform of a concentrated stator coil will be made using Fig. 6.4*a* and *b*.

When the current is in the direction shown in Fig. 6.4*a* a rectangular mmf waveform will be produced as shown in Fig. 6.4*b*. If the coil has N_s turns and carries current i_s the amplitude of the rectangular wave will be given by $N_s i_s/2$, where 2 is the number of poles. Stated in general form we can say that, if a P-pole concentrated winding consisting of

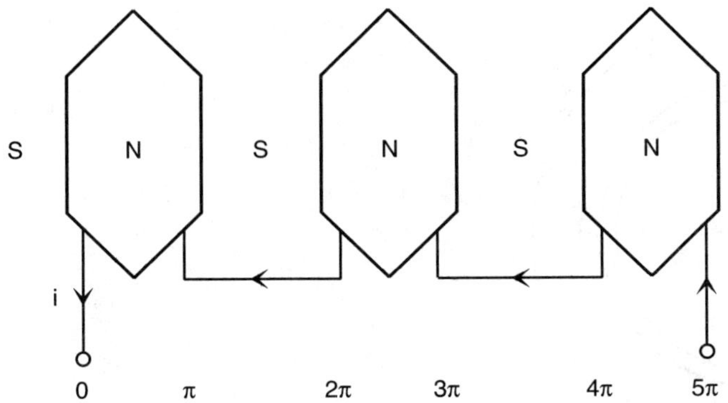

Figure 6.2 Half-coil winding of a six-pole induction motor.

N_s turns per pole were being considered, the peak amplitude of the rectangular mmf waveform would be

$$F_{sg} = \frac{N_s i_s}{P} \qquad \text{At} \tag{6.1}$$

The best way is to describe the rectangular mmf waveform by utilizing a Fourier series of space harmonics:

$$F_{sg} = \frac{4N_s i_s}{\pi P}\left(\sin\theta_{el} + \frac{1}{3}\sin 3\theta_{el} + \frac{1}{5}\sin 5\theta_{el} + \cdots\right] \tag{6.2}$$

The peak amplitude of a fundamental component of a rectangular wave is given by

$$F_{sg} = \frac{4N_s i_s}{\pi P} \qquad \text{At} \tag{6.3}$$

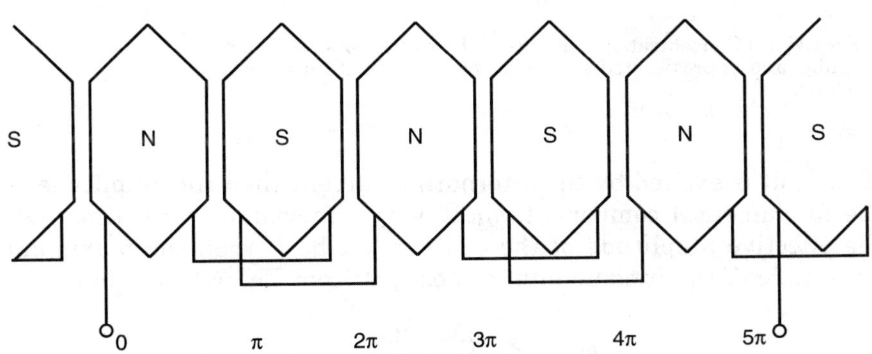

Figure 6.3 Whole-coil winding of six-pole induction motor.

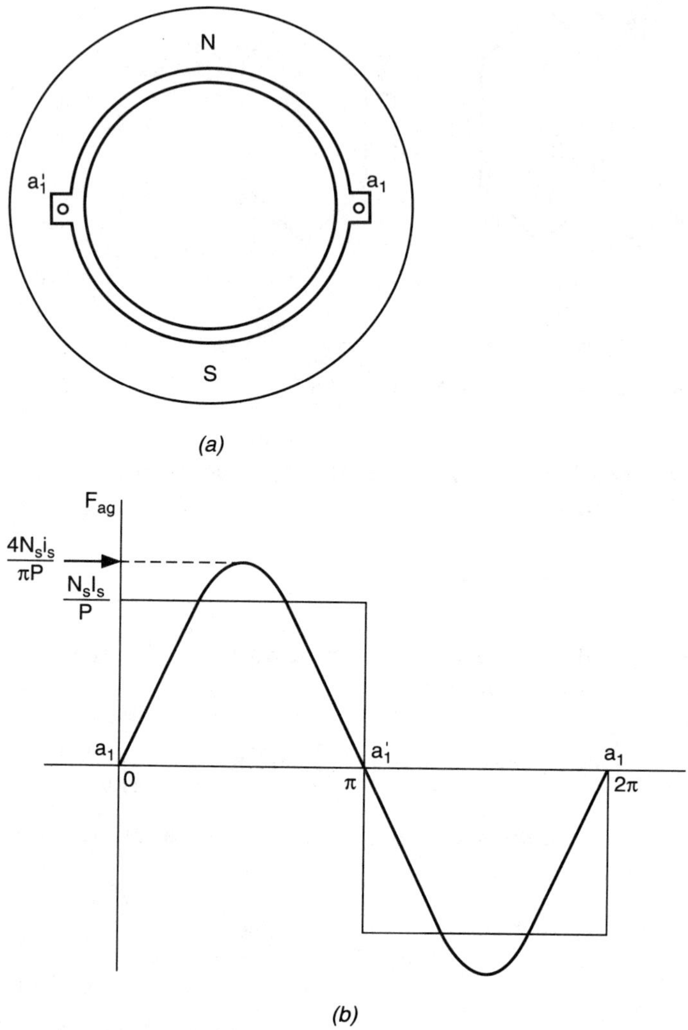

Figure 6.4 (*a*) Concentrated windings in the stator slots. (*b*) The rectangular and approximated sinusoidal mmf waves of an induction machine.

If the coil is excited by the alternating current then the amplitude of the fundamental component would vary sinusoidally with time. Let the effective amplitude of the coil current be I_s then the maximum amplitude of the fundamental component from Eq. (6.2) would be

$$F_{sg,\max} = \frac{4N_s\sqrt{2}i_s}{\pi P} \qquad \text{At} \tag{6.4}$$

6.2.2 Three-phase induction motor

The three-phase induction motor consists of a cylindrical laminated structure with slots in both stator and rotor. In the stator there are three identical symmetrically spaced windings placed uniformly in the slots. These uniformly distributed windings enable the sinusoidal mmf to be produced in the air gap. The rotor also has three identical symmetrical phase windings like the stator.

The three-phase stator excitation produces a rotating magnetic field in the air gap of an induction motor. This field rotates at a synchronous speed, $n_s = 120f/P$, causing the voltage to be induced in the conductors. The induced voltages give rise to rotor currents which interact in the air gap to produce a torque.

6.2.3 Slip and frequency of rotor currents

The *slip* is defined as the fraction of synchronous speed n_s and the actual speed of the rotor n. The slip is given by

$$s = \frac{n_s - n}{n_s} \tag{6.5}$$

At standstill the rotating magnetic field produced by the stator has the same relative speed with respect to rotor and stator windings. At this moment the frequency of the rotor currents f_r is the same as the frequency of the stator f_s. The frequency of the rotor f_r is zero when the motor is rotating at synchronous speed. At this time there is no relative motion between the rotating field and the rotor. At any other speeds, the rotor frequency is proportional to the slip s. A P-pole machine rotating at n_s rpm has the frequency of the current as

$$f_r = \frac{Pn_r}{120} = \frac{P(n_s - n)}{120} \tag{6.6}$$

The induction motor relative speed n_r between the rotor conductors and the rotating field produced by the stator is given by

$$n_r = n_s - n \tag{6.7}$$

From Eqs. (6.5) and (6.6) we get

$$n_s - n = sn_s = \frac{s(120f_s)}{P} \tag{6.8}$$

Substituting Eq. (6.8) into Eq. (6.6) the slip frequency becomes

$$f_r = \left(\frac{P}{120}\right)\left[\frac{s(120f_s)}{P}\right] = sf_s \tag{6.9}$$

6.3 The Induction Motor Equivalent Circuit

6.3.1 The stator equivalent circuit of the induction motor

As stated in Sec. 6.1 the stator of the induction motor is always connected to the ac grid, while the rotor voltage and current are reduced by induction. In this case the induction motor behaves like a transformer with an air gap and a variable resistance in the secondary. Therefore the stator side of the induction motor corresponds to the primary side of the transformer, as shown in Fig. 6.5. Just like the transformer, the stator current I_1 of the induction motor is made of two components: the load current I_2 and the excitation current I_ϕ. The excitation current, on the other hand, also has two components, that is, the core current I_c and the magnetization current I_m, as shown by Eqs. (6.10) and (6.11) and in Fig. 6.5.

$$I_1 = I_2 + I_f \tag{6.10}$$

and

$$I_\phi = I_c + I_m \tag{6.11}$$

From Fig. 6.5 the stator terminal voltage is given by

$$V_1 = (R_1 + jX_1)I_1 + E_1 \tag{6.12}$$

6.3.2 The rotor equivalent circuit of the induction motor

Defining X_2 as the rotor reactance at standstill, that is, at unity slip ($s = 1$), and using Eq. (6.9), the actual rotor impedance is expressed as

$$Z_2 = R_2 + jsX_2 \tag{6.13}$$

where X_2 is the reactance at standstill.

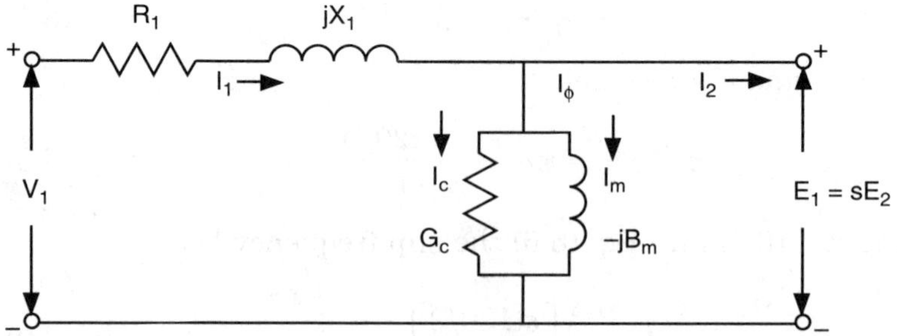

Figure 6.5 The stator equivalent circuit of an induction motor.

The voltage equation for one phase of the actual rotor circuit becomes

$$E_1 = sE_2 = (R_2 + jsX_2)I_2 \tag{6.14}$$

Dividing Eq. (6.14) by slip s we have

$$E_2 = \left(\frac{R_2}{s} + jX_2\right) I_2 \tag{6.15}$$

Equation (6.15) describes the secondary circuit of a rotor of the induction motor with a constant reactance and a variable resistance. The secondary impedance of each phase of the rotor referred to the stator side is equal to

$$Z_2 = \frac{E_2}{I_2} = \frac{R_2}{s} + jX_2 \tag{6.16}$$

Figure 6.6 shows the equivalent circuit of an induction motor.

6.3.3 Mechanical power and torque by using Thevenin's theorem

The use of Thevenin's theorem greatly simplifies the determination of mechanical power and torque of the induction motor. To determine the Thevenin voltage or equivalent voltage, and impedance or equivalent impedance, external to points a and b in Fig. 6.7 the following procedure will be followed:

First Thevenin's impedance Z_{TH} will be found by short-circuiting the voltage source. Then the two impedances are connected in parallel as

$$Z_{TH} = Z_{eq} = \frac{jX_m(R_1 + jX_1)}{R_1 + j(X_1 + X_m)}$$

$$= -\frac{X_1X_m}{R_1 + j(X_1 + X_m)} + j\frac{R_1X_m}{R_1 + j(X_1 + X_m)} \tag{6.17}$$

Thevenin's voltage is found by using the voltage divider rule on the magnetizing reactance.

Redrawing the circuit of Fig. 6.7 into the circuit of Fig. 6.8, we find that the real power is converted only in the resistive elements R_1 and R_2/s. The armature (stator) losses per phase are equal to

$$P_1 = I_1^2R_1 \tag{6.18}$$

The power converted at resistance R_2/s is the total power per phase

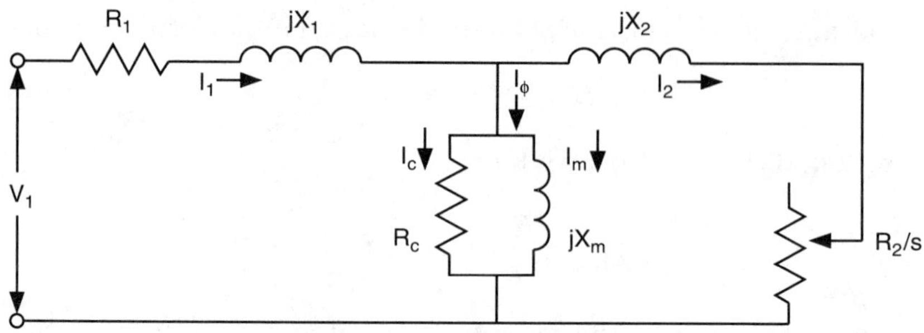

Figure 6.6 Equivalent circuit of an induction motor.

converted in the rotor. This is the power per phase transferred across the air gap.

$$P_g = I_2^2 \frac{R_2}{s} \tag{6.19}$$

The rotor winding losses per phase are given by

$$P_{\text{rotor}} = I_2^2 R_2 \tag{6.20}$$

Therefore the mechanical power across the gap is given by

$$P_m = P_g - P_r = I_2^2 R_2 \left(\frac{1 - s}{s}\right) \tag{6.21}$$

The rotor resistance R_2/s can be treated as the sum of two resistances:

$$\frac{R_2}{s} = R_2 + R_2 \left(\frac{1 - s}{s}\right) \tag{6.22}$$

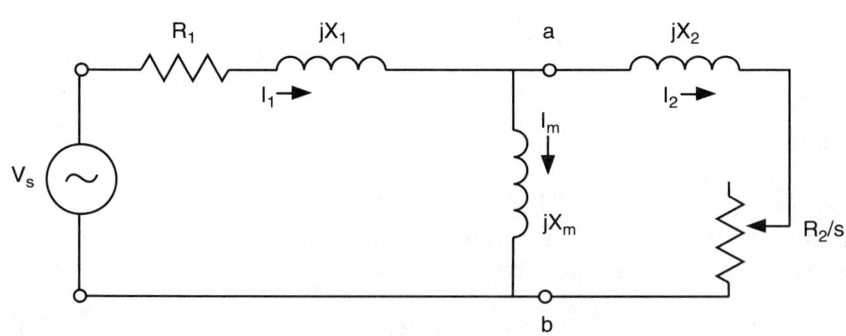

Figure 6.7 Equivalent circuit for the induction motor determination of mechanical power and torque.

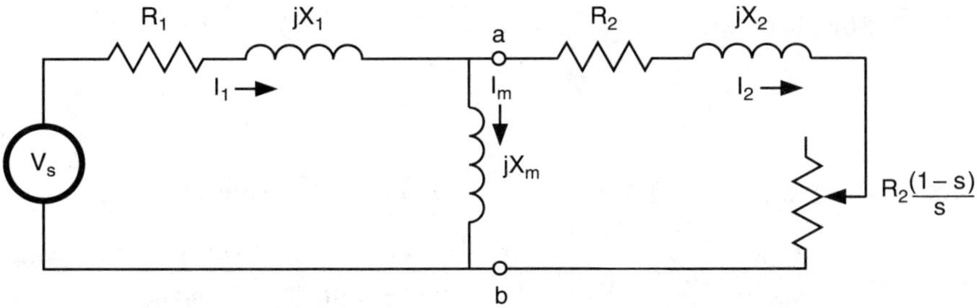

Figure 6.8 Modified equivalent circuit of an induction motor.

On the right-hand side of Eq. (6.22), R_2 determines the rotor losses while $R_2[(1 - s)/s]$ determines mechanical power converted. Figure 6.8 is the modified equivalent circuit, which includes the two resistances found in Eq. (6.22).

Figure 6.9 shows the induction motor equivalent circuit simplified by Thevenin's theorem. To determine I_2 from Fig. 6.9 we make use of Kirchhoff's voltage law, that is,

$$I_2 = \frac{V_{\text{eq},1}}{[R_{\text{eq},1} + (R_2/s)] + j(X_{\text{eq},1} + X_2)} \quad \text{A} \tag{6.23}$$

The mechanical power developed in an m-phase induction motor is equal to

$$P_m = mI_2^2 \frac{R_2}{s}(1 - s) \tag{6.24}$$

Therefore the developed torque is equal to

$$T_m = \frac{P_m}{\omega_m} = \frac{mI_2^2}{\omega_m}\left(\frac{R_2}{s}\right)(1 - s) \tag{6.25}$$

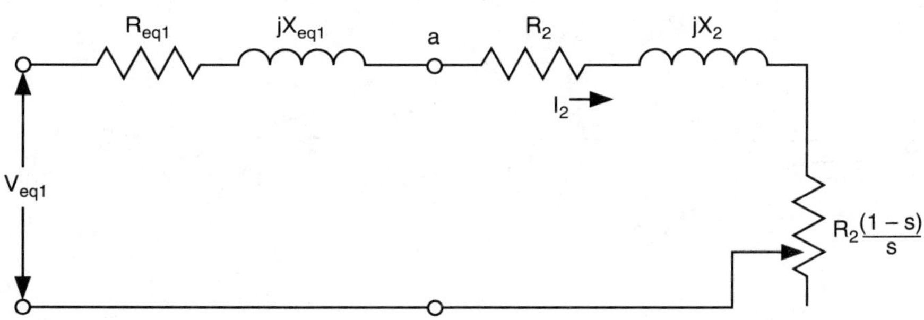

Figure 6.9 Simplified equivalent circuit of induction motor using Thevenin's theorem.

Let the angular velocity ω_m be equal to

$$\omega_m = \frac{2\pi n}{60} = \frac{2\pi(1 - s)n_s}{60} \tag{6.26}$$

Substituting Eqs. (6.26) and (6.21) into Eq. (6.25) yields

$$T_m = \frac{P_m}{\omega_m} = mI_2^2 R_2 \left(\frac{1 - s}{s}\right)\left[\frac{60}{2\pi(1 - s)n_s}\right] = \frac{60mR_2I_2^2}{2\pi s n_s} \tag{6.27}$$

Substituting Eq. (6.23) into Eq. (6.27), the final expression of the developed torque is

$$T_m = \frac{60mR_2}{2\pi s n_s}\left[\frac{V_{\text{eq},1}}{(R_{\text{eq},1} + R_2) + j(X_{\text{eq},1} + X_2)}\right]^2 \quad \text{N} \cdot \text{m} \tag{6.28}$$

In Eq. (6.28) the torque is expressed as a function of slip. The torque is positive for all slip greater than zero ($s > 0$). This is the speed region of the motor operation. When the slip is less than zero ($s < 0$), the torque is negative and the machine operates in the generator region. The torque magnitude approaches zero when the slip is equal to zero or infinity, as shown in Fig. 6.10. To find the maximum torque Eq. (6.28) must be differentiated with respect to slip and then equated to zero:

$$\frac{dT_m}{ds} = 0 \tag{6.29}$$

Example 6.1 A 15-hp, 230-V, 60-Hz, three-phase wound rotor induction motor has a rated speed of 1750 rpm. The parameters of the equivalent circuit are

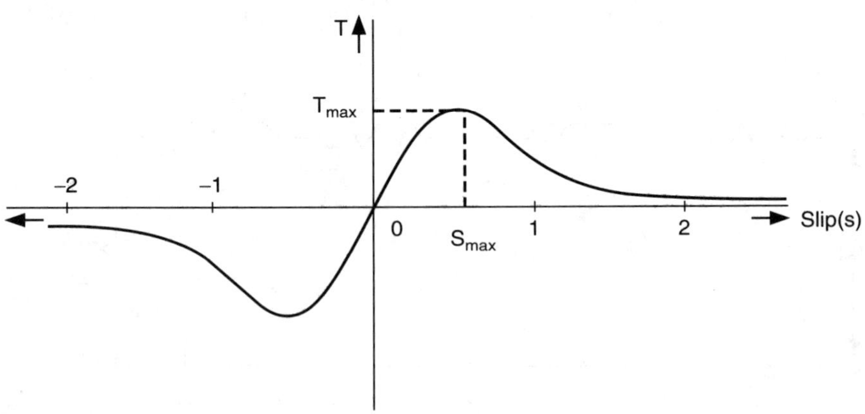

Figure 6.10 Torque versus slip of an induction motor.

$R_1 = 0.26\ \Omega$, $R_2 = 0.32\ \Omega$, $X_1 = 0.832\ \Omega$, $X_2 = 0.671\ \Omega$, and $X_m = 13.12\ \Omega$. Calculate the torque and the line current at rated speed, if the synchronous speed is 1800 rpm.

solution For rated speed, the slip is given by Eq. (6.5).

$$s = \frac{n_s - n}{n_s} = \frac{1800 - 1750}{1800} = 0.0278$$

Using Eqs. (6.25), (6.26), and (6.28) we get

$$T_m = \frac{mR_2}{s\omega_m}\left\{\frac{V_{\text{eq},1}}{[R_{\text{eq},1} + (R_2/s)] + j(X_{\text{eq},1} + X_2)}\right\}^2$$

$$R_{\text{eq},1} = -\frac{X_1 X_m}{R_1 + j(X_1 + X_m)} = 0.7824\underline{/-268.93°}\ \Omega$$

$$X_{\text{eq},1} = \frac{R_1 X_m}{R_1 + j(X_1 + X_m)} = 0.2445\underline{/1.012°}\ \Omega$$

$$V_{\text{eq},1} = j\frac{X_m V_s}{R_1 + j(X_1 + X_m)} = 124.85\underline{/1.06°}\ \text{V}$$

Therefore,

$$T_m = 21.2673\underline{/-12.28}\ \text{N}\cdot\text{m}$$

$$I_2 = \frac{V_{\text{eq},1}}{[R_{\text{eq},1} + (R_2/s)] + j(X_{\text{eq},1} + X_2)} = 10.7744\underline{/-6.14°}\ \text{A}$$

Example 6.2 A three-phase, Y-connected, 220-V, 4-pole, 60-Hz induction motor has stator leakage impedance of $(1.5 + j3)\ \Omega$ and rotor leakage impedance referred to the stator at standstill of $(1.5 + j3)\ \Omega$. The magnetizing impedance is $j35\ \Omega$ and the core loss impedance is $250\ \Omega$. All impedances are in ohms per phase. Using Fig. 6.6 compute the input current I_1, torque T, output power P_{out}, and efficiency η of a slip of $s = 0.06$.

solution At $s = 0.06$,

$$Z_2 = \frac{R_2}{s} + jX_2 = \frac{1.5}{0.06} + j3 = 25.18\underline{/6.84°}$$

$$Z_f = \frac{(j35)(250)}{250 + j35} = 34.66\underline{/82.03°}$$

$$Z_{\text{in}} = 1.5 + j3 + \left[\frac{(25.18\underline{/6.84°})(34.66\underline{/82.03°})}{25 + j3 + 4.806 + j34.325}\right] = 21.32\underline{/41.42°}\ \Omega$$

The input current

$$I_1 = \frac{220}{\sqrt{3}(21.32\underline{/41.42°})} = 5.96\underline{/41.42°}\ \text{A}$$

$$I_2 = \frac{(4.81 + j34.32)(5.96\underline{/41.43°})}{4.81 + j34.32 + 25 + j3} = 4.325\underline{/-10.77°}\ \text{A}$$

$$I_\phi = I_1 - I_2 = 5.96\underline{/-41.42^\circ} - 4.325\underline{/-10.77^\circ} = 3.143\underline{/-85.98^\circ}\text{ A}$$

Rotor input power

$$P_g = T\omega_m = m|I_2^2|\frac{R_2}{s} = 3(4.325)^2\left(\frac{1.5}{0.06}\right) = 1402.92\text{ W}$$

$$\omega_m = \frac{2\pi n}{60} = 60\pi\text{ rad/s}$$

Torque

$$T = \frac{P_g}{\omega_m} = \frac{1402.92}{60\pi} = 7.4427\text{ N}\cdot\text{m}$$

$$P_{\text{out}} = P_g(1 - s) = 1402.92(1 - 0.06) = 1318.74\text{ W}$$

$$P_{\text{in}} = \sqrt{3}(220)(5.96)\cos(-41.42^\circ) = 1703.026\text{ W}$$

Efficiency

$$\eta = \frac{P_{\text{out}}}{P_{\text{in}}} \times 100\% = \frac{1318.74}{1703.026} \times 100\% = 77.435\%$$

Example 6.3 The per-phase constants of a three-phase, 600-V, 60-Hz, 4-pole, Y-connected wound rotor induction motor are $R_1 = 0.75\ \Omega$, $X_1 = X_2 = 2.0\ \Omega$, $R_2 = 0.85\ \Omega$, $X_m = 50\ \Omega$. Neglect core losses.

1. Calculate the slip at which the maximum developed torque occurs.
2. Find the value of the maximum torque.
3. Specify the range of speed for a stable operation of the motor.

solution Using Fig. 6.7 and Eqs. (6.24) and (6.25), we have

$$P_d = T\omega_m = (1 - s)\omega_s T = P_g(1 - s)$$

or

$$T = \frac{P_g}{\omega_s} = \left(\frac{T_2^2}{\omega_s}\right)\left(\frac{R_2}{s}\right)$$

$$R_{\text{TH}} + jX_{\text{TH}} = \frac{jX_m(R_1 + jX_1)}{R_1 + j(X_1 + X_m)} = \frac{j50(0.75 + j2)}{0.75 + j(2 + 50)} = 2.05\underline{/158.6^\circ}$$

1. $s = \dfrac{R_2}{\sqrt{(R_{\text{TH}})^2 + j(X_{\text{TH}} + X_2)^2}}$

From Eq. (6.29) we get

$$\frac{dT}{ds} = 0$$

$$R_{\text{TH}}^2 - \left(\frac{R_2}{s}\right)^2 + (X_{\text{TH}} + X_2)^2 = 0$$

$$s = \frac{R_2}{\sqrt{(R_{\text{TH}})^2 + (X_{\text{TH}} + X_2)^2}} = \frac{0.85}{\sqrt{(0.75)^2 + (1.252)^2}} = 0.582$$

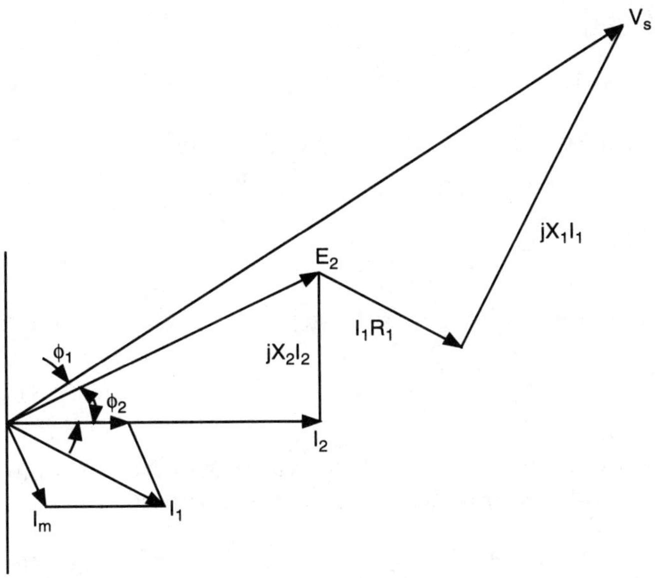

Figure 6.11 Phasor diagram for a three-phase induction motor.

2. $$T_m = \frac{V_{\mathrm{TH}}^2}{2\omega_s\sqrt{(R_{\mathrm{TH}})^2 + (X_{\mathrm{TH}} + X_2)^2}}$$

$$V = \frac{jX_m V_t}{R_1 + j(X_m + X_2)} = \frac{600(50\angle 90^\circ)}{0.75 + j52} = 576.86\angle -0.83^\circ$$

$$T_m = \frac{50\angle 90^\circ}{2[2\pi(1800)/2](52.005)} = 91.81\ \mathrm{N \cdot m}$$

3. Stable range of operation is $0 < s < 0.582$.

6.3.4 The phasor diagram of a three-phase induction motor

The equivalent circuit of an induction motor shown in Fig. 6.7 represents a one-phase diagram of either a squirrel cage rotor or a wound rotor motor. A complete phasor diagram of an induction motor has been drawn in Fig. 6.11.

6.4 Three-Phase Induction Motor Performance from the Equivalent Circuit

The performance characteristics of a three-phase induction motor include the variations of speed, losses, current, starting torque, and maximum torque. All the performance calculations are made per phase basis and assuming balanced operations of the induction motor. Using the equivalent circuit of the induction motor (Fig. 6.8) and the power flow diagram in Fig. 6.12, the power can be analyzed from stator to rotor.

The power P_g transferred through the air gap from the stator is the difference between the electrical input power P_{in} and the stator copper loss. Therefore,

$$P_g = P_{in} - I_1^2 R_1 \tag{6.30}$$

This power is dissipated in resistance R_2/s as

$$P_g = (I_2^2)\left(\frac{R_2}{s}\right) \tag{6.31}$$

If the rotor copper loss is subtracted from the power transferred through the air gap, P_g, the developed mechanical power is obtained:

$$P_m = P_g - I_2^2 R_2 \tag{6.32}$$

Substituting Eq. (6.31) into Eq. (6.32) the developed mechanical power is found to be

$$P_m = P_g(1 - s) \tag{6.33}$$

The developed mechanical power is the power appearing or absorbed by the load resistance, $R_2[(1 - s)/s]$. To obtain the mechanical power output for the induction motor, the rotational losses must be subtracted from the developed mechanical power yielding

$$P_{out} = P_m - P_{rot} = T_0 \omega_m \tag{6.34}$$

where T_0 is the torque and ω_m is the angular velocity. The efficiency of the induction motor is then given by

$$\eta = \frac{P_{out}}{P_{in}} \tag{6.35}$$

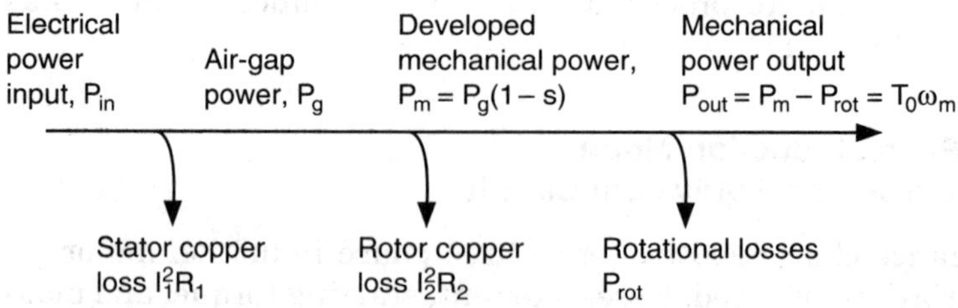

Figure 6.12 The power flow diagram of the induction motor.

6.5 No-Load Test

Rated balanced voltage at rated frequency is applied to the stator of the motor and the motor is allowed to run on no-load. Measurements of input power P_0, line to line voltage V_0, and line current I_0 are taken and recorded. At no-load, the slip of the induction motor is nearly zero ($s = 0$). During no-load test the region to the right of the shunt branch of Fig. 6.13*a* is open, and a further simplification of Fig. 6.13*a* gives Fig. 6.13*b*. Figure 6.13*b* facilitates the evaluation of the reactances and resistances of the induction motor since, at no-load, a low value of rotor frequency and negligible rotor core loss are experienced. To evaluate the motor parameters at no-load test, we get

$$R_0 = R_1 + R_m = \frac{P_0}{I_0^2} \tag{6.36}$$

$$Z_0 = \frac{V_0}{I_0} \tag{6.37}$$

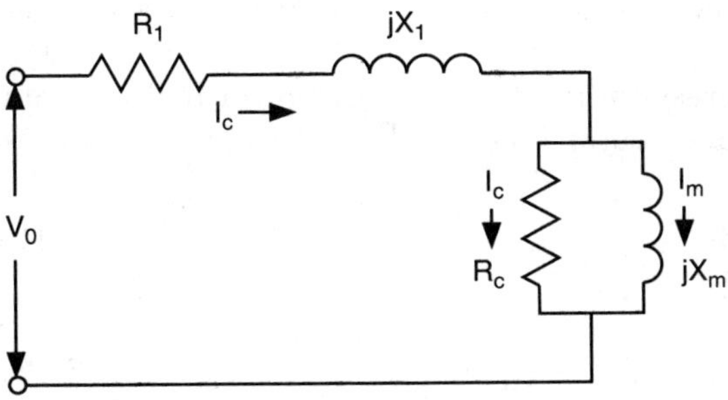

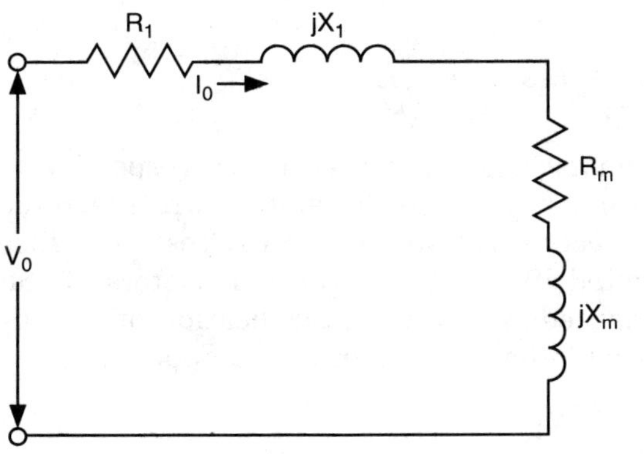

Figure 6.13 Per-phase equivalent circuits of induction motor (*a*) and (*b*), no-load test.

$$X_0 = X_1 + X_m = \sqrt{Z_0^2 - R_0^2} \tag{6.38}$$

$$\text{No-load power factor} = \cos \Psi = \frac{P_0}{V_0 I_0} \tag{6.39}$$

where R_1 is the stator resistance taken as the dc value.

The rotational losses are made up of the sum of windage, friction, and core losses. The losses are found by subtracting the stator copper loss from the no-load power input, yielding:

$$P_{\text{rot}} = P_0 - 3I_0^2 R_1 \tag{6.40}$$

6.6 Locked Rotor Test

In this test the rotor of the induction motor is blocked so that the slip is equal to unity ($s = 1$). A reduced voltage is applied to the stator of the induction motor, so that the rated current flows through the stator windings. The values of input power P_L, line to line voltage V_L, and the line current I_L are observed and recorded. The iron losses are neglected.

The equivalent circuit corresponding to the locked rotor test is given in Fig. 6.14*a* and *b*. The locked rotor test results are

$$Z_L = \frac{V_L}{I_L} \tag{6.41}$$

$$R_L = R_1 + R_2' = \frac{P_L}{I_L^2} \tag{6.42}$$

$$X_L = j(X_1 + X_2') = \sqrt{Z_L^2 - R_L^2} \tag{6.43}$$

$$\text{Power factor (pf)} = \cos \Psi = \frac{P_L}{V_L I_L} \tag{6.44}$$

where R_2 and X_2 are equivalent resistance and reactance, respectively, which must be calculated from Fig. 6.14*b*. R_2 and X_2 are the rotor resistance and reactance, respectively. Using the IEEE Test Code List of empirical proportions quoted in Table 6.1, the parameters of the induction motor can be computed. When the classification of the induction motor is unknown, the following expression is applied:

$$X_1 = X_2 = 0.5X_L \tag{6.45}$$

Let R_2 be the equivalent resistance of the induction motor. The equivalent resistance R_2 is smaller than the rotor resistance R_2', and

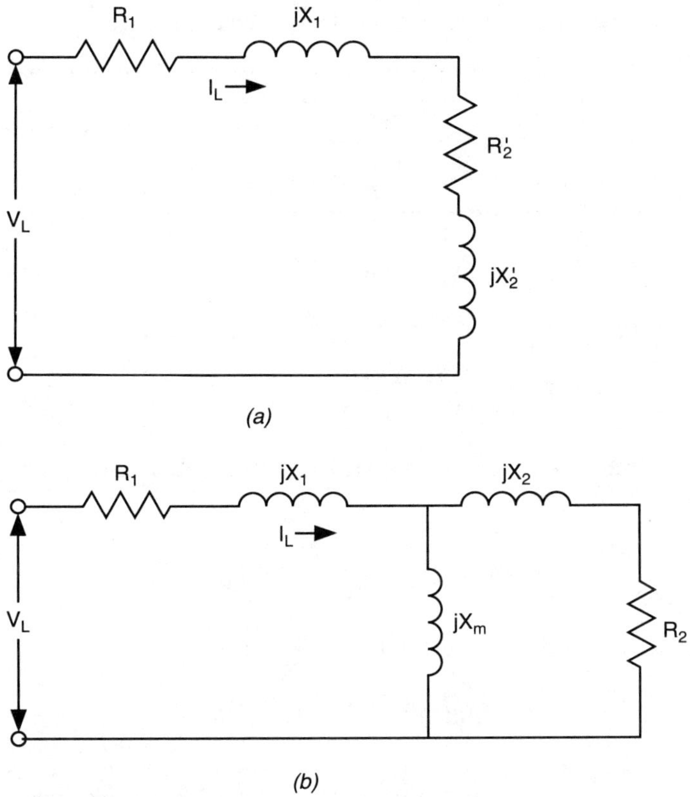

Figure 6.14 Per-phase equivalent circuits of induction.

$R_2/s \gg X_1 + X_m$ in running range has a correspondingly greater effect on the performance of the induction motor. The equivalent circuits of Fig. 6.14*a* and *b* applies to locked rotor conditions when the slip is equal to unity, therefore,

$$R_2' + jX_2' = \left[\frac{jX_m(R_2 + jX_2)}{R_2 + j(X_2 + X_m)}\right]\left[\frac{R_2 - j(X_2 + X_m)}{R_2 - j(X_2 + X_m)}\right]$$

$$= \frac{R_2 X_2^2}{R_2^2 + (X_2 + X_m)^2} + j\,\frac{X_m R_2^2 + X_2(X_2 + X_m)}{R_2^2 + (X_2 + X_m)^2} \tag{6.46}$$

Therefore the equivalent circuit resistance R_2 equals

$$R_2' = \frac{R_2 X_m^2}{R_2^2 + (X_2 + X_m)^2} \tag{6.47}$$

TABLE 6.1 Empirical Proportions of Three-Phase Induction Motor Leakage Reactances

Type of motor	Class A	Class B	Class C	Class D	Wound rotor
X_1	$0.5X_L$	$0.4X_L$	$0.3X_L$	$0.5X_L$	$0.5X_L$
X_2	$0.5X_L$	$0.6X_L$	$0.7X_L$	$0.5X_L$	$0.5X_L$

Since $R_2 \ll X_2 + X_m$, Eq. (6.47) becomes

$$R_2' = \frac{R_2 X_m^2}{(X_2 + X_m)^2} \tag{6.48}$$

For the rotor resistance referred to the stator, after substituting Eq. (6.48) into Eq. (6.42), we get

$$R_L = R_1 + R_2' = R_1 + \frac{R_2 X_m^2}{(X_2 + X_m)^2} \tag{6.49}$$

Rearranging Eq. (6.49) and solving for R_2 we have

$$R_2 = \frac{(R_L - R_1)(X_2 + X_m)}{X_m^2} \tag{6.50}$$

Example 6.4 The following test results were obtained on a 3-kW, three-phase, 220-V, 4-pole, 60-Hz squirrel cage induction motor (design A).

No load test: 220 V line to line, 5.61 A, 635 W, three-phase power.

Locked rotor test at 60 Hz: 47.25 V line to line, 12 A, three-phase, 560 W.

The dc resistance of the stator is $R_1 = 0.976\ \Omega$/phase. Determine the no-load rotational losses and the equivalent circuit constants.

solution From the no-load test data and Eqs. (6.36) and (6.37), we get

$$Z_0 = \frac{V_0}{\sqrt{3}I_0} = \frac{220}{\sqrt{3}(5.61)} = 22.64\ \Omega$$

$$R_0 = \frac{P_0}{3I_0^2} = \frac{635}{3(5.61)^2} = 6.73\ \Omega$$

$$X_0 = \sqrt{Z_0^2 - R_0^2} = \sqrt{(22.64)^2 - (6.73)^2} = 21.62\ \Omega$$

From the locked rotor test equations (6.41), (6.42), and (6.43), we get

$$Z_L = \frac{V_L}{\sqrt{3}I_L} = \frac{47.25}{\sqrt{3}(12)} = 2.273\ \Omega$$

$$R_L = \frac{P_L}{3I_L^2} = \frac{560}{3(12)^2} = 1.296\ \Omega$$

The locked rotor reactance is

$$X_L = \sqrt{Z_L^2 - R_L^2} = \sqrt{(2.273)^2 - (1.296)^2} = 1.867\ \Omega$$

From Table 6.1

$$X_1 = X_2 = 0.5X_L = (0.5)(1.867) = 0.934\ \Omega$$

From Eqs. (6.38) and (6.50) we obtain

$$X_0 = X_1 + X_m$$

$$X_m = X_0 - X_1 = 21.62 - 0.934 = 20.68\ \text{W}$$

$$R_2 = \frac{(R_L - R_1)(X_2 + X_m)}{X_m^2}$$

$$= \frac{(1.296 - 0.976)(0.934 - 20.68)}{(20.68)^2} = 0.0162\ \Omega$$

The no-load rotational losses are given by

$$P_{\text{rot}} = P_0 - 3I_0^2R_1 = 635 - 3(5.61)^2(0.976) = 542.84\ \text{W}$$

6.7 The Induction Generator

Any three-phase induction motor can be operated as an induction generator. As a motor the induction machine runs at a speed slightly below

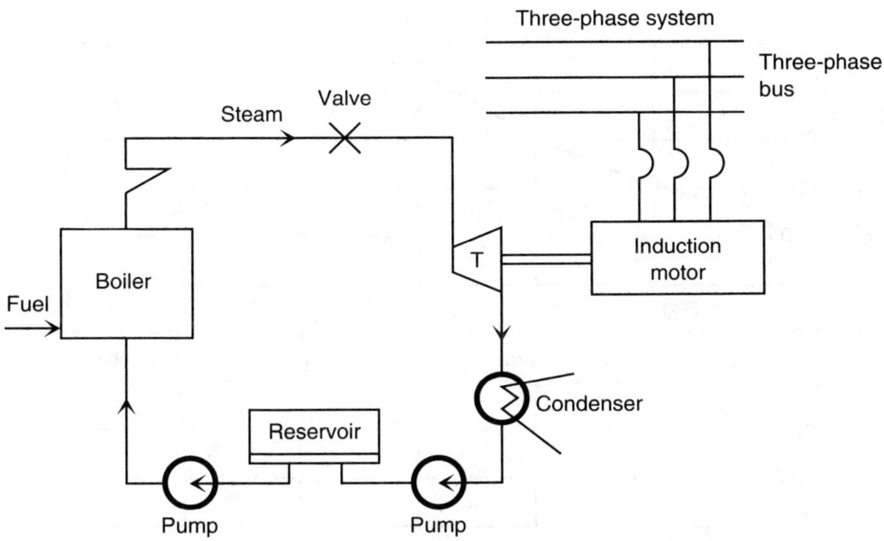

Figure 6.15 An induction motor connected to a three-phase power system.

its synchronous speed. The main application for the induction generator is in cogeneration applications for a plant which has a waste heat available and has an existing induction motor of about 1000 hp or larger. Figure 6.15 shows an induction motor connected to a three-phase system, with its shaft mechanically coupled to the steam turbine.

By closing the circuit breaker, the motor is started at full voltage, at the same time the steam is prevented from entering the turbine by closing the steam valve. The induction motor accelerates and drives the turbine at a speed somewhat less than the synchronous speed of the rotating stator flux. Gradually opening the turbine valve causes a gradual buildup of torque adding to that developed by the induction motor, resulting in an increase in rotor speed. When the speed of the turbine-motor set reaches synchronous speed of the stator the slip becomes zero ($s = 0$), R_2/s becomes infinite, the rotor current becomes zero, and no motor torque is developed. At slip equals zero the induction machine is neither a motor or a generator. The machine is floating on the bus. The exciting current is the only one that supplies the rotating magnetic field and the iron losses.

The prime movers (such as turbines, diesels, and water turbines) can also supply active power. Therefore, when connected to an energized system, the magnetizing current I_M is supplied by the power system whether the machine operates as a motor or as a generator. Increasing the rotor speed above synchronous by increasing the energy input to the turbine will cause the slip to become negative reversing the direction of the air-gap power as shown in Fig. 6.16. Thus, instead

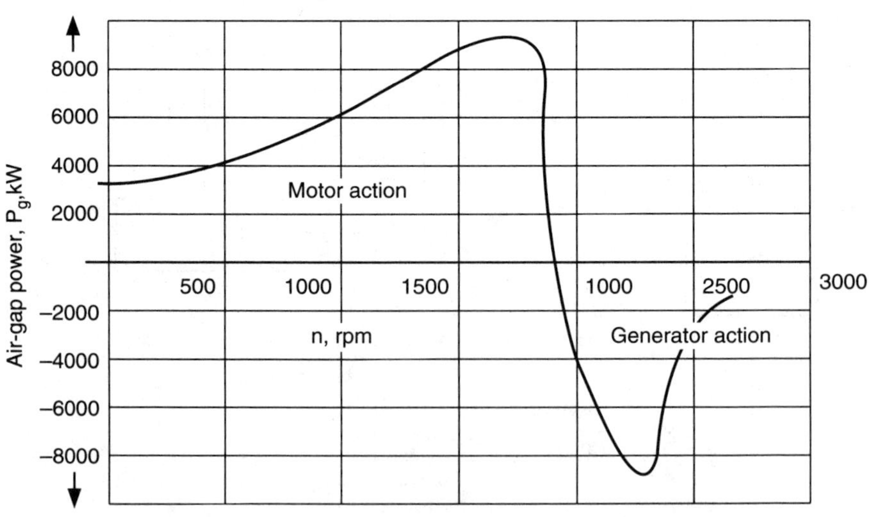

Figure 6.16 Air-gap power versus motor speed of the induction machine.

of active power being transferred across the air gap from the stator to the rotor, as in motor action, a negative slip causes the power to be transferred in the reverse direction from the rotor to the stator, and on into the distribution grid.

Figure 6.16 shows a plot of air gap power versus the motor speed of induction machine operation, as a motor below synchronous speed and as a generator above synchronous speed. When operating as an induction generator, the interaction of the magnetic flux of the stator and the magnetic flux of the rotor produce a countertorque in opposition to the driving torque of the prime mover. Increasing the prime mover speed increases electrical power supply to the distribution system by the induction generator causing an increase in countertorque.

The countertorque increases with the increase in speed until it attains a maximum value called the pushover torque. The pushover torque, when operating as an induction motor, is shown in Fig. 6.16. The increase of the prime mover speed beyond its pushover point causes the power to decrease. This decreases the load (countertorque) on the prime mover causing a rapid increase in speed. The resultant overspeed can damage both the induction machine and the prime mover. Automatic closure of the turbine valve is necessary to prevent damaging overspeed when overloading past the pushover point. A similar effect occurs when an induction generator is loaded, and the breaker connecting the generator to the power system is tripped. The sudden loss of load causes the turbine to overspeed. To protect against damage, accidental overspeed NEMA specifications require squirrel-cage and wound-rotor motors to be so constructed that in an emergency they will be able to withstand overspeed without mechanical injury. The respective overspeeds are listed in Table 6.2 [NEMA, 1987].

Like induction motors, the induction generator absorbs reactive power from the system to which it is connected. The utility companies accept the supply of reactive power to the operator of the motor, probably because it is a necessary part of the earnings from selling the power and energy required by the motor.

The utility companies are not so keen about supplying reactive power to an induction generator when they are buying the power and energy

TABLE 6.2 Allowable Emergency Overspeed of Squirrel-Cage and Wound-Rotor Motors

Horsepower	Synchronous speed, in rpm	Percent above synchronous speed
<200	1201 and over	25
	1200 and below	50
250–500 inclusive	1801 and over	20

it produces. The supply of reactive power through the inductive reactance of the power system causes a drop in system voltage. Except in large ratings, the induction generator's reactive power requirements should not cause a serious voltage problem to the utility companies. For instance, the allowable induction generator size is based on a steady-state system voltage dropoff, about 1½ percent for a distance of a mile from the substation. A 2000-hp motor with a 12-kV supply can be used for an average-size substation and an average circuit conductor size and configuration. The following calculations will be made in order to illustrate the determination of the maximum acceptable induction generator at various locations on the utility system based on a 0.015 per unit voltage drop during generator operation.

Assume the maximum tolerable step change in voltage, such as that associated with capacitor switching, is 0.03 per unit. The generator power factor is equal to 0.8. This means that for every kilowatt produced, 0.75 of KVAR is consumed.

Example Given the circuit in Fig. 6.17 with the following per-unit values. The base values are 10 MVA and 12 kV and the inductive reactance per mile is $j0.7$ Ω. For a 12-kV circuit determine various sizes of the induction generators.

Per-unit calculation bases: 100 MVA, 12 kV.

Using Eqs. (7.16) and (7.13) we get the transformer reactance and line reactance for 1, 2, and 3 miles from the substation. For 1 mile:

$$\text{pu } X_{tr} = j0.07(100/15) = j0.4667$$

$$\text{pu } X_L = j0.7\left[\frac{100{,}000}{12^2(1000)}\right] = j0.4861$$

For 2 miles:

$$\text{pu } X_L = j0.7(2)\left[\frac{100{,}000}{12^2(1000)}\right] = j0.9722$$

For 3 miles:

$$\text{pu } X_L = j0.7(3)\left[\frac{100{,}000}{12^2(1000)}\right] = j0.4583$$

$$\text{pu Voltage drop} = \text{KVAR}(X)$$

$$\text{Total } X = j(X_s + X_{tr} + X_L)$$

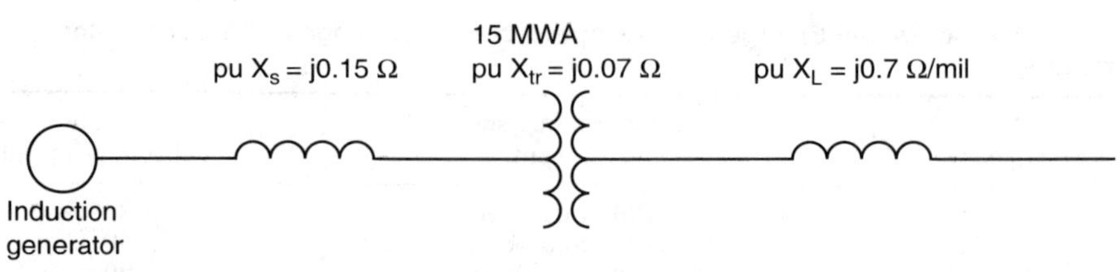

Figure 6.17 One-line diagram of a section of a power system.

TABLE 6.3 Determination of Sizes of Induction Generators

Distance, miles	Total reactance	KVAR that will cause 0.015 voltage (KVAR = $0.015/jX$)	kW = KVAR/0.75	Kilowatts	Horsepower
1	1.1028	0.0136	0.0181	1813.56	2431
2	1.5889	0.0094	0.0125	1250	1675.6
3	2.075	0.0072	0.0096	960	1286.9

Therefore,

$$\text{pu } X_{(1\text{ mile})} = j(0.15 + 0.4667 + 0.4861) = j1.1028$$

$$\text{pu } X_{(2\text{ miles})} = j(0.15 + 0.4667 + 0.9722) = j1.5889$$

$$\text{pu } X_{(3\text{ miles})} = j(0.15 + 0.4667 + 1.4583) = j2.075$$

$$\text{pu KVAR} = \left(\frac{0.015 \text{ per unit voltage drop}}{jX}\right) = \frac{0.015}{j1.1028} = 0.0136\underline{/90^\circ}$$

$$\text{pu kW} = \left(\frac{|\text{KVAR}|}{0.75}\right) = \frac{0.0136}{0.75} = 0.0181$$

or

$$\text{kW} = 0.0181(100{,}000) = 1813.3$$

Tabulating the results for 1, 2, and 3 miles we get Table 6.3.

In Table 6.3 the sizes of induction generators have been determined for 1, 2, and 3 miles from the substation. The other characteristic of the induction generator that the utility companies see as a problem is its ability to operate self-excited. Normally, the generator's voltage and frequency are determined by the system frequency and the system voltage at the point of connection. If the generator was to be separated from the main system but a portion of the system connected to its terminals, the following conditions could prevail. The kilowatt load would be greater than generator capacity and the frequency would be decreased. If the kilowatt load is less than the generator capacity, the frequency would increase. If the reactive power, on the other hand, is greater than the generator's excitation requirements, the frequency would increase and the voltage would increase. When the reactive power source is less than the generator's excitation requirements, the voltage decreases. The induction generators are protected with voltage, frequency, current, negative sequence, and differential relays.

Problems

1 A 220-V, three-phase, 4-pole, Y-connected, 60-Hz induction motor has the following parameters: $R_1 = 0.21\ \Omega$, $R_2' = 0.12\ \Omega$, $X_1 = 0.45\ \Omega$, $X_2' = 0.21\ \Omega$, $X_m = 25.0\ \Omega$. The total iron and mechanical losses are 400 W. For a slip of 3 percent, find the input current, output power, output torque, and efficiency.

2 A 400-V, 60-Hz, three-phase, Y-connected induction motor with slip of 0.05 has the following per-phase equivalent circuit parameters: $R_1 = 2R_2' = 0.2$, $X_1 = 2.5X_2' = 0.5$, $X_m = 20\ \Omega$. Develop a Thevenin equivalent voltage and impedance. Using these values calculate the output torque, when the core loss or rotational loss is 350 W.

3 A three-phase, 60-Hz, 4-pole induction motor runs at 1710 rpm. If the power crossing the air gap is 120 kW, what is the rotor copper loss?

The motor has core and mechanical losses (at 1710 rpm) of 1.7 kW and 2.7 kW, respectively, and the stator copper loss is 3 kW. Find the motor efficiency and output torque.

4 A 220-V, 15-hp, 4-pole, 60-Hz, Y-connected induction motor has a full-load slip of 5 percent. Find the following:

a The synchronous speed of the motor
b The speed of the motor at rated load
c The frequency of the motor at the rated load
d The shaft torque of this motor at the rated load

5 The equivalent circuit parameters of a 100-W, 100-V, 60-Hz, two-pole servomotor are $R_1 = 0.288\ \Omega$, $X_1 = 0.6$ W, $R_2 = 0.849\ \Omega$, $X_2 = 0.6\ \Omega$, and $X_m = 95.5\ \Omega$. For 100-V balanced stator excitation, $s = 0.5$ and core, friction, and windage losses equal 0.736 W. Calculate the followings:

a The synchronous speed of the field
b The mechanical speed of the motor
c The frequency of the rotor currents
d The input current and power input
e The developed power and torque
f The output power and torque
g The efficiency

6 A Y-connected, three-phase, 6-pole, 60-Hz, 440-V induction motor has 75 A and 50 hp. The following test data were obtained:

1. DC test: The dc resistance is 0.21 Ω.
2. No-load test at 60 Hz: $I_0 = 2.5$ A, $P_0 = 1700$ W ($P_1 = 900$ W and $P_2 = 800$ W).
3. Blocked rotor test at 60 Hz for an applied voltage of 40 V, the following measurements were recorded: $I_L = 10$ A, $P_L = 1700$ W ($P_1 = 450$ W and $P_2 = 1250$ W).

Calculate the rotational losses of the motor and the parameters of the equivalent circuit for the normal running conditions, that is, for frequency $f = 60$ Hz and $s = 0.03$.

7 The results of the no-load and blocked rotor tests on a three-phase, Y-connected induction motor are as follows:

1. No-load test: $V_0 = 230$, $P_0 = 2000$ W, $I_0 = 30$ A, friction and windage loss = 500 W.
2. Blocked rotor test: $V_L = 40$ V, $P_L = 2500$ W, $I_L = 60$ A.

The stator resistance is 0.12 Ω. Calculate the parameters of the approximate equivalent circuit.

8 A 400-V, 20-hp, 60-Hz, 4-pole, Y-connected induction motor has the following impedances in ohms per phase referred to the stator circuit: $R_1 = 0.561$ Ω, $R_2 = 0.310$ Ω, $X_1 = 1.09$ Ω, $X_2 = 0.481$ Ω, $X_m = 25.3$ Ω. The rotational losses are 1720 W and are assumed to be constant. For a rotor slip of 3 percent at rated voltage and rated frequency, find the motor's:

a Speed
b Stator current
c Power factor
d Power converted and power output
e Induced torque and output torque
f Efficiency

9 A 220-V, 4-pole, Y-connected, 20-hp design class B induction motor is tested with the following results:

No-load: 220 V, 20.0 A, 1400 W, 60 Hz

Blocked rotor: 25 V, 65 A, 2800 W, 15 Hz

Dc test: 12.5 V, 63 A

Calculate the equivalent circuit of this motor.

10 A three-phase, Y-connected, 400-V, 25-hp, 60-Hz, 4-pole induction motor has the following equivalent circuit constants in ohms per phase referred to the stator: $R_1 = 0.21$ Ω, $R_2 = 0.19$ Ω, $X_1 = 1.12$ Ω, $X_2 = 1.09$ Ω, $X_m = 35$ Ω. The total friction, windage, and core losses may be assumed constant at 1400 W. The motor is connected directly to a 400-V source. Calculate the speed, shaft torque output in newton-meters, power in horsepower, efficiency and terminal power factor for slips of 1 and 2 percent.

Bibliography

Brown, D., and E. P. Hamilton III: *Electromechanical Energy Conversion,* MacMillan, New York, 1984.

Chapman, S. J.: *Electric Machinery Fundamentals,* McGraw-Hill, New York, 1985.

Elgerd, O. I.: *Basic Electric Power Engineering,* Addison-Wesley, Reading, Mass., 1977.

Hunt, W. Jr., and R. Stein: *Electric Power Systems—Transformers and Rotating Machines,* Van Nostrand Reinhold, New York, 1979.

Kingsley, C., Jr., and A. E. Fitzgerald: *Electric Machinery,* McGraw-Hill, New York, 1980.

Mablekos, Van E.: *Electric Machine Theory for Power Engineers,* Harper and Row, New York, 1980.

McLean, P. G.: *Electromagnetic and Electromechanical Machines,* Ellis Horwood, Chichester, U.K., 1984.

Matsh, L. W., and J. D. Mogan: *Electromagnetic and Electromechanical Machines,* Harper and Row, New York, 1986.

Nasar, S. A., and L. E. Unnewehr: *Electromechanics and Electric Machines,* John Wiley, New York, 1983.

National Electric Manufacturers Association (NEMA) Motors and Generators, Publication No. MG1-1987.

Sarma, M. S.: *Electric Machines,* Wm. C. Brown, Dubuque, Ia., 1985.

Chapter

7

The Electric Power System Network

7.1 Power System Representation

7.1.1 One-line diagram

A balanced three-phase power system is always represented for calculation purposes as a one (single)-line diagram. In the one-line diagram, all interconnections and locations for generators, transformers, transmission lines, static loads, dynamic loads (motors), circuit breakers, reactors, and switches are as shown in Fig. 7.1. The normal procedure is to show only those elements needed to solve a particular problem. For instance, circuit breakers and relays are included in the one-line diagram if stability of the power under transient conditions has to be studied. Calculations are performed on a single-phase circuit composed of one of the three lines.

7.1.2 Impedance and reactance diagrams

The performance of a power system under load conditions or upon occurrence of a fault can be calculated by representing the three-phase system as a one-line diagram, as shown in Fig. 7.2. From a one-line diagram, an impedance diagram is drawn, as shown in Fig. 7.3. In the one-line diagram in Fig. 7.2 two buses 1 and 2 are shown. The bus representation in a one-line diagram has the same meaning as the node in a circuit diagram. Figure 7.3 shows the equivalent circuit of various elements of a power system. When the power system is balanced, there is no current to ground; therefore, the limiting impedances are omitted on the impedance diagram.

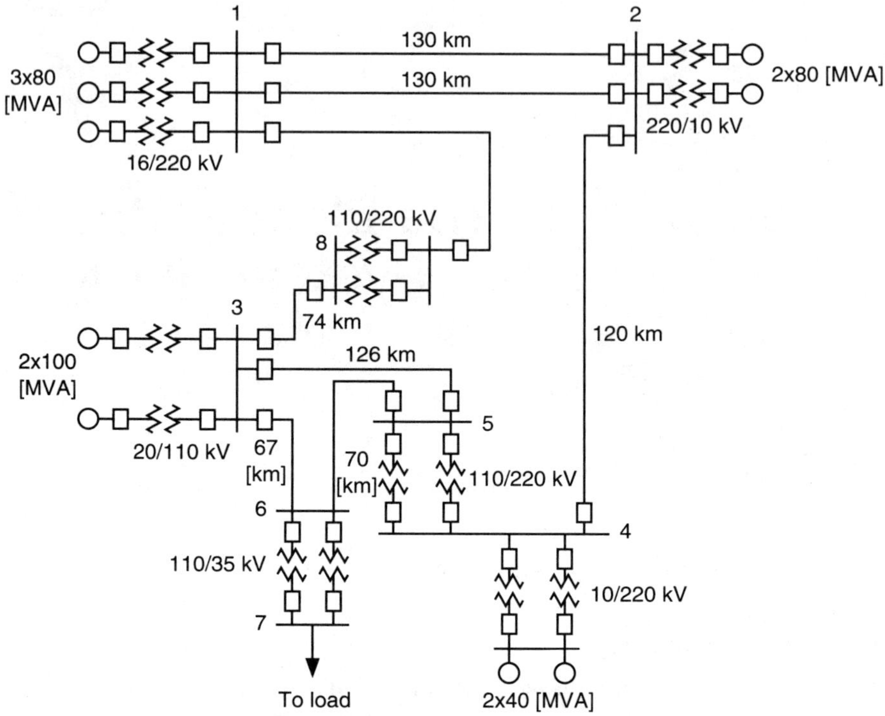

Figure 7.1 One-line diagram of electric power system.

The excitation current I_f of the transformer is usually very small compared to full-load current, therefore the shunt admittance is neglected. For fault calculations the impedance diagram is further simplified by omitting the capacitors of the transmission line.

7.2 The Per-Unit Representation (System)

Power system calculations are usually normalized or scaled. The method of normalization is known as per-unit representation and is defined as a ratio of the quantity to its base value as a decimal. Since we deal with quantities of large magnitudes, such as kilowatts, mega-

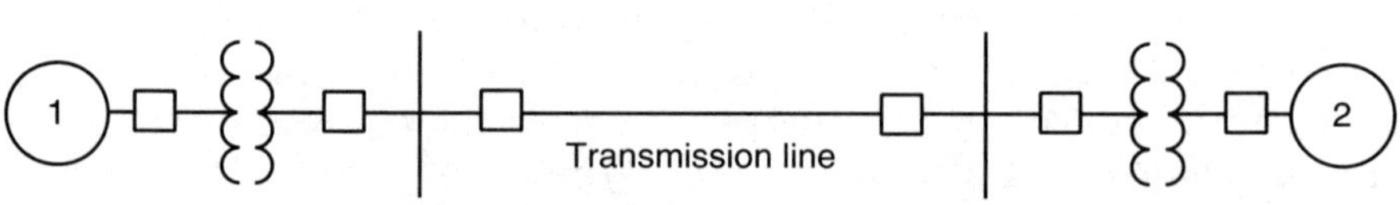

Figure 7.2 One (single)-line diagram of a power system with two generators.

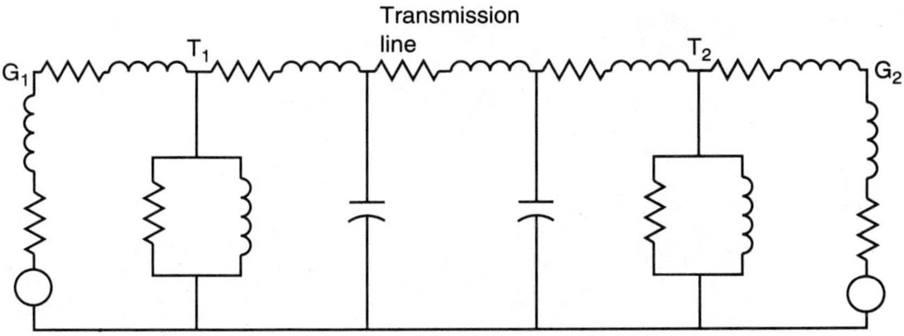

Figure 7.3 Impedance diagram of a one-line diagram, Fig. 7.2.

watts, kilovolts, etc., it is preferable to carry out all calculations in a per-unit system. When the actual values are needed, the inverse calculations are performed. Such calculations will be shown in the following sections.

The voltage, current, kilovolt-amperes, and impedance are all related. Therefore selecting any two of them as base values, the values of the remaining two can be calculated. In a single-phase system, for instance, a voltage-amperes base (VA_{base}) and a voltage base (V_{base}) may be chosen as bases. These values become base values throughout the system. In the formulas developed below, the terms *voltage* refers the voltage to neutral, *current* refers to the line current, and *kilovolt-amperes* refers to the kilovoltamperes per phase.

Let us first calculate the base values of current, impedance, and admittance by using the selected base values of the VA_{base} and V_{base}:

$$I_{base} = \frac{VA_{base}}{V_{base}} \tag{7.1}$$

$$Z_{base} = \frac{V_{base}}{I_{base}} \tag{7.2}$$

$$Y_{base} = \frac{I_{base}}{V_{base}} \tag{7.3}$$

Substituting Eq. (7.1) into Eq. (7.2) we have

$$Z_{base} = \frac{V_{base}}{I_{base}} = \frac{V_{base}^2}{VA_{base}} \tag{7.4}$$

or

$$Z_{base} = \frac{(kV_{base})^2}{MVA_{base}} \tag{7.5}$$

The basic per-unit scaling equation is

$$\text{Per-unit (pu) value} = \frac{\text{actual value}}{\text{base value}} \tag{7.6}$$

Therefore,

$$\text{pu}Z = \frac{Z_{\text{actual}}}{Z_{\text{base}}} = \frac{Z_{\text{actual}}}{(\text{kV}_{\text{base}})^2/\text{MVA}_{\text{base}}} = Z_{\text{actual}}\left[\frac{\text{MVA}_{\text{base}}}{(\text{kV}_{\text{base}})^2}\right] \tag{7.7}$$

For a three-phase system,

$$\text{VA}_{\text{base}} = \sqrt{3}\ \text{V}_{\text{base}}I_{\text{base}} \tag{7.8}$$

$$I_{\text{base}} = \frac{\text{VA}_{\text{base}}}{\text{V}_{\text{base}}} = \frac{\text{V}_{\text{base}}I_{\text{base}}}{\sqrt{3}\ \text{V}_{\text{base}}} \tag{7.9}$$

$$Z_{\text{base}} = \frac{\text{V}_{\text{base}}}{3I_{\text{base}}} = \frac{\text{V}^2_{\text{base}}}{\text{V}_{\text{base}}I_{\text{base}}} \tag{7.10}$$

or

$$Z_{\text{base}} = \frac{\text{kV}^2_{\text{base}}}{\text{MV}_{\text{base}}} \tag{7.11}$$

The per-unit value is

$$\text{pu}Z = Z_{\text{actual}}\frac{\text{MVA}_{\text{base}}}{(\text{kV}_{\text{base}})^2} \tag{7.12}$$

Equations (7.7) and (7.12) give the per-unit values of impedances for single-phase and three-phase systems, respectively; it is seen that both equations are identical.

The per-unit values of electrical devices in power systems are generally given in terms of their own kVA and voltage ratings. It is customary to express all impedances in any part of the system on the same impedance base when doing computations. We are interested here in developing a procedure to convert per-unit impedances from one base to another. Suppose an impedance with actual value (Z_{actual}) is referred to base 1 as per-unit value (puZ_1) and it is desired to change the puZ_1 to a new per-unit value (puZ_2), using Eq. (7.12) we get

$$\text{pu}Z_1 = Z_{\text{actual}}\frac{\text{MVA}_1}{(\text{kV})^2} \tag{7.13}$$

and

$$\text{pu}Z_2 = Z_{\text{actual}}\frac{\text{MVA}_2}{(\text{kV}_2)^2} \tag{7.14}$$

Dividing Eq. (7.13) by Eq. (7.14) we get

$$\frac{\text{pu}Z_1}{\text{pu}Z_2} = \left[\left(\frac{\text{MVA}_1}{\text{MVA}_2}\right)\left(\frac{\text{kV}_2}{\text{kV}_1}\right)^2\right] \tag{7.15}$$

Therefore

$$\text{pu}Z_2 = \text{pu}Z_1 \left[\left(\frac{\text{MVA}_2}{\text{MVA}_1}\right)\left(\frac{\text{kV}_1}{\text{kV}_2}\right)^2\right] \tag{7.16}$$

Example 7.1 Draw an impedance diagram for the power system shown as a one-line diagram in Fig. 7.4. Show all the impedances in per-unit values, neglect all resistances. Use a base of 30 MVA and 138 kV for 80-Ω and 20-Ω lines. The ratings of generators and transformers are

Generator 1	20 MVA	18 kV	$X'' = 0.2$ pu
Generator 2	30 MVA	18 kV	$X'' = 0.2$ pu
Generator 3	30 MVA	20 kV	$X'' = 0.2$ pu
Transformer 1 Single-phase units	10 MVA	138/18 kV	$X = 20\%$
Transformer 2	25 MVA	138/18 kV	$X = 10\%$
Transformer 3	35 MVA	138/20 kV	$X = 10\%$

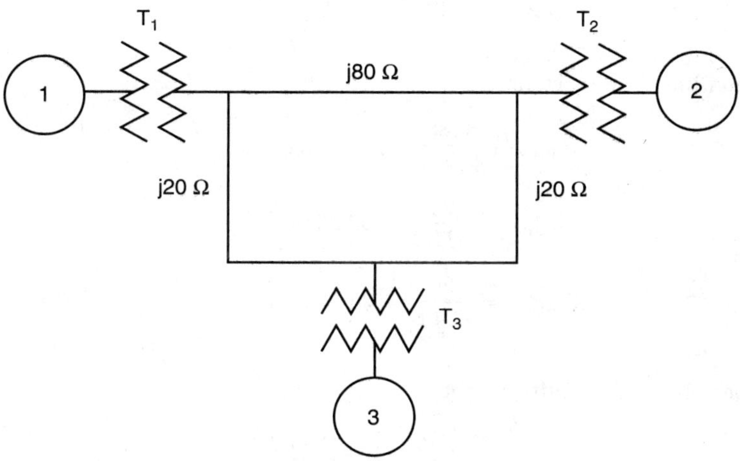

Figure 7.4 One-line diagram for Example 7.1.

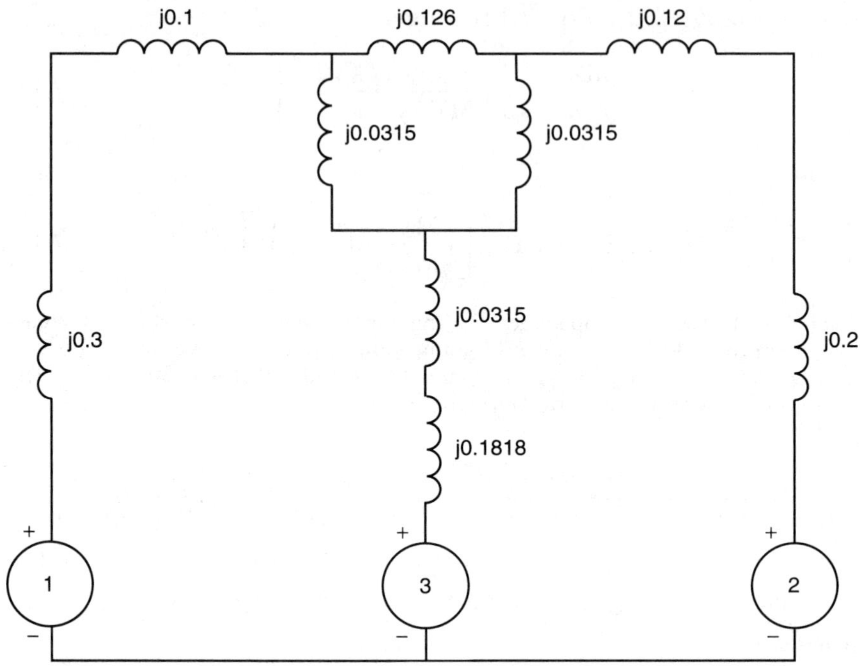

Figure 7.5 Impedance diagram of Example 7.1.

solution

Generator 1 $\quad \text{pu } X'' = 0.2(30/20) = 0.3 \text{ pu}$

Generator 2 $\quad \text{pu } X'' = 0.2(30/30) = 0.2 \text{ pu}$

Generator 3 $\quad \text{pu } X'' = 0.2(30/30) = 0.2 \text{ pu}$

Transmission lines
Transmission line base reactance:

$$X_{1(\text{base})} = \frac{(138)^2}{30} = 634.8\ \Omega$$

$$X_{11(\text{base})} = \frac{20}{634.8} = 0.126 \text{ pu}$$

$$X_{12(\text{base})} = \frac{20}{634.8} = 0.0315 \text{ pu}$$

Transformers
For three single-phase units we get

$$T_1 = 0.1(30/30) = 0.1 \text{ pu}$$

$$3(10 \text{ MVA}) = \sqrt{3}(138/18) \text{ kV}$$

$$30 \text{ MVA} = {}^{239}\!/_{18} \text{ kV}$$

$$T_2 = 0.1({}^{30}\!/_{25}) = 0.12 \text{ pu}$$

$$T_3 = 0.1({}^{30}\!/_{35}) = 0.0315 \text{ pu}$$

7.3 The Electric Power System

7.3.1 Introduction

It is not possible at the present time to build power stations (plants) near the consumers' location because of environmental, economical, and reliability problems. As a result of these problems, power stations are built in remote areas (nuclear and hydroelectric plants). The energy produced by such plants is transmitted over high-voltage (high-tension) transmission lines for long distances to consumers via the subtransmission systems and distribution systems.

The locations of power stations are dictated by the following factors: fuel availability (coal), environmental problems (pollution), the availability of adequate water head (hydroelectric plants), and enough cooling water for steam plants. In this section the radial and ring (loop) power systems and transmission lines will be discussed. To be complete, the transmission lines discussion will include the mechanical and electrical designs.

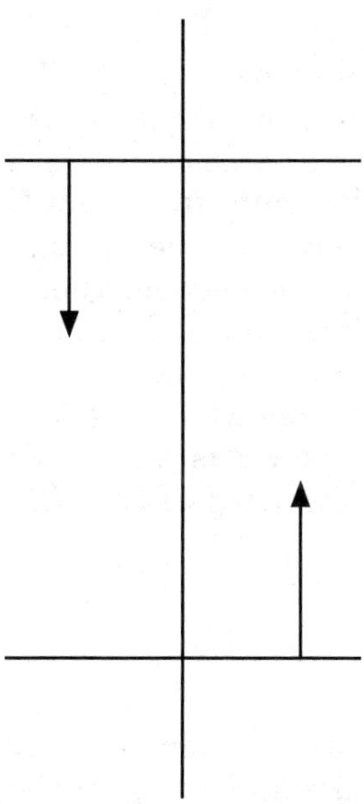

Figure 7.6 A schematic diagram of a radial system shown in Fig. 7.2.

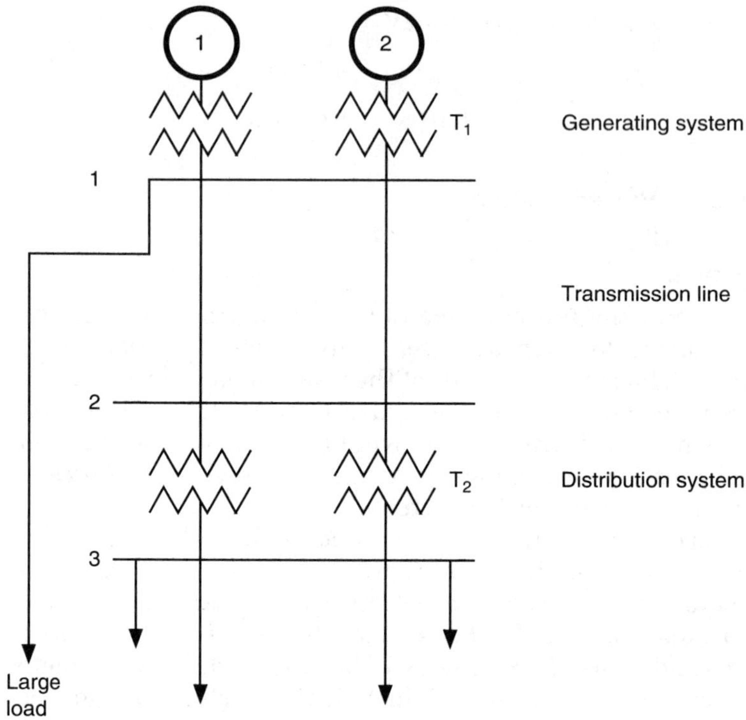

Figure 7.7 The power system showing main divisions.

7.3.2 Radial and ring (loop) connections

All power systems are connected either in radial as shown in Figs. 7.6 and 7.7 or in ring (loop) as shown in Fig. 7.1. The loop structure has more advantages than the radial structure. When a fault occurs in a radial structure, the entire system will have to be shut off. If a fault occurs in a loop system, the faulted part is isolated from the system. The healthy lines of the system will share excessive load between them.

The power system is mainly subdivided into three major divisions, namely, the generating system, the transmission system, and the distribution system. These divisions are schematically shown in Fig. 7.7. When large loads are required by the customers, power is tapped directly from the transmission grid (buses 5, 7, 8) as shown in Fig. 7.1.

7.4 Transmission Line Parameters

7.4.1 Power line resistance

Resistance affects the line impedance and thus transmission capacity due to the voltage drop along the line. If a dc current flowing in the homogeneous conductor is uniformly distributed over the cross-sectional area, the resistance of the conductor at 20°C is given by the following expression:

$$R_{dc} = \rho \frac{l}{A} \qquad \Omega/\text{m}^2 \tag{7.17}$$

where ρ = resistivity, in Ω/m
l = length of the conductor, in m
A = cross-sectional area, in m^2

The resistivity of metals increases proportionally with temperature within the temperature limits encountered on transmission lines. In general, the resistivity is given by the following equation:

$$\rho = \rho_{20}[1 - \alpha(t - 20°)] \tag{7.18}$$

where t is the actual temperature, in °C, and α is the temperature coefficient of the resistance.

The temperature coefficients for copper and aluminum are given as

For copper: $$\alpha = 0.0034 - 0.004 \left(\frac{1}{°\text{C}}\right)$$

For aluminum: $$\alpha = 0.0032 - 0.0056 \left(\frac{1}{°\text{C}}\right)$$

The dc resistance approaches ac resistance at a frequency of 60 Hz or under. The effective resistance measured in ohms per phase and meter is obtained from the formula:

$$R = R_{dc}(1 + \alpha\Delta t)K_1K_2 \tag{7.19}$$

where α = temperature coefficient of resistance
K_1, K_2 = constants due to stranded conductors
Δt = change in temperature

7.4.2 The power line inductance

In Sec. 2.3 a detailed discussion on self- and mutual inductances were dealt with, and their basic equations developed. The main discussion here is to determine the inductance inside and outside the conductor due to internal and external fluxes. The same inductance will also be calculated using geometric mean distance (GMD) and geometric mean radius (GMR).

7.4.3 The inductance of a conductor due to internal flux

The internal inductance can be computed by taking a ratio of flux linkages to the current as given by Eq. (2.3) (see Fig. 7.8). The internal flux links only a fraction of the total current. Assuming uniform current density, the mmf around a closed loop is given by Eq. (2.5)

$$\oint \mathbf{H} \cdot d\mathbf{l} = I \tag{7.20}$$

where $\mathbf{H}$ = magnetic field intensity
$\mathbf{l}$ = distance along the path
I = enclosed current

From Fig. 7.8 at a distance of x meters from the center the magnetic field H_x is constant. Therefore integrating through a closed path we get the fractional current I_x,

$$\oint H_x \, dl = 2\pi x(H_x) = I_x \tag{7.21}$$

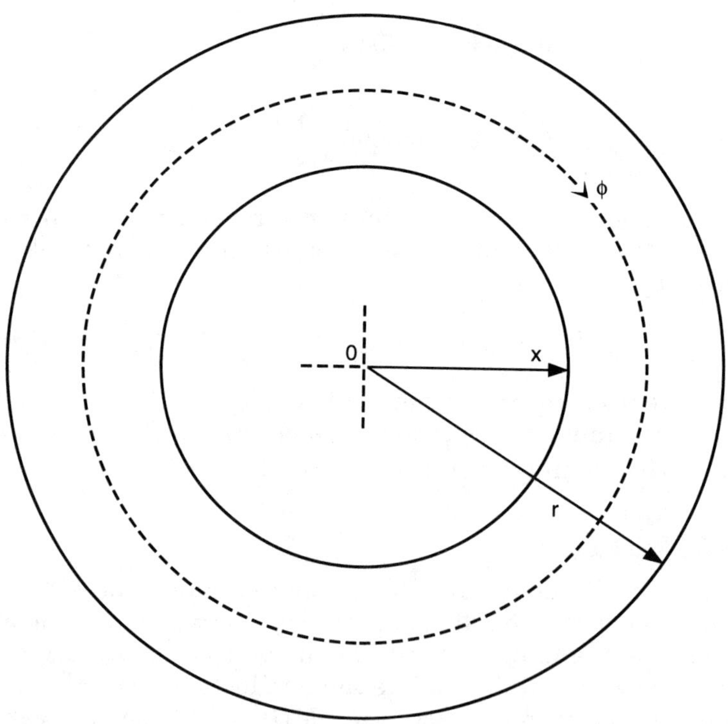

Figure 7.8 Cylindrical conductor with a cross section A.

The flux per unit length is given by

$$d\phi_x = B_x A_x = mH_x\,dx = \mu\,\frac{I_x}{2\pi x}\,dx \tag{7.22}$$

Since I_x is a fraction of the total current, then

$$I_x = \left(\frac{\pi x^2}{\pi r^2}\right) I = \frac{X^2}{r^2}\,I \tag{7.23}$$

The flux, $d\phi_x$, links only a fraction of the total current in the conductor and is given by

$$d\lambda_{\text{int}} = \left(\frac{\pi x^2}{\pi r^2}\right) d\phi_x = \frac{\mu X^3 I}{2\pi r^4}\,dx \tag{7.24}$$

Integrating flux linkages between 0 and r, the result is

$$\lambda_{\text{int}} = \int \frac{\mu x^3 I}{2\pi r^2}\,dx = \frac{\mu_0 \mu_r I}{8\pi} = \frac{1}{2} \times 10^{-7}\,I \tag{7.25}$$

From Eq. (2.31) the total internal inductance is given by

$$L_{\text{int}} = \frac{\mu_0 \mu_r I}{8\pi I} = \frac{1}{2} \times 10^{-7}\,I \tag{7.26}$$

where μ_r is the relative permeability, and μ_0 ($= 4\pi \times 10^{-7}$ H/m) is the permeability of free space.

7.4.4 Inductance of a conductor due to external flux

The external inductance is found by taking points external to conductor (see Fig. 7.9). Let dx be an infinitesimal tubular thickness external to the conductor. The flux density at point x with a magnetic field intensity of H_x is

$$B_x = \mu H_x = \frac{\mu I}{2\pi x} \tag{7.27}$$

The external flux linkage is

$$d\lambda_{\text{ext}} = \frac{\mu I}{2\pi x}\,dx \tag{7.28}$$

Integrating Eq. (7.27) from $x = a$ to $x = b$ we get

$$d\lambda_{\text{ext}} = \frac{\mu l}{2\pi}\int_a^b \frac{dx}{x} = \frac{\mu_0 l}{2\pi}\ln\frac{b}{a} \tag{7.29}$$

where $\mu = \mu_0\,\mu_r$ is the permeability.

In Eq. (7.29) the flux linkage $d\lambda$ per meter is numerically equal to flux $d\phi$, because the flux external to the conductor links all the current in the conductor only once. Substituting the value of μ_0 in Eq. (7.29) and replacing $a = D_1$ and $b = D_2$, Eq. (7.29) becomes

$$\lambda_{\text{ext}} = \frac{4\pi \times 10^{-7}}{2\pi}(l)\left(\ln\frac{D_2}{D_1}\right) = (2 \times 10^{-7})(l)\left(\ln\frac{D_2}{D_1}\right) \tag{7.30}$$

The inductance due to total flux linkages is

$$L = (2 \times 10^{-7})\left(\ln\frac{D_2}{D_1}\right) \tag{7.31}$$

The total flux linkage and inductance are given by the following equations:

$$\lambda_{\text{total}} = \lambda_{\text{int}} + \lambda_{\text{ex}} = (2 \times 10^{-7})\left(\ln\frac{D_2}{e^{-0.25}D_1}\right) \tag{7.32}$$

and

$$L = (2 \times 10^{-7})\left(\ln\frac{D_2}{e^{-0.25}D_1}\right) \tag{7.33}$$

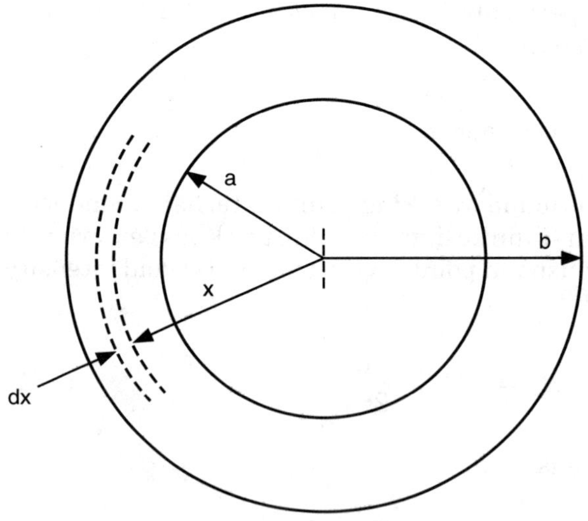

Figure 7.9 A cylindrical conductor showing external points between a and b.

7.4.5 Line inductance using geometric mean distance

A common method of calculating the inductance of a power line is the use of geometric mean distance (GMD). To determine the GMD, the spacing of conductors in the transmission towers must be known. Conductors are placed in such positions that no flashover is allowed to take place between them. The diameter of the conductor and its spacing affect the inductance, capacitance, and corona inception voltage. The conductors can be either equilaterally or unsymmetrically spaced. Equilateral spacing means that the distances D between the conductors are equal. Unsymmetrical spacing means that the distances between the conductors are not equal. In a three-phase line the flux linkages and inductance of each phase are not the same and this results in an unbalanced system. The balance is restored by transposing the conductors at regular intervals or at the switching stations. Figure 7.10 shows the transposition method.

The geometric mean distance of the three-phase transmission line is given by

$$D_{\text{eq}} = \sqrt[3]{D_{ab}D_{bc}D_{ca}} \tag{7.34}$$

Substituting Eq. (7.34) into Eq. (7.33) and replacing $D_1 = r$ we get the inductance of the power line.

$$L = (2 \times 10^{-7}l)\left(n\ \frac{\sqrt[3]{D_{ab}D_{bc}D_{ca}}}{e^{-0.25}r}\right)$$

$$= (2 \times 10^{-7})\left(\ln \frac{\sqrt[3]{D_{ab}D_{bc}D_{ca}}}{r'}\right) \tag{7.35}$$

where $r' = e^{-0.25}r = 0.779r$ is the geometric mean radius (GMR).

Example 7.2 A three-phase, 60-Hz power transmission line has its conductors arranged in an unequilateral triangle. The distance between the conductors are 3 m, 3.5 m, and 4.2 m, as shown in Fig. 7.11. The conductor's diameter is 0.365 cm. Calculate the inductance and the inductive reactance.

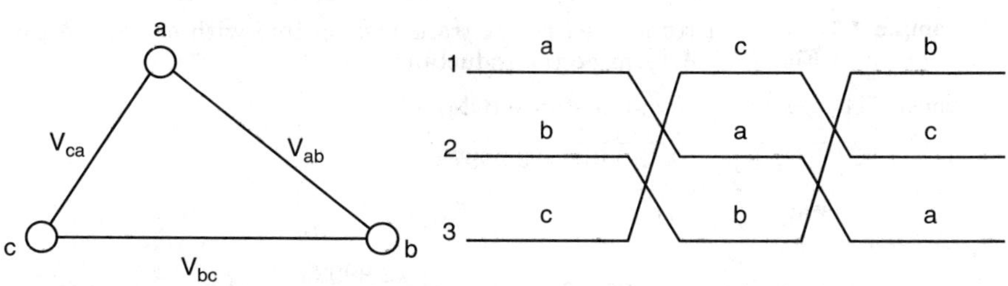

Figure 7.10 Transposed three-phase power lines.

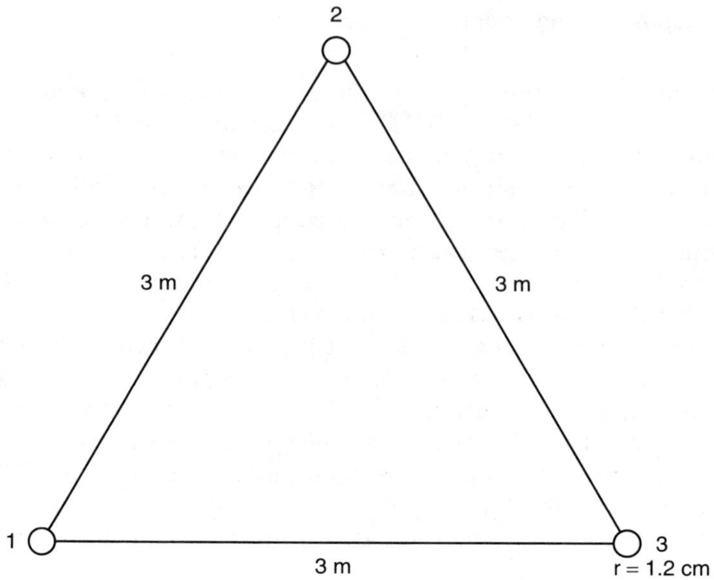

Figure 7.11 Example 7.2.

solution Using Eqs. (7.34) and (7.35) we have

$$D_{eq} = \sqrt[3]{D_{12}D_{23}D_{31}} = \sqrt[3]{3(3.5)(4.2)} = 3.5286 \text{ m}$$

$$r' = e^{-0.25}r = e^{-0.25}\left(\frac{0.365}{2}\right) = 0.1421 \text{ cm} = 0.00142 \text{ m}$$

The inductance and inductive reactance are

$$L = (2 \times 10^{-7})\left(\ln \frac{D_{eq}}{r'}\right) = (2 \times 10^{-7})\left(\ln \frac{3.5286}{0.00142}\right)$$

$$= 15.636 \times 10^{-7} \text{ H/m}$$

and $$X_L = \omega L = 2\pi fL = (2\pi 60)(15.636 \times 10^{-7})$$

$$= 0.5895 \times 10^{-7}\ \Omega$$

Example 7.3 Given a transposed power transmission line with a cross section as shown in Fig. 7.12, determine the inductance per phase.

solution The geometric mean distance (GMD)

$$D_{eq} = \sqrt[3]{D_{12}D_{23}D_{31}} = \sqrt[3]{3(3)(3)} = 2.9997 \text{ m}$$

$$r' = e^{-0.25}(0.012) = 0.0093 \text{ m}$$

$$L = (2 \times 10^{-7})\left(\ln \frac{D_{eq}}{r'}\right) = (2 \times 10^{-7})\left(\ln \frac{2.9997}{0.0093}\right) = 11.55 \times 10^{-7} \text{ H/m}$$

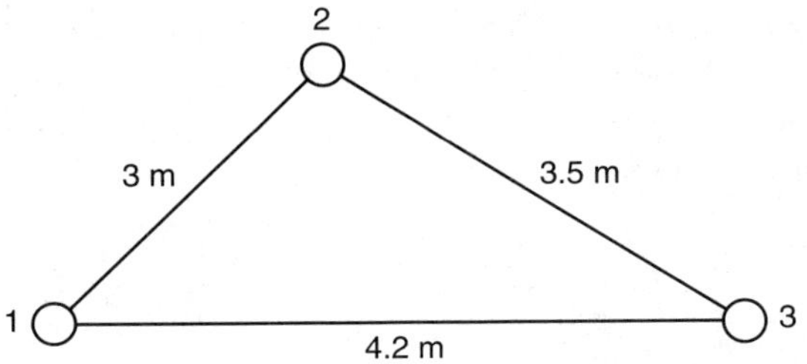

Figure 7.12 A cross section of a transmission line for Example 7.3.

7.4.6 The power line capacitance

The line resistance and inductance constitute the elements that make up the series impedance of the power transmission line. The capacitance forms a shunt or parallel admittance of the line. The series elements of which the inductance is dominant set a limit to the current that flows through the line and therefore determine the power that can be transmitted. The shunt elements of which capacitance is dominant represent a leakage path of the line currents. The leakage currents are proportional to the line voltage. The importance of the shunt elements increases with the increase of magnitude of the operating voltage.

Line capacitance of a single-phase line. The electric field between two parallel power lines 1 and 2 can be determined by using Fig. 7.13. The voltage V_{12} between conductors 1 and 2 is calculated by first assuming that only conductor 1 is charged and then conductor 2 is charged.

$$V_{12} = \frac{Q_1}{2\pi\varepsilon}\left[\ln\left(\frac{D}{r}\right) - \ln\left(\frac{r}{D}\right)\right]$$

$$= \frac{Q}{2\pi\varepsilon}\ln\left(\frac{D^2}{r^2}\right) = \frac{Q_1}{\pi\varepsilon}\ln\left(\frac{D}{r}\right) \quad \text{V} \qquad (7.36)$$

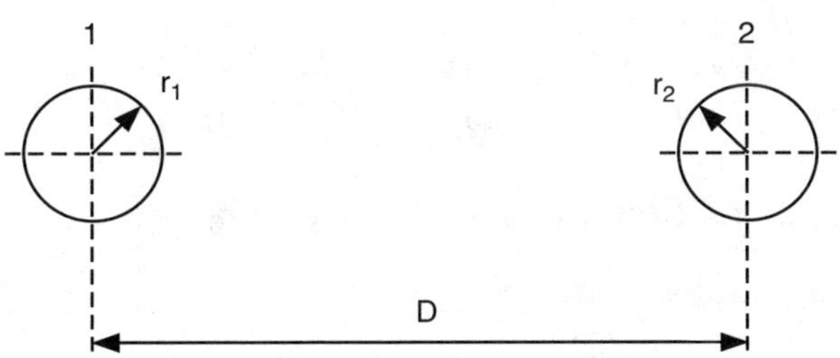

Figure 7.13 Two parallel conductors of a single-phase line.

where ε is the permittivity constant of the dielectric medium. Since the radii of the two conductors are equal and their charges are also equal in magnitude but opposite in polarity, we have $r_1 = r_2 = r$ and $Q_1 = -Q_2$. Equation (7.36) becomes

$$V_{12} = \frac{Q_1}{2\pi\varepsilon} \ln\left(\frac{D}{r}\right) - \ln\left(\frac{r}{D}\right)$$

$$= \frac{Q_1}{2\pi\varepsilon} \ln\left(\frac{D^2}{r^2}\right) = \frac{Q_1}{\pi\varepsilon} \ln\left(\frac{D}{r}\right) \quad \text{V} \tag{7.37}$$

The capacitance between the two conductors of a power line is

$$C_{12} = \frac{Q_1}{V_{12}} = \frac{\pi\varepsilon}{\ln (D/r)} \quad \text{F/m} \tag{7.38}$$

Capacitance of a three-phase power line with equal spacing. Equation (7.36) gives the voltage between two conductors of a single-phase power line. The voltage V_{12} was found by taking into account the charges in conductors 1 and 2, thus

$$V_{12} = \frac{1}{2\pi\varepsilon} \left[Q_1 \ln\left(\frac{D}{r_1}\right) + Q_2 \ln\left(\frac{r_2}{D}\right) \right] \tag{7.39}$$

The voltage in conductor 3 due to the charge Q_3 is

$$V_{12} = \frac{Q_3}{2\pi\varepsilon} \ln\left(\frac{D}{D}\right) \tag{7.40}$$

Summing up Eqs. (7.39) and (7.40) we get the voltage of conductor 1 in terms of the other two conductors.

$$V_{12} = \frac{1}{2\pi\varepsilon} \left[Q_1 \ln\left(\frac{D}{r_1}\right) + Q_2 \ln\left(\frac{r_2}{D}\right) + Q_3 \ln\left(\frac{D}{D}\right) \right] \tag{7.41}$$

and

$$V_{13} = \frac{1}{2\pi\varepsilon} \left[Q_1 \ln\left(\frac{D}{r}\right) + Q_2 \ln\left(\frac{D}{D}\right) + Q_3 \ln\left(\frac{r_3}{D}\right) \right] \tag{7.42}$$

Summing Eqs. (7.41) and (7.42) and letting $r_1 = r_2 = r_3 = r$ and $\ln (D/D) = 0$, we get

$$V_{12} + V_{13} = \frac{1}{2\pi\varepsilon} \left[2Q_1 \ln\left(\frac{D}{r}\right) + (Q_2 + Q_3) \ln\left(\frac{r}{D}\right) \right] \tag{7.43}$$

Substituting $Q_2 + Q_3 = -Q_1$ into Eq. (7.43) we have

$$V_{12} + V_{13} = \frac{1}{2\pi\varepsilon}\left[2Q_1 \ln\left(\frac{D}{r}\right) - Q_1 \ln\left(\frac{r}{D}\right)\right]$$

$$= \frac{3Q_1}{2\pi\varepsilon} \ln\left(\frac{D}{r}\right) \tag{7.44}$$

From Fig. 7.14*b*

$$X_1 = V_{n2} \cos 60° = 0.5V_{n2} \tag{7.45}$$

$$X_2 = V_{\text{in}} + X_1 \tag{7.46}$$

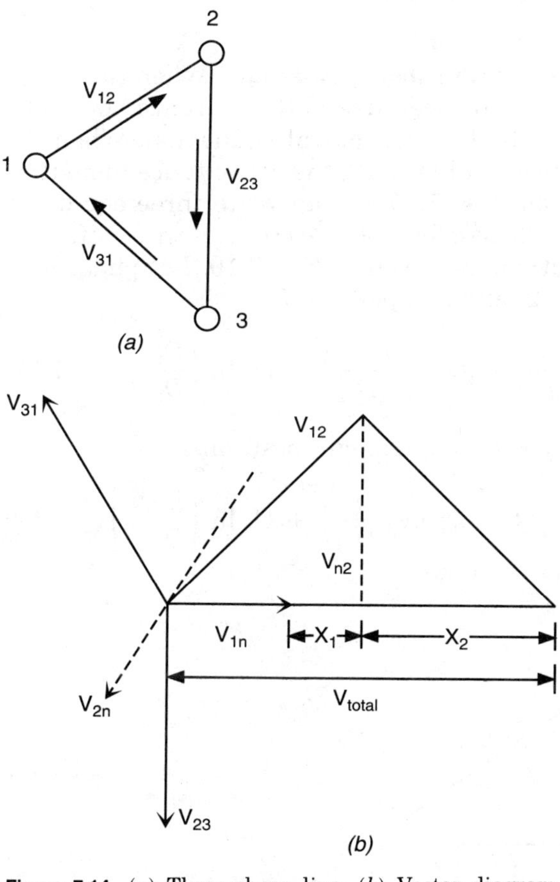

Figure 7.14 (*a*) Three-phase line. (*b*) Vector diagram showing the voltage to neutral.

But, $V_{in} = V_{2n} = V_{3n}$

$$V_{total} = V_{in} + 0.5V_{in} + V_{in} + 0.5V_{in} = 3V_{in} \tag{7.47}$$

Therefore summing up vectorially from Fig. 7.14*b* we get

$$V_{12} + V_{13} = 3V_{in} \tag{7.48}$$

Therefore

$$V_{in} = \frac{1}{3}(V_{12} + V_{13}) = \frac{Q_1}{2\pi\varepsilon} \ln\left(\frac{D}{r}\right) \tag{7.49}$$

Capacitance to the neutral is

$$C_{in} = \frac{Q_1}{V_{in}} = \frac{2\pi\varepsilon}{\ln (D/r)} \quad \text{F/m} \tag{7.50}$$

Capacitance of a three-phase line with unequal spacing. When the three-phase power lines are not equally spaced, and the lines are not transposed, the capacitance of each phase to neutral is unequal. When the lines are transposed, as shown in Fig. 7.10, the capacitance to neutral of each phase is equal. From Fig. 7.15 we can write three equations for any of the three phases. Phase *a* for instance occupies three different positions during transposition, as shown in Fig. 7.10. Let phase *a* be in position 1, *b* in position 2, and *c* in position 3.

$$V_{12} = \frac{1}{2\pi\varepsilon}\left[Q_a \ln\left(\frac{D_{12}}{r}\right) + Q_b \ln\left(\frac{r}{D_{12}}\right) + Q_c \ln\left(\frac{D_{23}}{D_{31}}\right)\right] \tag{7.51}$$

With *a* in position 2, *b* in position 3, and *c* in position 1

$$V_{12} = \frac{1}{2\pi\varepsilon}\left[Q_a \ln\left(\frac{D_{23}}{r}\right) + Q_b \ln\left(\frac{r}{D_{23}}\right) + Q_c \ln\left(\frac{D_{31}}{D_{12}}\right)\right] \tag{7.52}$$

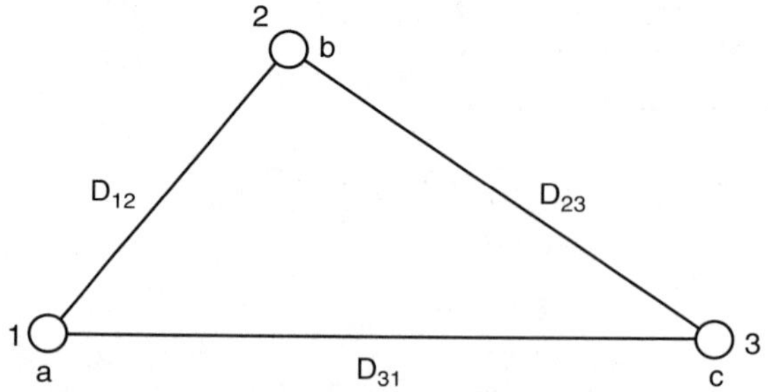

Figure 7.15 Three-phase lines unequally spaced.

With a in position 3, b in position 1, and c in position 2

$$V_{12} = \frac{1}{2\pi\varepsilon}\left[Q_a \ln\left(\frac{D_{31}}{r}\right) + Q_b \ln\left(\frac{r}{D_{31}}\right) + Q_c \ln\left(\frac{D_{12}}{D_{23}}\right)\right] \tag{7.53}$$

The average value of V_{12} is found by adding Eqs. (7.51) to (7.53) and then dividing by 3 as given by Eq. (7.54):

$$\begin{aligned} V_{12} &= \frac{1}{2\pi\varepsilon}\left[Q_a \ln\left(\frac{D_{12}D_{23}D_{31}}{r^3}\right)\right. \\ &\quad \left. + Q_b \ln\left(\frac{r^3}{D_{12}D_{23}D_{31}}\right) + Q_c \ln\left(\frac{D_{12}D_{23}D_{31}}{D_{12}D_{23}D_{31}}\right)\right] \\ &= \frac{1}{6\pi\varepsilon}\left[Q_a \ln\left(\frac{D_{eq}}{r}\right) + Q_b \ln\left(\frac{r}{D_{eq}}\right)\right] \end{aligned} \tag{7.54}$$

where $D_{eq} = \sqrt[3]{D_{12}D_{23}D_{31}}$ is the geometric mean distance (GMD).

To find the average voltage drop from conductor a to c, we use the same procedure and find the value of V_{13} as

$$\begin{aligned} V_{13} &= \frac{1}{6\pi\varepsilon}\left[Q_a \ln\left(\frac{D_{eq}}{r}\right)\right. \\ &\quad \left. + Q_b \ln\left(\frac{D_{12}D_{23}D_{31}}{D_{12}D_{23}D_{31}}\right) + Q_c \ln\left(\frac{r}{D_{eq}}\right)\right] \\ &= \frac{1}{6\pi\varepsilon}\left[Q_a \ln\left(\frac{D_{eq}}{r}\right) + Q_c \ln\left(\frac{r}{D_{eq}}\right)\right] \end{aligned} \tag{7.55}$$

The voltage to the neutral is found by using Eq. (7.48) as

$$\begin{aligned} 3V_{in} &= V_{12} + V_{13} \\ &= \frac{1}{2\pi\varepsilon}\left[2Q_a \ln\left(\frac{D_{eq}}{r}\right) + Q_b \ln\left(\frac{r}{D_{eq}}\right) + Q_c \ln\left(\frac{r}{D_{eq}}\right)\right] \end{aligned} \tag{7.56}$$

In a balanced three-phase operation, the charges are

$$Q_a = -(Q_b + Q_c) \tag{7.57}$$

Substituting Eq. (7.57) into Eq. (7.56) we get

$$3V_{in} = \frac{3}{2\pi\varepsilon} \ln\left(\frac{D_{eq}}{r}\right) \qquad \text{V} \tag{7.58}$$

Therefore the capacitance to neutral is

$$C_{\text{in}} = \frac{Q_a}{V_{\text{in}}} = \frac{2\pi\varepsilon}{\ln(D_{\text{eq}}/r)} \qquad \mu\text{F} \tag{7.59}$$

7.5 Power Transmission Line Calculations

7.5.1 Mechanical design

Before discussing methods of calculating line constants, it is desirable at this stage to study briefly the mechanical design of power lines. In mechanical design two most important parameters will be discussed: first, the material of the conductor and second the tensile strength or breaking load of the conductor. The conductors can be hard-drawn like copper, cadmium, copper, aluminum, and aluminum alloy or aluminum conductor steel reinforced (ACSR). The hard-drawn conductors are used for low voltages and the ACSR is used for high voltages. In stranded lines of the aluminum conductor steel reinforced (ACSR), the aluminum is used for conducting the current while the steel core provides mechanical strength. The choice of proper types of towers or poles, and the calculation of sag are very important parameters to the design engineer. The sag calculation is very essential and depends on the topography of the terrain, condition of wind, ice, and temperature. Sag and tension will be calculated using the parabola and the catenary methods.

7.5.2 Parabola

Figure 7.16 shows a transmission line strung between two fixed points A and B, which are on the same level. The equation of a parabola is

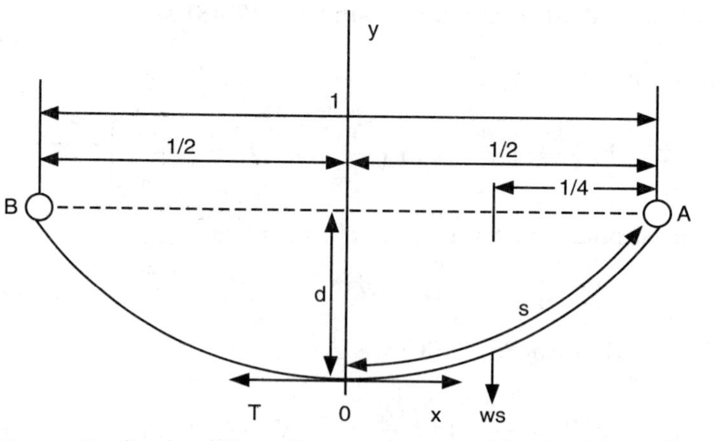

Figure 7.16 Overhead line with supports at equal heights (A and B).

given by

$$y = ax^2 \tag{7.60}$$

In Fig. 7.16, l is the span (the horizontal distance between the two poles in meters), d is the sag at midspan, and s is the conductor distance from the pole to 0. Let $y = d$ and $x = ½$ be substituted in Eq. (7.60) as

$$d = a\left(\frac{l}{2}\right)^2 \tag{7.61}$$

$$a = \frac{4d}{l^2} \tag{7.62}$$

Substituting Eq. (7.62) into Eq. (7.60) we have

$$y = \left(\frac{4d}{l^2}\right) x^2 \tag{7.63}$$

In most power transmission lines, the sag is small compared with the span ($d < l$) so that the curved length of the conductor is approximately equal to the span. The conductor length s is equal to half the span $s = l/2$. The weight of the half-span (ws) is acting at $l/4$ meters from point A.

Taking moments about point A we have

$$Td = (ws)\left(\frac{l}{4}\right) = w\frac{l^2}{8} \tag{7.64}$$

Therefore the tension is

$$T = \left(\frac{wl^2}{8d}\right) \qquad \text{N} \cdot \text{m} \tag{7.65}$$

and sag is

$$d = \frac{wl^2}{8T} \qquad \text{m} \tag{7.66}$$

The transmission lines possess minimum clearance height for the line above ground. If this clearance height is added to the height of sag, the height of the insulator support at points A and B to the ground can be found. If during calculations of sag, wind, ice and temperature have to be included, the sag equation becomes

$$d = \frac{F_r l^2}{8T} \tag{7.67}$$

where $F_r = [F_w^2 + (w + w_i)^2]^{1/2}$ (7.68)

F_w = wind force

w_i = weight of ice

7.5.3 Catenary

The formulas, which are discussed here, are used for very high voltage lines whenever sag > span/10 approximately. Figure 7.17*a* shows an unknown height $D0 = C(m)$, the horizontal tension $T_H = wc$, and point P where the weight of the half-span is acting. With the help of Fig. 7.17*b* the actual tension T can be computed as

$$T = \sqrt{T_H^2 + (ws)^2} = w\sqrt{c^2 + s^2} \tag{7.69}$$

The length of the conductor from one support to the midspan is

$$s = c \tan \theta \qquad \text{m} \tag{7.70}$$

At point $P(x, y)$ we get

$$\frac{dy}{dx} = \frac{ws}{T_H} \tag{7.71}$$

and

$$ds = \sqrt{dx^2 + dy^2} \tag{7.72}$$

We also know that

$$\frac{dx}{ds} = \cos \theta = \frac{T_H}{T} = \frac{c}{\sqrt{c^2 + s^2}} \tag{7.73}$$

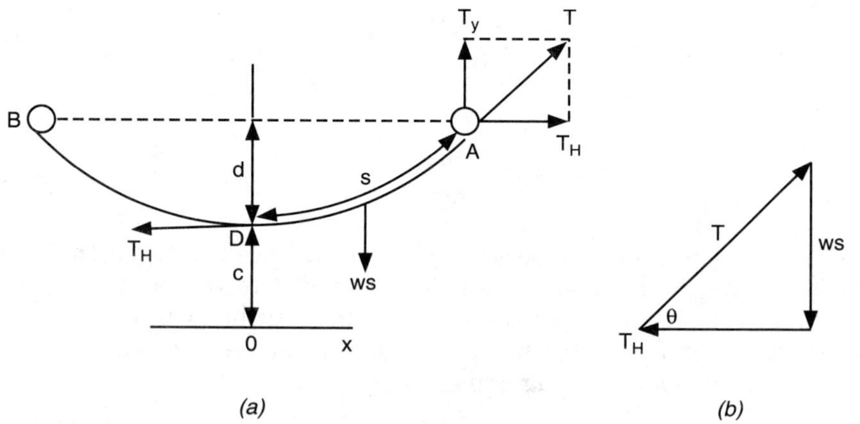

Figure 7.17 (*a*) Power transmission lines at equal level (*A* and *B*). (*b*) Forces on the transmission line.

Integrating Eq. (7.73) with respect to s we have

$$x = \int \left(\frac{c}{c^2 + s^2}\right) ds = c \sinh^{-1}\left(\frac{s}{c}\right) + K \tag{7.74}$$

Taking the initial conditions for $x = 0, \theta = 0$ and $\cos\theta = 1$, Eqs. (7.70), (7.73), and (7.74) become

$$s = 0 \qquad y = c^2 + s^2 \qquad \text{and} \qquad \sinh^{-1}\frac{s}{c} = 0 \tag{7.75}$$

With the values obtained in expressions (7.75), $k = 0$; therefore Eq. (7.74) becomes

$$\frac{x}{c} = \sinh^{-1}\left(\frac{s}{c}\right) \tag{7.76}$$

and

$$\frac{s}{c} = \sinh\left(\frac{x}{c}\right) \tag{7.77}$$

From Fig. 7.17*b* we also get

$$\frac{s}{c} = \tan\theta = \frac{dy}{dx} = \sinh\left(\frac{x}{c}\right) \tag{7.78}$$

Integrating Eq. (7.78) with respect to x we have

$$y = \int \sinh\left(\frac{x}{c}\right) dx = c\cosh\left(\frac{x}{c}\right) + N \tag{7.79}$$

To eliminate N from Eq. (7.79), let $D0 = c(m)$ in Fig. 7.17*a*. For $x = 0$, Eq. (7.79) becomes

$$y = c0 + N = 0 \tag{7.80}$$

Therefore Eq. (7.79) becomes

$$y = c\cosh\left(\frac{x}{c}\right) \tag{7.81}$$

Squaring Eqs. (7.77) and (7.81) and subtracting Eq. (7.78) from Eq. (7.81) we get

$$y^2 - s^2 = c^2\left[\cosh^2\left(\frac{x}{c}\right) - \sinh^2\left(\frac{x}{c}\right)\right] = c^2 \tag{7.82}$$

Hence

$$y = \sqrt{c^2 + s^2} \tag{7.83}$$

From Eq. (7.69), the actual tension (T) at any point (x, y) in the conductor is given by

$$T = wy \tag{7.84}$$

The tension in the conductor becomes maximum at the supports where $y = c + d$ and minimum tension occurs at the midspan, where $y = c$; therefore

$$y = c + d = c^2 + s^2 \tag{7.85}$$

Therefore

$$c = \frac{s^2 - d^2}{2d} \tag{7.86}$$

$$T_{\max} = wy = w(c + d) = w\left(d + \frac{s^2 - d^2}{2d}\right) \quad \text{N} \cdot \text{m} \tag{7.87}$$

and $$T_m = wc = w\left(\frac{s^2 - d^2}{2d}\right) \tag{7.88}$$

7.5.4 Supports at different levels

Occasionally we have to calculate sag when the supports are at different levels as shown in Fig. 7.18. Using the parabolic method we have

$$d_2 = w\,\frac{(1 - q)^2}{2T} \tag{7.89}$$

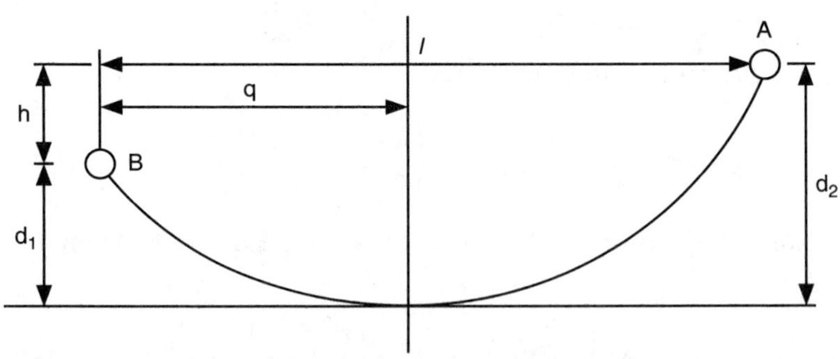

Figure 7.18 Transmission line with supports at different levels.

and
$$d_1 = w\frac{q^2}{2T} \tag{7.90}$$

Hence

$$h = d_2 - d_1 = \frac{w}{2T}(l^2 - 2lq + q^2 - q^2) = \frac{wl}{2T}(1 - 2q) \tag{7.91}$$

Therefore,

$$q = \frac{1}{2}\left(1 - \frac{2Th}{w}\right) \tag{7.92}$$

Example 7.4 A power transmission line conductor hangs in the form of a catenary. The distance from the lowest point of the catenary to the ground is 150 m. The span of the catenary is 310 m and the conductor weighs 5.5 N · m. Calculate

1. The length of the conductor
2. The sag
3. Maximum and minimum values of tension using the catenary and the parabola methods

solution

1. From Eq. (7.81) we get

$$y = c\cosh\left(\frac{x}{c}\right) = 150\cosh\left(\frac{155}{150}\right) = 150\cosh 1.033$$

$$\cosh 1.033 = \left(1 + \frac{1.033}{2}\right)^2 = 1582$$

$$y = 150(1.5827) = 237.407 \text{ m}$$

From Eq. (7.85)

$$y = \sqrt[3]{c^2 + s^2}$$

$$y^2 = c^2 + s^2$$

$$s = \sqrt[3]{y^2 - c^2} = \sqrt[3]{237.407^2 - 150^2} = 23.512 = 183.629 \text{ m}$$

2. Sag:

$$y = c + d$$

$$d = y - c = 237.407 - 150 = 87.407 \text{ m}$$

3. $T_{\max} = w(c + d) = 55(150 + 87.407) = 13{,}057.3 \text{ N}$

$T_{\min} = wc = 55(150) = 8250.0 \text{ N}$

From Eq. (7.65), the tension using the parabola method is

$$T = \frac{wl^2}{8d} = \frac{55(310)^2}{8(87.407)} = 7558.7 \text{ N}$$

7.5.5 Electrical design of power transmission lines (mathematical model)

The calculations of voltages, currents, and power flow in a power transmission line requires the development of the mathematical models of the line. The models must take into account the series resistance, inductance, and shunt capacitance of each phase. The power transmission line is commonly subdivided into three according to their lengths:

1. The short transmission line is about 80 km (50 miles).
2. The medium transmission line is between 80 km and 240 km (50 miles to 150 miles).
3. The long transmission line is above 240 km (150 miles).

For short and medium lines the parameters are lumped, and the shunt capacitance is usually neglected. For a long line parameters are uniformly distributed along the entire length of the line.

The short power transmission lines. Figure 7.19*a* shows an equivalent circuit of a short transmission line, and its phasor diagram is shown in Fig. 7.19*b*. Using the two figures, the following equations for voltages and currents can be derived:

$$\begin{aligned} I_s &= I_R \\ V_s &= V_R + ZI_R \end{aligned} \tag{7.93}$$

where I_s and V_s are sending end current and sending end voltage, and I_R and V_R are received end current and receiving end voltage.

In a generalized form, Eqs. (7.92) and (7.93) can be expressed in matrix form:

$$\begin{bmatrix} V_s \\ I_s \end{bmatrix} = \begin{bmatrix} A & B \\ C & D \end{bmatrix} \begin{bmatrix} V_R \\ I_R \end{bmatrix} \tag{7.94}$$

where the arbitrary line constants are $A = 1$, $B = Z$, $C = 0$, and $D = 1$.

The variation of power factor of the load causes changes on the receiving voltage. This variation is better understood when discussing the voltage regulation on short power lines. The voltage regulation of a power line is defined as the rise in voltage at the receiving end

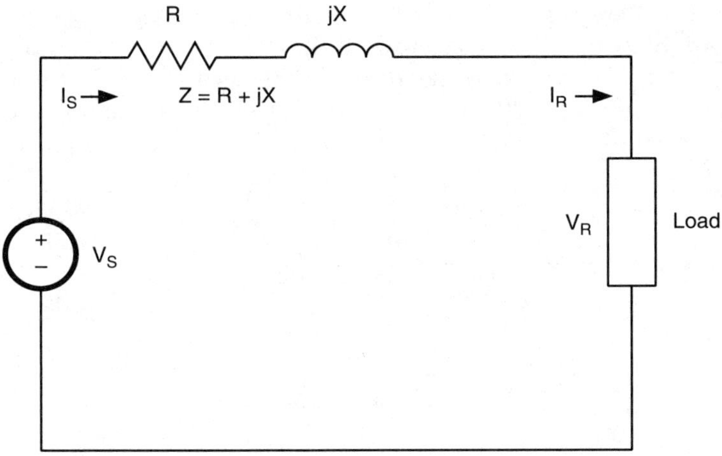

(a)

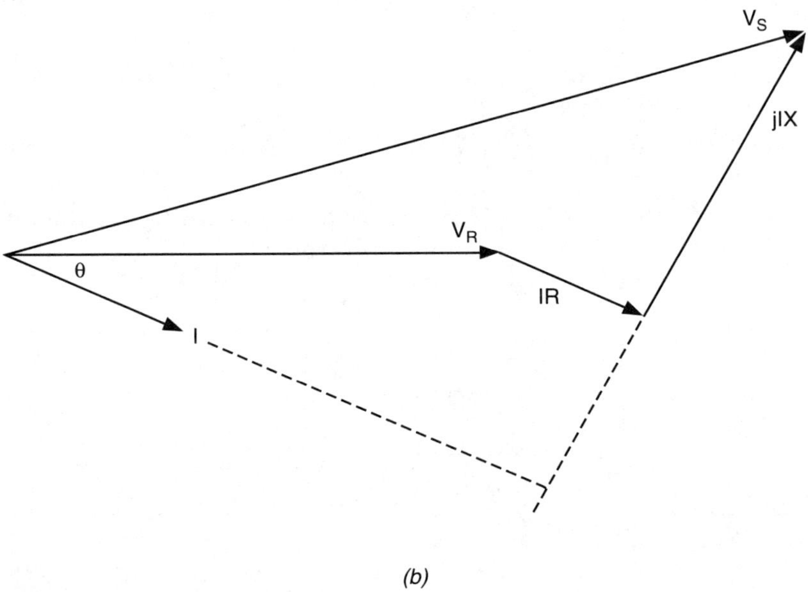

(b)

Figure 7.19 (*a*) Short transmission line equivalent circuit. (*b*) Phasor diagram of equivalent circuit.

expressed as per unit of full-load voltage. This happens when full-load at a specific power factor is moved while the sending end voltage is kept constant. The voltage regulation is

$$\text{Percentage voltage regulation (VR)} = \frac{|V_s| - |V_R|}{|V_R|} \times 100\% \quad (7.95)$$

The medium power transmission line. The medium line can be represented by nominal Π or *T* circuits. In nominal Π circuits the shunt capacitance is split into two equal parts while the series impedance is lumped as shown in Fig. 7.20. Referring to Fig. 7.20 and using the nodal method at nodes *a* and *b*, we can derive the following equations:

At node *a*

$$I_s = V_s\left(\frac{Y}{2}\right) + \frac{V_s}{Z} - \frac{V_R}{Z} \tag{7.96}$$

and at node *b*

$$I_R = \frac{V_R - V_s}{Z} + V_R\left(\frac{Y}{2}\right) \tag{7.97}$$

Simplifying Eq. (7.97) we get

$$V_s = \left(1 + \frac{YZ}{2}\right)V_R - ZI_R \tag{7.98}$$

Substituting the value of V_s from Eq. (7.98) into Eq. (7.96) we have

$$I_s = \left(1 + \frac{YZ}{4}\right)YV_R - \left(1 + \frac{YZ}{2}\right)I_R \tag{7.99}$$

Putting Eqs. (7.98) and (7.99) into matrix form, the values of the arbitrary constants *A*, *B*, *C*, and *D* can be extracted.

$$\begin{bmatrix} V_s \\ I_s \end{bmatrix} = \begin{bmatrix} 1 + \dfrac{YZ}{2} & -Z \\ Y + \dfrac{Y^2Z}{4} & -\left(1 + \dfrac{YZ}{2}\right) \end{bmatrix} \begin{bmatrix} V_R \\ I_R \end{bmatrix} \tag{7.100}$$

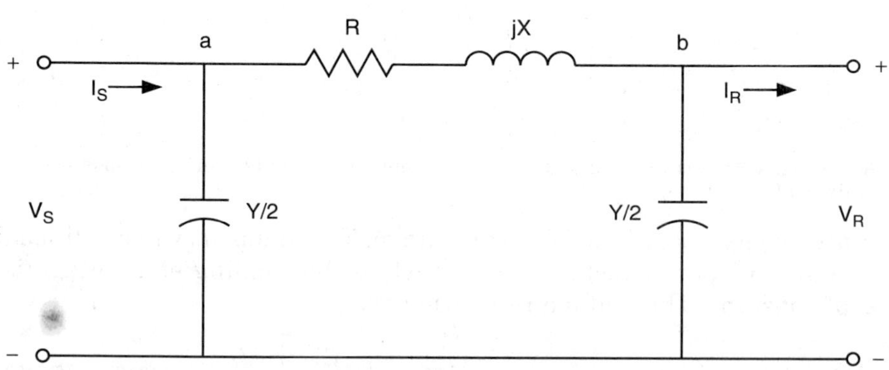

Figure 7.20 Nominal Π circuit of a medium power transmission line.

Therefore,

$$A = 1 + \frac{YZ}{2} \qquad C = Y + \frac{Y^2 Z}{4}$$

$$B = -Z \qquad D = -\left(1 + \frac{YZ}{2}\right)$$

Similar equations can be derived by using a nominal T circuit, as shown in Fig. 7.21. Applying Kirchhoff's voltage law (KVL) to Fig. 7.21 we get

$$V_s = I_s \frac{Z}{2} + I_s \left(\frac{1}{Y}\right) - I_R \left(\frac{1}{Y}\right) \tag{7.101}$$

and

$$V_R = \frac{I_R}{Z/2} + \frac{I_R}{Y} - \frac{I_s}{Y} \tag{7.102}$$

Simplifying Eqs. (7.101) and (7.102) we have

$$I_s = YV_R + \left(1 + \frac{YZ}{2}\right) I_R \tag{7.103}$$

and

$$V_s = \left(1 + \frac{YZ}{2}\right) V_R + \left(1 + \frac{YZ}{4}\right) ZI_R \tag{7.104}$$

Putting the above equations in matrix form and extracting the line constants we get

$$\begin{bmatrix} V_s \\ I_s \end{bmatrix} = \begin{bmatrix} 1 + \dfrac{YZ}{2} & Z + \dfrac{YZ^2}{4} \\ Y & 1 + \dfrac{YZ}{2} \end{bmatrix} \begin{bmatrix} V_R \\ I_R \end{bmatrix} \tag{7.105}$$

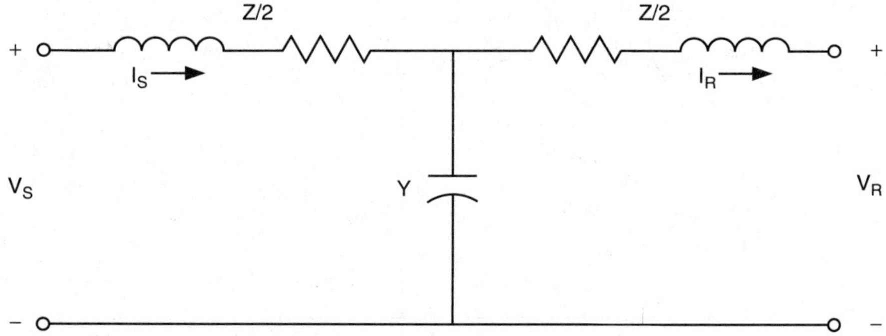

Figure 7.21 Nominal T equivalent circuit of a medium power transmission line.

Therefore,

$$A = 1 + \frac{YZ}{2} \qquad C = Y$$

$$B = Z + \frac{YZ^2}{4} \qquad D = 1 + \frac{YZ}{2}$$

Example 7.5 A medium 60-Hz, three-phase power transmission line is 160 km long. The total series impedance is $Z = 2.38 + j58.81\ \Omega$ and a shunt admittance is $Y = 7.122 \times 10^{-4}$ S. The line delivers 750 MW at 500 kV, with 0.95 power factor lagging. Determine the sending end voltage and sending end current (V_s, I_s) by

$$\text{Nominal } V_{12} = \frac{Q_1}{2\pi\varepsilon}\left[\ln\left(\frac{D}{r}\right) - \ln\left(\frac{r}{D}\right)\right]$$

$$= \frac{Q}{2\pi\varepsilon}\ln\left(\frac{D^2}{r^2}\right) = \frac{Q_1}{\pi\varepsilon}\ln\left(\frac{D}{r}\right) \quad \text{V} \qquad (7.36)$$

1. The nominal Π circuit
2. The nominal T circuit

solution

$$I_R = \frac{750{,}000}{\sqrt{3}(500)(0.95)} = 911.60 \text{ A}$$

1. Using Eqs. (7.98) and (7.99) and Fig. 7.22 the values of the sending end voltage and sending end current are found by

$$V_s = \left(1 + \frac{YZ}{2}\right) V_R - ZI_R$$

$$Z = 2.38 + j58.81 = 58.858\angle 87.68^\circ\ \Omega$$

$$Y = 7.122 \times 10^{-4} \text{ S}$$

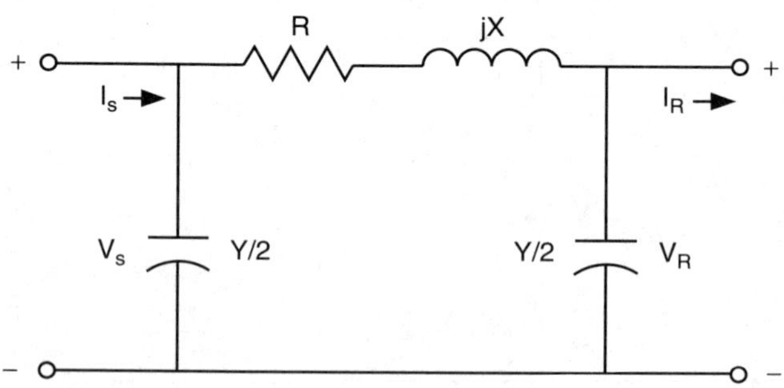

Figure 7.22 Nominal Π circuit for Example 7.5.

$$V_R = \frac{500{,}000}{\sqrt{3}} = 288{,}675.13 \text{ V}$$

$$V_s = \left(1 + \frac{YZ}{2}\right) V_R - ZI_R = 284{,}954.58\underline{/1.262°} \text{ V}$$

$$I_s = \left(Y + \frac{Y^2Z}{4}\right) V_R - \left(1 + \frac{YZ}{2}\right) I_R = 198.433\underline{/59.174°} \text{ A}$$

2. Using Fig. 7.23 and Eq. (7.105), the sending end voltage and sending end current are

$$V_s = 1 + \left(\frac{YZ}{2}\right) V_R + \left(Z + \frac{YZ^2}{4}\right) I_R = 284{,}927.09\underline{/1.25°} \text{ V}$$

$$I_s = YV_R + \left(1 + \frac{YZ}{2}\right) I_R = 200\underline{/59.42°} \text{ A}$$

The long power transmission line. The equivalent circuit of a power transmission line is found by taking the impedance and capacitance uniformly distributed along the entire length of the power line. The entire line is divided into infinitesimal segments, as shown in Fig. 7.24. Applying the differential equation method to one segment and then summing up for the entire line, the voltages and currents of the line can be found.

Defining the parameters of Fig. 7.24 we have:

$V_{(x)}$ is the phasor voltage at location x on the line

$I_{(x)}$ is the phasor current at location x on the line

$z = R + j\omega L$ is the series impedance per unit length, in Ω/m

$y = j\omega C$ is the shunt admittance per unit length, in S/m

x is the position along the line, measured from the receiving end toward the left in meters

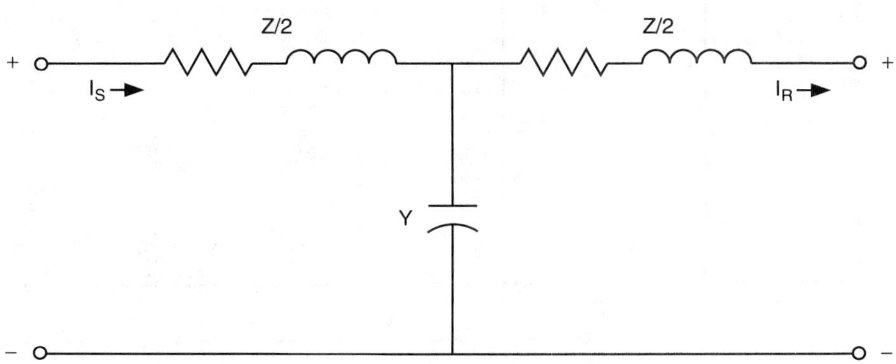

Figure 7.23 Nominal T circuit for Example 7.5.

From Fig. 7.24 the following equations can be derived:

$$dV = zI\,dx \tag{7.106}$$

$$dI = yV\,dx \tag{7.107}$$

Differentiating Eqs. (7.106) and (7.107)

$$\frac{d^2V}{dx^2} = z\,\frac{dI}{dx} \tag{7.108}$$

$$\frac{d^2I}{dx^2} = y\,\frac{dV}{dx} \tag{7.109}$$

Substituting the value of dI/dx and dV/dx into Eqs. (7.108) and (7.109) we have

$$\frac{d^2V}{dx^2} = zyV \tag{7.110}$$

and

$$\frac{d^2I}{dx^2} = zyI \tag{7.111}$$

Using Laplace transform on Eq. (7.110), we get

$$s^2V(s) - sV(x_0) - V'(x_0) - zyV(s) = 0 \tag{7.112}$$

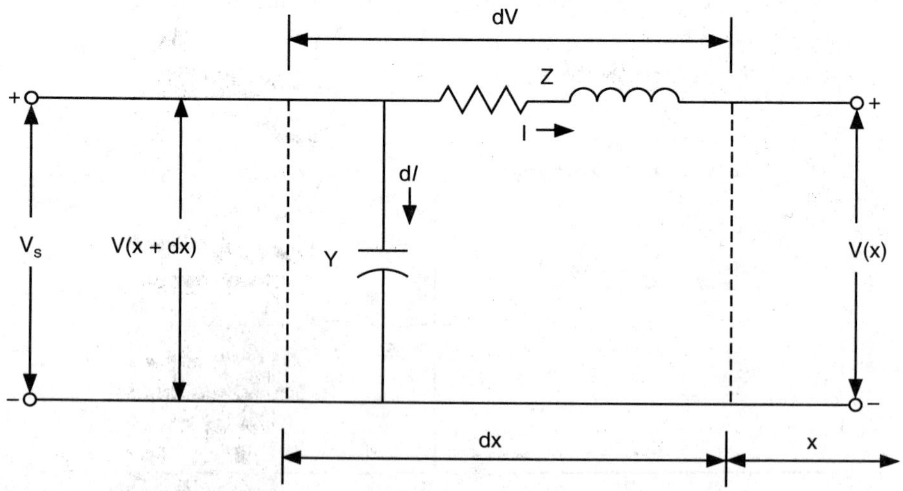

Figure 7.24 Three-phase long line with distributed parameters.

From the initial conditions $x = 0$, $V = V_R$ and $I = I_R$, we also know that

$$V'_{(x_0)} = \left.\frac{dV}{dx}\right|_{x=0} = ZI|_{x=0} = ZI_R \tag{7.113}$$

Substituting Eq. (7.113) into Eq. (7.112) and applying Laplace inverse transform, we get

$$\begin{aligned} s^2V(s) - sV_R - zI_R - zyV(s) &= 0 \\ V(s) &= \left(\frac{s}{s^2 - zy}\right) V_R + \left(\frac{I}{s^2 - zy}\right) ZI_R \end{aligned} \tag{7.114}$$

Therefore,

$$V_{(x)} = [\cosh(\sqrt{zy})x]V_R + \left[\sqrt{\frac{z}{y}}\sinh(\sqrt{zy})x\right] I_R \tag{7.115}$$

Similarly for $I_{(x)}$ from Eq. (7.111), we get

$$I_{(x)} = [\cosh(\sqrt{zy})x]I_R + \left[\sqrt{\frac{z}{y}}\sinh(\sqrt{zy})x\right] V_R \tag{7.116}$$

where $\sqrt{zy} = \gamma$ is the propagation constant, and $\sqrt{z/y} = Z_0$ is the characteristic impedance of the line.

Substituting the propagation constant and characteristic impedance into Eqs. (7.115) and (7.116) we have

$$V(x) = (\cosh \gamma x)V_R + (Z_0 \sinh \gamma x)I_R \tag{7.117}$$

and

$$I(x) = (Y_0 \sinh \gamma x)V_R + (\cosh \gamma x)I_R \tag{7.118}$$

Putting Eqs. (7.117) and (7.118) in matrix form we have

$$\begin{bmatrix} V(x) \\ I(x) \end{bmatrix} = \begin{bmatrix} \cosh \gamma x & Z_0 \sinh \gamma x \\ Y_0 \sinh \gamma x & \cosh \gamma x \end{bmatrix} \begin{bmatrix} V_R \\ I_R \end{bmatrix} \tag{7.119}$$

From Eq. (7.119) the line constants A, B, C, and D can be found to be

$$\begin{aligned} A = D &= \cosh \gamma x \qquad C = Y_0 \sinh \gamma x \\ B &= Z_0 \sinh \gamma x \end{aligned} \tag{7.120}$$

Equation (7.119) can be computed by expressing the hyperbolic functions in terms of exponential functions:

$$\sinh \gamma x = \frac{e^{\gamma x} - e^{-\gamma x}}{2} = \gamma x + \frac{(\gamma x)^3}{3!} \qquad (7.121)$$

$$\cosh \gamma x = \frac{e^{\gamma x} + e^{-\gamma x}}{2} = I + \frac{(\gamma x)^2}{2!} \qquad (7.122)$$

In the above equations, the propagation constant is a complex quantity. It is made up of a real part and an imaginary part as

$$\gamma = a + jb \qquad (7.123)$$

where α = attenuation constant, nepers/m
β = phase constant, rad/m

Substituting Eqs. (7.121) and (7.122) into Eqs. (7.115) and (7.116) we get

$$V_{(x)} = \frac{1}{2}(e^{\gamma x} + e^{-\gamma x})V_R + \frac{Z_0}{2}(e^{\gamma x} - e^{-\gamma x})I_R$$

$$= \frac{1}{2}e^{\gamma x}(V_R + Z_0 I_R) + \frac{1}{2}e^{-\gamma x}(V_R - Z_0 I_R) \qquad (7.124)$$

$$I_{(x)} = \frac{Y_0}{2}(e^{\gamma x} - e^{-\gamma x})V_R + \frac{1}{2}(e^{\gamma x} + e^{-\gamma x})I_R$$

$$= \frac{1}{2}e^{\gamma x}(Y_0 V_R + I_R) + \frac{1}{2}e^{-\gamma x}(I_R - Y_0 V_R) \qquad (7.125)$$

Substituting $\gamma = \alpha + j\beta$ in Eqs. (7.124) and (7.125) and setting $x = l$, the sending end voltage V_s and the sending end current I_x are

$$V_s = \tfrac{1}{2}e^{(\alpha + j\beta)l}(V_R + Z_0 I_R) + \tfrac{1}{2}e^{-(\alpha + j\beta)l}(V_R - Z_0 I_R) \qquad (7.126)$$

and

$$I_s = \tfrac{1}{2}e^{(\alpha + j\beta)l}(Y_0 V_R + I_R) + \tfrac{1}{2}e^{-(\alpha + j\beta)l}(I_R - Y_0 V_R) \qquad (7.127)$$

From Eq. (7.126) we observe that $\tfrac{1}{2}e^{(\alpha + j\beta)l}(V_R + Z_0 I_R)$ is known as the incident wave (incident voltage), and $\tfrac{1}{2}e^{-(\alpha + j\beta)l}(V_R - Z_0 I_R)$ is known as the reflected wave (reflected voltage). $e^{\alpha l}$ and $e^{-\alpha l}$ are known as magnitude operators. These operators change in magnitude as the value of l changes and $e^{j\beta l}$ and $e^{-j\beta l}$ are known as phase angle operators. They shift the phase angle as the value of l changes.

From the above discussion, it is now possible to define an infinite line. An infinite line (flat line) is one which is terminated in its characteristic impedance (Z_0) in which case the reflected wave is eliminated for both V_s and I_s. Therefore, for the flat line, Eqs. (7.126) and (7.127) are reduced to

$$V_s = Z_0 I_R e^{(\alpha + j\beta)l} \tag{7.128}$$

and

$$I_s = I_R e^{(\alpha + j\beta)l} \tag{7.129}$$

To determine the wavelength or distance x on the line there must be a phase shift of 2π radians for incident and reflected waves. This occurs when

$$e^{j\beta l} = e^{j2\pi} \tag{7.130}$$

$$\lambda = \frac{2\pi}{\beta} \tag{7.131}$$

The long transmission line equivalent circuits. Using A, B, C, and D constants in hyperbolic form we can obtain accurate long line equivalents for the steady-state circuit. Using the equivalent Π circuit, the A and B values from Eq. (7.100) can be written as

$$A = 1 + \frac{Z_{\Pi} Y_{\Pi}}{2} \tag{7.132}$$

and

$$B = Z_P \tag{7.133}$$

The A, B constants are also found from Eq. (7.119) as

$$A = \cosh \gamma l \tag{7.134}$$

$$B = Z \sinh \gamma l \tag{7.135}$$

Equating Eqs. (7.132) and (7.134) and Eqs. (7.133) and (7.135) we get

$$1 + \frac{Z_{\Pi} Y_{\Pi}}{2} = \cosh \gamma l \tag{7.136}$$

and

$$Z_P = Z_0 \sinh \gamma l \tag{7.137}$$

Substituting Eq. (7.137) into Eq. (7.136) and simplifying we have

$$1 + \frac{Z_0 \sinh \gamma l Y_{\Pi}}{2} = \cosh \gamma l$$

$$\frac{Y_{\Pi}}{2} = \frac{\cosh \gamma l^2 - l \sinh \gamma l}{Z_0} = \left(\frac{l}{Z_0}\right) \tan \left(\frac{\gamma l}{2}\right) \tag{7.138}$$

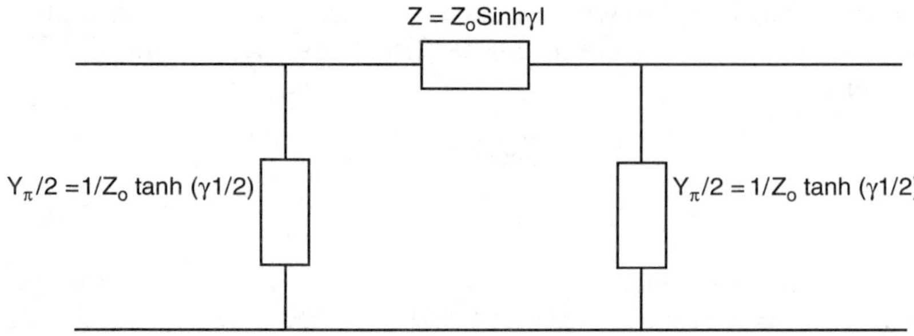

Figure 7.25 The long line equivalent Π circuit.

The equivalent to Eq. (7.138) is given in Fig. 7.25. An equivalent *T* circuit can also be found using the same procedure.

Example 7.6 A three-phase, 400-km, 60-Hz transmission line has a series impedance of $0.15 + j0.78\ \Omega/\text{km}$ and shunt admittance of $j5.0 \times 10^{-6}$ S/km. Determine the *A*, *B*, *C*, *D* constants using

1. The medium line equations
2. The long line equations

Observe the differences between results of part (1) and part (2).

solution For medium line use the nominal Π circuit (Fig. 7.22) and for long line use the Π equivalent circuit.

$$z = 0.15 + j0.78 = 0.7943\underline{/79.11^\circ}\ \Omega/\text{km}$$

$$y = j5 \times 10^{-6} = 5 \times 10^{-6}\underline{/90^\circ}\ \text{S/km}$$

$$Z = zl = 317.72\underline{/79.11^\circ}\ [\Omega]$$

$$\gamma = \sqrt{zy} = 1.9929 \times 10^{-3}\underline{/84.56^\circ}$$

$$\gamma l = 0.79772\underline{/84.56^\circ}$$

$$Z_0 = \sqrt{\frac{z}{y}} = 0.7942\underline{/79.11^\circ}(\sqrt{5 \times 10^{-6}\underline{/90^\circ}}) = 0.3986 \times 10^{-3}\underline{/-5.44^\circ}$$

$$Y_0 = \sqrt{\frac{y}{z}} = 2.509 \times 10^{-3}\underline{/5.44^\circ}$$

1. Nominal Π circuit for a medium line.

$$A = D = 1 + \frac{YZ}{2} = 0.6906\underline{/4.984^\circ}$$

$$B = Z = 317.72\underline{/79.11^\circ}$$

$$C = Y + \frac{Y^2Z}{4} = 16.8907 \times 10^{-4}\underline{/92.04^\circ}$$

2. Π equivalent circuit for a long line.

$$A = D = \cosh \gamma l = 1 + \frac{(\gamma l)^2}{2!} = 0.6905\underline{/4.984^\circ}$$

$$B = Z_0 \sinh \gamma l = 284.79\underline{/80.38^\circ}$$

$$C = Y_0 \sinh \gamma l = 17.927 \times 10^{-4}\underline{/91.28^\circ}$$

Comparison:

Nominal Π circuit	Π equivalent circuit
$A = D = 0.6906\underline{/4.984^\circ}$	$A = D = 0.6905\underline{/4.984^\circ}$
$B = 317.72\underline{/79.11^\circ}$	$B = 284.79\underline{/80.38^\circ}$
$C = 16.89 \times 10^{-4}\underline{/92.04^\circ}$	$C = 17.93 \times 10^{-4}\underline{/91.28^\circ}$

7.5.6 Reactive power compensation in power transmission lines

Reactive power or voltampere reactive (VARs) are defined and discussed in detail in Appendix A. Lagging VARs flowing into an inductance are defined as positive and those flowing into a capacitance are defined as negative VARs. A synchronous machine, whether delivering or generating power, generates VARs when overexcited and delivers VARs to supply the magnetizing currents of transformers, induction motors, etc., when underexcited. At a moment in time, the VAR demand on the grid network is fixed in magnitude and location by the voltage at the consumer's terminals. The relation between VAR demand and the voltage drop along a short line shows that the major cause of the voltage drop is the lagging reactive current flowing in the lines series inductance, and this statement is especially valid when inductive reactance of the line is greater than its resistance.

Power lines are limited in their ability to deliver power. The two important limits are the thermal effects and stability. Reactive lagging currents flowing in the conductors cause I^2R losses and heat is generated. This loss of power reduces transmission efficiency, and heat generated causes the temperature to rise. Since line temperature of overhead power lines must be kept within safe limits to prevent excessive line sag between the transmission towers (see Sec. 7.5 on sag calculations), and to prevent irreversible stretching, this imposes limitations on the maximum safe current a line can carry. Therefore, the purpose of the grid network is to transmit power with the reactive power transmission reduced to the minimum possible. For a given reactive power transmission along a line the per-unit voltage drop is high if the line impedance is high. The reverse is true for short lengths of heavy 500-kV line transmitting large amounts of power. The voltage relative to nominal voltage at all the bus bars of the grid system is called the voltage profile. Since the circuit breakers, transformers, and line insulation for the 500-kV system are operating close to the present limits of design, the maximum permitted voltage is 525 kV (i.e., 5 percent). The 500-kV lines may have to transfer load with a voltage drop less than 10 percent and hence the load power factor must be near unity or leading. This is especially true for longer lines (>150 mi or 240 km). The 345-kV and lower voltage systems normally operate within about ±10 percent. Reactive power is normally obtained from generators connected to the system. These have rated power factors varying from about 0.8 to 0.9 lagging. During maximum on the grid with generators at or near rated power, there is a limit to VARs which they can generate. During minimum load on the grid, with the generators underexcited but operating near rated power, there is a limit to the VARs they can absorb since their load angle (rotor angle) should not exceed about 70°. Therefore, it is sometimes necessary to provide additional equipment to generate or absorb VARs. The shunt capacitor has the same effect as an overexcited synchronous condenser. A synchronous condenser is a synchronous motor running idle with an overexcited field which is used in a power system solely for power factor correction or for control of reactive power flow.

Such a machine in a larger size may be more economical than static capacitors. The V-curves for zero mechanical power (Fig. 4.28*a*) approximately represents the synchronous condenser operation. When rotational losses and armature resistance are neglected, the real power of the synchronous condenser is zero with an associated zero torque angle, and the reactive power output is given by Eq. (4.62). At a time of maximum load the VARs absorbed by the series inductance of the line is greater than the VARs generated by the shunt capacitance of the line and $V_S > V_R$ especially for lagging power factor load, where V_S is the sending end voltage of the line and V_R is the receiving end voltage of the line. In extreme cases it may be necessary for the line to import VARs at both ends, that is, the delivered load is at a leading power factor in order to reduce the voltage drop. At times of light load the line generates VARs and if these exceed the consumer's demand, the excess VARs must be absorbed by operating the generators underexcited. There is also a tendency for the system voltage to rise ($V_R > V_S$). Switching some lines out of service and reducing the grid voltage might ease the problem. When the load at a receiving end of a line is equivalent to that taken by the characteristic impedance of the line, Z_0, the line is said to be operating at its natural load, as discussed on page 223. At natural load the VARs absorbed by the series inductance equal those generated by the shunt capacitance and $V_S = V_R$.

Static compensation. Figure 7.26 shows static compensation equipment using a shunt-connected inductor and capacitors. This equipment is often connected at the receiving end of the line to the tertiary delta winding of the transformer. The capacitors are used when the load demand is heavy. During maximum system load, when the system voltage is liable to be low, the VAR output of these capacitors is a minimum. The inductor, which is used when the system load is low, can be a coreless type, a gapped-core type, a saturable reactor, or a transducer. The coreless type is linear and is used for the larger ratings and voltages (e.g., 100 MVA at 500 kV). The gapped-core type is designed to be linear up to a voltage of about 1.3 per unit. Both these types will absorb VARs proportional to the square of the terminal voltage. For slow control an on-load tap changer can be fitted to the reactor or to an auxiliary transformer.

It is possible to use a tapped saturable reactor which saturates at selected voltages within the range ±10 percent of rated voltage. When saturated, the VARs absorbed by the reactor change rapidly for small change in voltage. The saturable reactors are used in parallel with capacitors. The voltage versus generated VARs characteristic of a capacitor is a straight line through the origin.

The voltage versus delivered VARs characteristic of a saturable reactor is similar to the magnetization curve of a transformer. When saturable reactors and capacitors operate in parallel, at low voltage, the equipment generates VARs while at high voltage it delivers VARs.

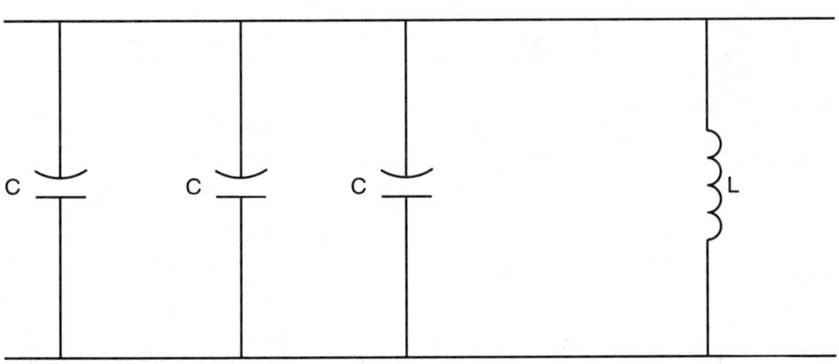

Figure 7.26 Static reactive power compensation.

It is appropriate to mention series capacitors which are connected, in series, with the line and they generate VARs proportional to the square of the current in them. Their primary use is to reduce the series impedance of the line, by canceling about 50 percent of the lines series inductance, in order to raise the power limit and to improve the system stability. The effect of a series capacitor on the system voltage is shown in Fig. 7.27. For a load of lagging power factor $V_{R2} > V_{R1}$, the voltage drop between V_S and V_{R2} is less than the voltage drop between V_S and V_{R1}. A problem with series capacitors is that when they carry fault current the voltage across them generally exceeds the voltage rating of the capacitor. Series capacitors are, therefore, shunted by arc gaps, set to flash over when the fault current is about three times the rated

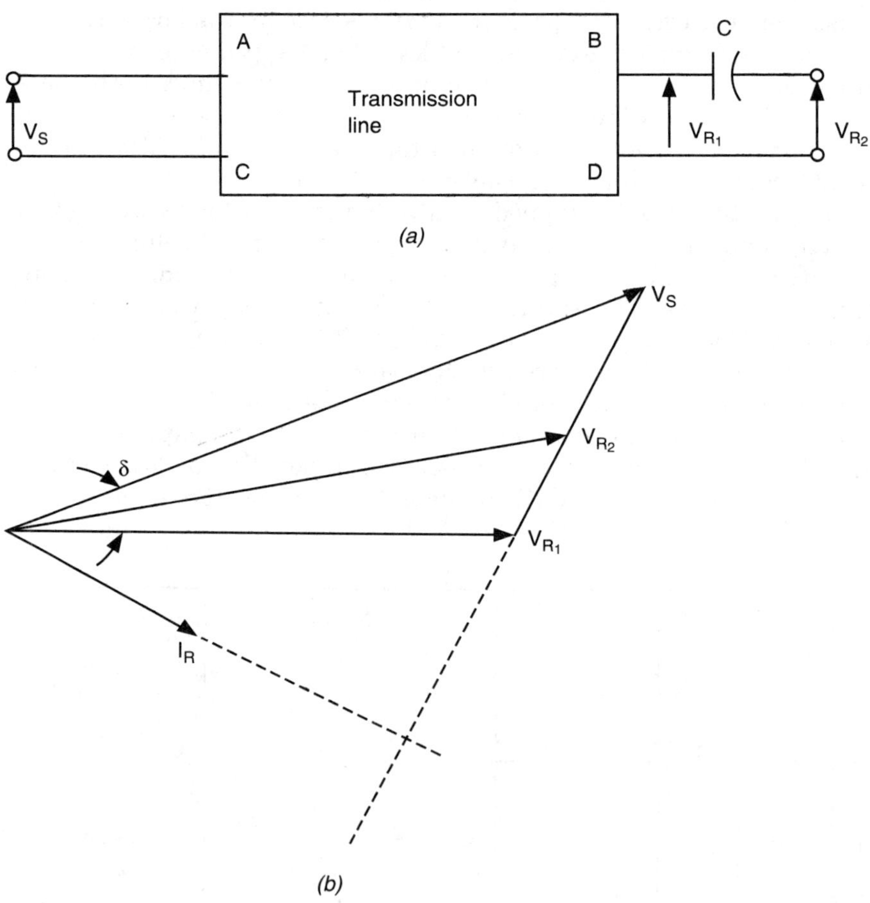

Figure 7.27 (*a*) Series capacitor in series with the line with constant I_R. (*b*) Phasor diagram of (*a*).

current of the capacitor, in series with a relay which closes a normally open circuit breaker shunted across the capacitor. Alternatively the arc gap can be made self-extinguishing when the voltage across the capacitor has fallen to a safe value. Figure 7.27 shows a series capacitor and its phasor diagram.

Natural load: Surge impedance load. A line is said to be operating at its natural load when the effective impedance of the delivered load equals the characteristic or surge impedance of the line, that is, $V_R/I_R = Z_R = Z_0$. Assuming that the line is loss-free, the characteristic impedance is a pure resistance and this will be emphasized by using Z_0 rather than a vector $\mathbf{Z}_0$.

For the attenuation $\alpha = 0$ and the propagation constant $\gamma = j\beta$ the hyperbolic functions become

$$\sinh j\beta l = j \sin \beta l \tag{7.139}$$

$$\cosh j\beta l = \cos \beta l \tag{7.140}$$

Substituting these values in Eqs. (7.115) and (7.118) we have

$$V_{(S)} = (\cos \beta l)V_R + (jZ_0 \sin \beta l)I_R \tag{7.141}$$

$$I_{(S)} = \left(\frac{1}{Z_0} j \sin \beta l\right) V_R + (\cos \beta l)I_R \tag{7.142}$$

where the quantities in parentheses are the A, B, C, and D constants for a loss-free line.

$$\begin{bmatrix} V_S \\ I_S \end{bmatrix} = \begin{bmatrix} \cos \beta l & jZ_0 \sin \beta l \\ \dfrac{1}{Z_0} j \sin \beta l & \cos \beta l \end{bmatrix} \begin{bmatrix} V_R \\ I_R \end{bmatrix} \tag{7.143}$$

If the loss-free line is operating at its natural load, then $V_R = Z_0 I_R$ and Eq. (7.143) becomes

$$V_{(S)} = V_R(\cos \beta l) + V_R(\sin \beta l) = V_R(\cos \beta l + j \sin \beta l)$$

$$= V_R e^{j\beta l} = V_R\underline{/\beta 1} \tag{7.144}$$

$$I_{(S)} = I_R(\cos \beta l + j \sin \beta l) = I_R e^{j\beta l} = I_R\underline{/\beta l} \tag{7.145}$$

Thus the voltage is constant in magnitude at all points along the line but V_S advances in phase relative to V_R by β radians per mile.

Problems

1 Given a 120-V sinusoidal voltage in series with a 4-Ω resistor, a 6-Ω inductor, and a 4-Ω capacitor, as shown in Fig. 7.28,
a Solve the problem in per unit on bases of $V_{base} = 120$ V and $S_{base} = 500$ VA.
b Draw an impedance diagram indicating the values per unit.

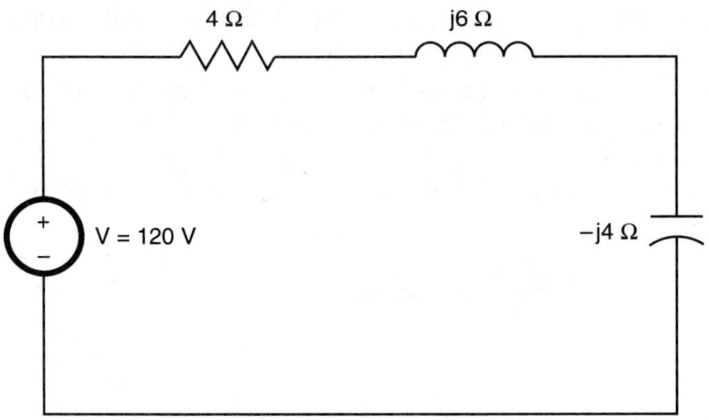

Figure 7.28 Problem 1.

2 Draw an impedance diagram of Fig. 7.2. Mark all the reactances in per unit. Choose the generator rating as the base in the generator circuit. The two 3-phase generators have the following ratings:

Generator 1	250 MVA	20.0 kV	$X'' = 20\%$
Generator 2	250 MVA	13.8 kV	$X'' = 20\%$

The two 3-phase transformers have the following ratings:

Transformer 1	100 MVA	220/20.0 kV
Transformer 2	150 MVA	220/13.8 kV

The leakage reactances are 10 percent each. Transmission line reactance is $j40\ \Omega$.

3 A transposed transmission line has a cross section as shown in Fig. 7.29. Determine the inductance and the inductive reactance per phase.

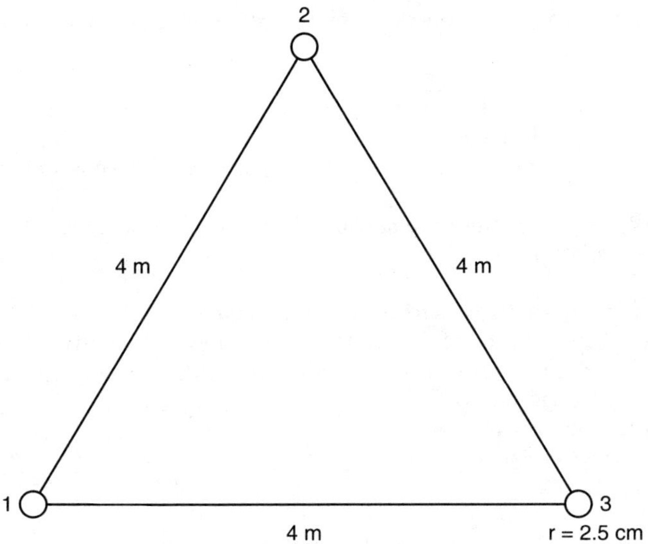

Figure 7.29 A cross section of a transmission line.

4 A three-phase, 60-Hz power transmission line has its inductors arranged in a unequilateral triangle. The distances between the conductors are 2.6 m, 3.1 m, and 3.6 m, as shown in Fig. 7.30. The diameter of the conductor is 0.365 cm. Determine the inductance and inductive reactance.

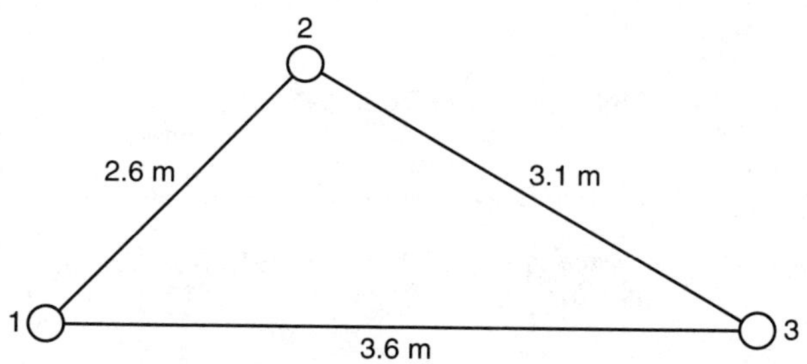

Figure 7.30 Three-phase transmission line.

5 A conductor hanging in the form of a catenary has a span of 160 m and a distance of 100 m above ground to the lowest point of the catenary. The conductor weight is 1.4 N/m. Determine

a The length of the conductor
b The sag
c Maximum and minimum values of tension using the
  (1) Catenary method
  (2) Parabolic method

6 A transmission line is supported from two towers at levels of 30 m and 60 m above the reference line. The horizontal distance between the towers is about 40 m. Determine the sag if the conductor weighs 5.6 N/m and the tension is 75 N.

7 A three-phase transmission line has the following characteristic: $R = 10^{-4}$ W/m, $L = 1.3 \times 10^{-6}$ H/m, $C = 9.1 \times 10^{-12}$ F/m, and $l = 300$ km. Using the Π equivalent model for a long line, calculate the A, B, C, D constants.

8 From Problem 7, calculate the propagation constant and characteristic impedance of the transmission line.

9 A three-phase transmission line with a series impedance of $30 + j210\ \Omega$ and a shunt capacitive admittance of $j0.001$ S is connected in T equivalent circuit. The receiving end has a load of $50 + j100\ \Omega$ per phase. The sending end voltage is 220 kV. Determine

a The sending end current per phase
b The sending end power per phase
c The receiving end voltage

10 A 130-kV transmission line has a series resistance and inductive reactance per phase per kilometer of 0.12 and 0.392 Ω, respectively. Find the magnitude of the sending end voltage when transmitting the full-line capacitance of $C = 9.1 \times 10^{-12}$ F/km. The receiving end power is 120 MVA at a power factor of 0.95 lagging. The receiving voltage is 130 kV for 200 km of the line.

Bibliography

Gross, C. A.: *Power System Analysis,* John Wiley, New York, 1979.

Neuenswander, J. R.: *Modern Power Systems,* International Textbook, Scranton, Pa., 1971.

Shultz, R. D., and R. A. Smith: *Introduction to Electric Power Engineering,* Harper and Row, New York, 1985.

Stevenson, W. D. Jr.: *Elements of Power System Analysis,* McGraw-Hill, New York, 1982.

Weedy, B. M.: *Electric Power System,* John Wiley, New York, 1979.

Chapter

8

Power System Fault Calculation

8.1 Introduction

A fault is said to occur on a power system when very high currents flow due to the partial or complete failure of the insulation at one or more points. Fault calculations involve finding the voltage and current distribution throughout the system during the fault. Very high current may flow as a consequence of lightning or switching surges, which puncture through or cause flashover across the surface of otherwise healthy insulation. It is important to determine the values of system voltages and currents during faulted conditions, so that protective devices may be set to minimize the harmful effects of such contingencies.

Symmetrical three-phase faults will be discussed first, then unsymmetrical faults such as single line to ground faults (SLG), double line to ground faults (DLG), and line to line faults (LL) will be discussed in detail. For unsymmetrical faults the method of symmetrical components will be used. When a fault occurs at a point in a power system, the corresponding fault MVA is referred to as the fault level at that point and, unless otherwise stated, will be taken to refer to a three-phase symmetrical short circuit. The per-unit system of units discussed in Sec. 7.2 will be used throughout this chapter.

8.2 Symmetrical Three-Phase Fault

In order to establish the basic calculation of a three-phase fault, first we need to calculate the initial current when the synchronous generator is short-circuited. We also need to evaluate what happens when an ac voltage is applied to the synchronous generator terminals as

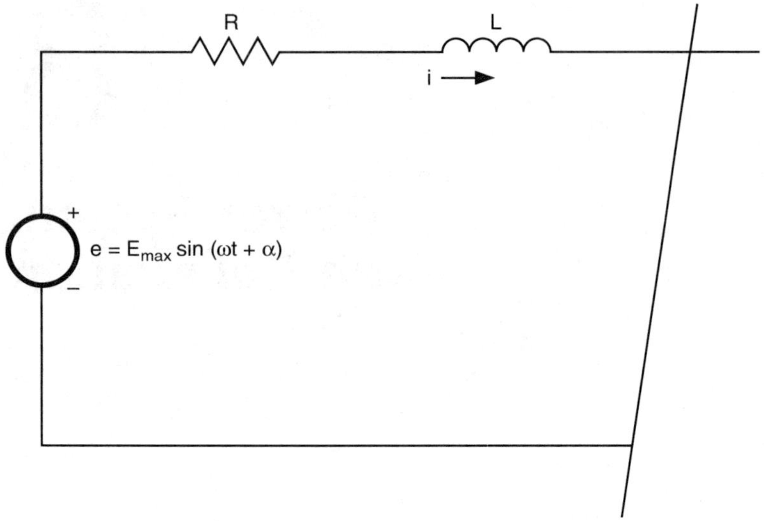

Figure 8.1 Synchronous generator with balanced fault.

shown in Fig. 8.1. Assuming a balanced three-phase fault on the terminals of the synchronous generator in Fig. 8.1 at a time zero with the generator initially on open circuit, the instantaneous voltage equation for the circuit is

$$L\frac{di}{dt} + R_L = E_{max} \sin(\omega t + \alpha)$$

$$= E_{max} (\sin \omega t \cos \alpha + \sin \alpha \cos \omega t) \tag{8.1}$$

There are many ways to arrive at a solution to Eq. (8.1). Here we are going to apply the Laplace transform to Eq. (8.1).

$$(Ls + R)I_{(s)} = E_{max}\left[\cos\alpha\left(\frac{\omega}{s^2+\omega^2}\right) + \sin\alpha\left(\frac{s}{s^2+\omega^2}\right)\right]$$

$$I_{(s)} = E_{max}\left[\frac{s(\sin\alpha) + \omega(\cos\alpha)}{(s^2+\omega^2)(Ls+R)}\right] \tag{8.2}$$

$$I_s = \frac{A}{s+R/L} + \frac{B}{s-j\omega} + \frac{B*}{s+j\omega}$$

$$= \left[\frac{(E_{max}/L)(\cos\alpha + s\sin\alpha)}{(s+R/L)(s^2+\omega^2)}\right] \tag{8.3}$$

$$I_{(t)} = A_e^{-t(R/L)} + 2|\mathrm{B}|\cos(\omega t + \underline{/B})u(t) \tag{8.4}$$

$$A = \left.\frac{(E_{max}/L)(\omega \cos \alpha + s \sin \alpha)}{s^2 + \omega^2}\right|_{s=-R/L}$$

$$= \frac{(E_{max}/L)(\omega \cos \alpha - R/L \sin \alpha)}{(-R/L)^2 + \omega^2}$$

$$\therefore A = \frac{(E_{max})(\omega L \cos \alpha - R \sin \alpha)}{R^2 + (\omega L)^2}$$

$$= \frac{(E_{max})(\omega L \cos \alpha - R \sin \alpha)}{(\sqrt{R^2 + (\omega L)^2})^2}$$

Therefore

$$A = \frac{E_{max} \sin \theta \cos \alpha - \cos \theta \sin \alpha)}{[\sqrt{R^2 + (\omega L)^2}]^2} \qquad |Z| = \sqrt{R^2 + (\omega L)^2}$$

$$A = \frac{E_{max}}{|Z|}[\sin(\theta - \alpha)] \qquad \text{or} \qquad A = \frac{E_{max}}{|Z|}[-\sin(\alpha - \theta)] \tag{8.5}$$

$$B = \left.\frac{(E_{max}/L)(\omega \cos \alpha + S \sin \alpha)}{(s + j\omega)(s + R/L)}\right|_{s=j\omega}$$

$$= \frac{(V_m/L)(\omega \cos \alpha + j\omega \sin \alpha)}{2j\omega(j\omega + R/L)}$$

$$= \frac{E_{max}(\cos \alpha + j \sin \alpha)}{2j(j\omega L + R)}$$

$$= \frac{(E_{max}/2)(\cos \alpha + j \sin \alpha)}{-\omega L + jR}$$

Therefore

$$|B| = \frac{E_{max}/2}{\sqrt{-\omega L^2 + R^2}} = \frac{E_{max}/2}{|Z|} \tag{8.6}$$

$$\underline{/B} = \tan^{-1}\left[\frac{(E_{max}/2) \sin \alpha}{(E_{max}/2) \cos \alpha} - \tan^{-1}\left(\frac{R}{-\omega L}\right)\right] \tag{8.7}$$

$$\tan^{-1}\left(\frac{R}{-\omega L}\right) = 90 + \tan^{-1}\left(\frac{\omega L}{R}\right) = 90 - \theta$$

Therefore, Eq. (8.7) becomes

$$\underline{/B} = \alpha - 90 - \theta \tag{8.8}$$

Therefore

$$i(t) = \frac{E_{\max}}{|Z|} [\cos(\omega t + \alpha - 90 - \theta) - \sin(\alpha - \theta)e^{-t(R/L)}] \tag{8.9}$$

Equation (8.9) becomes

$$i(t) = \frac{E_{\max}}{|Z|} [\sin(\omega t + \alpha - \theta) - \sin(\alpha - \theta)e^{-t(R/L)}] \quad \text{A} \tag{8.10}$$

From Eq. (8.10) the following observations can be made. The first term on the right-hand side varies sinusoidally with time, and the second term decays exponentially with the time constant of R/L seconds.

8.3 Symmetrical Three-Phase Fault on Synchronous Generators

For a symmetrical three-phase fault, short-circuit current calculations can be defined by observing the oscillograph of current in one phase of the synchronous generator undergoing the short circuit as shown in Fig. 8.3. In Fig. 8.3, $I'_{\max}$ is the transient current the first few cycles

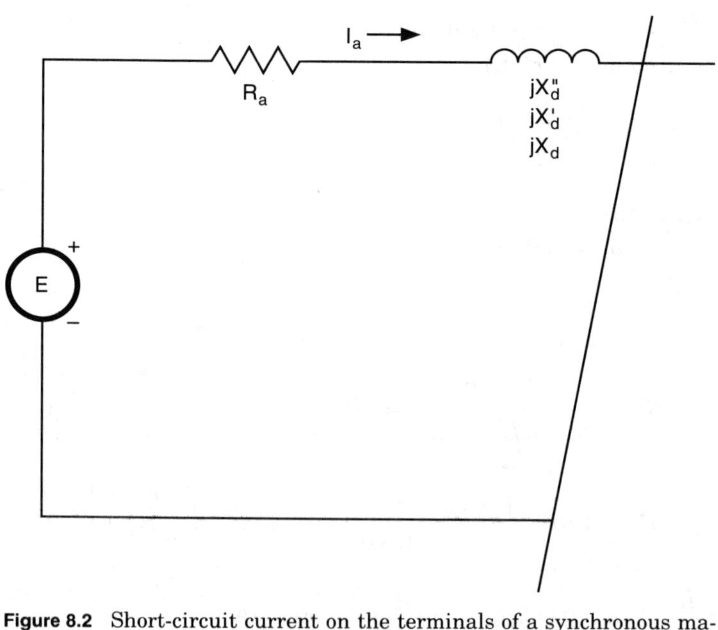

Figure 8.2 Short-circuit current on the terminals of a synchronous machine.

after the fault and $I''_{\max}$ is the subtransient current immediately after the occurrence of the fault. Figure 8.2 shows a synchronous reactance schematic for approximating short-circuit conditions where the choice of X''_d, X'_d, and X_d will depend upon the time at which the study is made. The no-load voltage of the generator divided by maximum current gives us the synchronous reactance. The same can be said for transient and subtransient reactances; thus

$$X''_d = \frac{E_{\max}}{I''_{a\ \max}} \tag{8.11}$$

$$X'_d = \frac{E_{\max}}{I''_{a\ \max}} \tag{8.12}$$

$$X_d = \frac{E_{\max}}{I''_{a\ \max}} \tag{8.13}$$

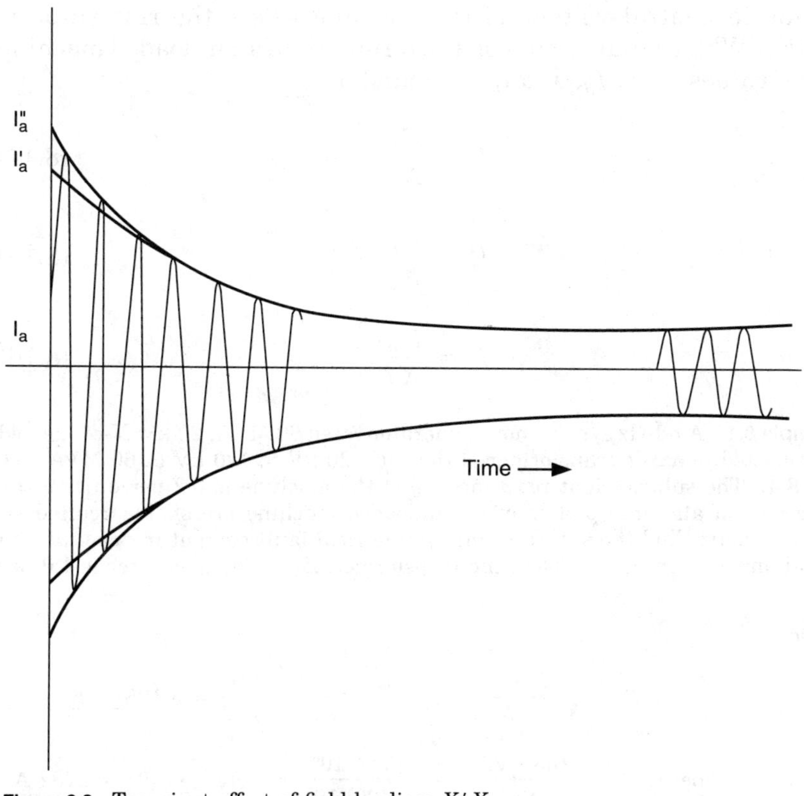

Figure 8.3 Transient effect of field loading, $X'_d X_d$.

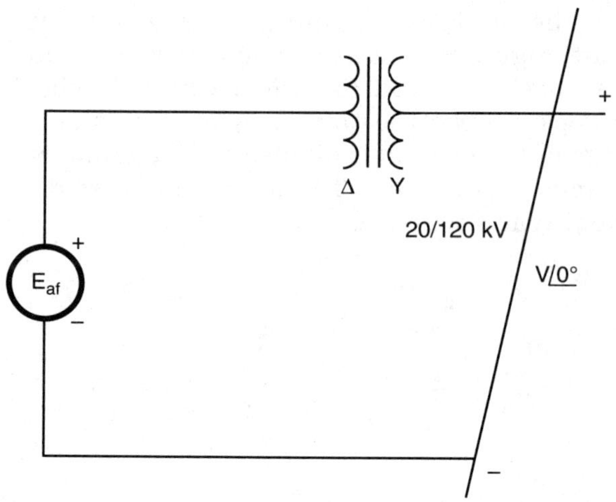

Figure 8.4 Example 8.1.

E_{max} is the same in Eqs. (8.11), (8.12), and (8.13) and this is the no-load line to neutral voltage of the machine where the rms value is called E_{af}. When a fault occurs on the terminals of an unloaded machine the rms values of I''_a, I'_a, and I_a are found as

$$I''_a = \frac{E_{af}}{X''_d} \tag{8.14}$$

$$I'_a = \frac{E_{af}}{X'_d} \tag{8.15}$$

$$I_a = \frac{E_{af}}{X_d} \tag{8.16}$$

Example 8.1 A 60-Hz synchronous generator rated 60 MVA, 20 kV, Y-connected is connected to a Δ-Y transformer of the ratio 20 kV to 120 kV at 60 MVA (see Fig. 8.4). The subtransient reactance X''_d of the machine is 0.2 per unit on the same base if all three phases of the unloaded machine are short-circuited simultaneously. Find the subtransient symmetrical fault current in per-unit and actual amperes on both sides of the transformer. E_{af} of the generator is 1.0 per unit.

solution

$$\text{pu } I''_f = \frac{E_{af}}{X''_d + X_{tr}} = \frac{1.0}{j0.2 + j0.12} = \frac{1.0}{j0.32} = 3.125\underline{/-90}$$

$$\text{Base ampere}_{(L\text{-}V)} = \frac{\text{Base VA}}{\sqrt{3}\ \text{Base } V} = \frac{60 \times 10^6}{\sqrt{3}(20)(10^3)} = 1.732 \times 10^3 = 1732 \text{ A}$$

$$\text{Base ampere}_{(\text{H-V})} = \frac{\text{Base VA}}{\sqrt{3}\ \text{Base } V} = \frac{60 \times 10^6}{\sqrt{3}(120 \times 10^3)} = 288.7 \text{ A}$$

$$\text{Actual } I_f'' \text{ on 20 kV} = \text{pu } I_f'' \times \text{Base } I = 3.125(1732) = 5412.5 \text{ A}$$

$$\text{Actual } I_f'' \text{ on 120 kV} = 3.125(288.7) = 902.1875 \text{ A}$$

Up to now, referring to Eqs. (8.14), (8.15) and (8.16), we have ignored the effects of prefault system loading. The prefault load currents are significant, and two approaches will be used in the solution of the symmetrical fault currents. Both methods will give identical results.

1. *The use of the Thevenin method at the point of fault.* Referring to Fig. 8.5 the Thevenin impedance and voltage external to f and n are

$$Z_{\text{TH}} = \frac{Z_1 Z_2}{Z_1 + Z_2} \tag{8.17}$$

$$I_1'' = \frac{E_{\text{TH}}}{Z_{\text{TH}}} \tag{8.18}$$

$$I_{f1}'' = \frac{Z_2}{Z_1 + Z_2} I_f'' \tag{8.19}$$

and

$$I_{f2}'' = \frac{Z_1}{Z_1 + Z_2} I_f'' \tag{8.20}$$

where I_f'' is the total fault current at point f and the branch currents I_{f1}'' and I_{f2}'' do not include prefault lead currents. Let the load current I_L be

$$I_1'' = I_{f1}'' + I_L \tag{8.21}$$

$$I_2'' = I_{f2}'' - I_L \tag{8.22}$$

2. The second method is the use of machine internal voltages behind subtransient reactances. When a prefault current I_L flows in the circuit three internal machine voltages are found; the voltage behind reactance E, the voltage behind transient reactance E', and the voltage behind the subtransient reactance E''. These internal voltages are

$$E'' = V_t + jI_L X_d'' \tag{8.23}$$

$$E' = V_t + jI_L X_d' \tag{8.24}$$

$$E = V_t + jI_L X_d \tag{8.25}$$

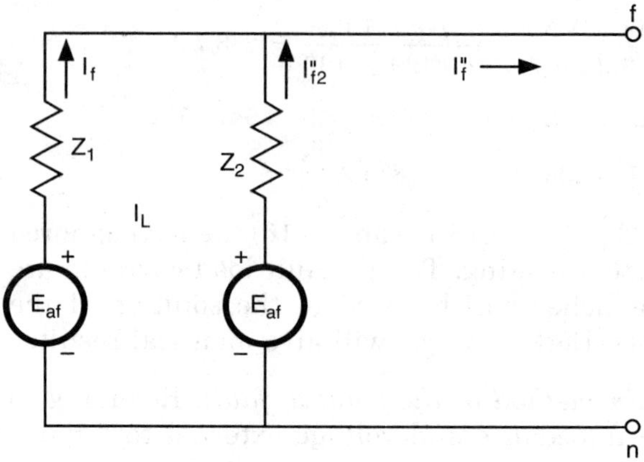

(a)

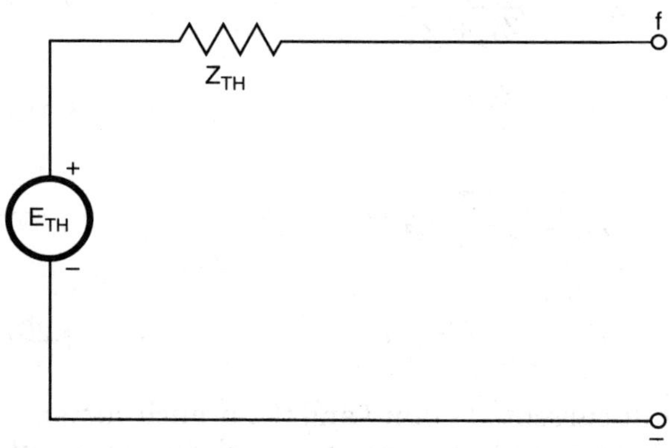

(b)

Figure 8.5 (*a*) The two-generator system. (*b*) The equivalent Thevenin voltage and Thevenin impedance of system (*a*).

8.4 Unbalanced Three-Phase Systems

8.4.1 Introduction

The electrical power system normally operates in a balanced three-phase sinusoidal steady-state load. When a fault occurs, the system voltages become unbalanced. A method of symmetrical components was introduced by C. L. Fortescure in 1918 and has become the most important tool for dealing with unbalanced polyphase circuits. The discovery proved that an unbalanced system of N phases can be resolved into systems of N systems of balanced phasors. He called these components "symmetrical components." Symmetrical components are also used in the analysis of rotating machinery with unbalanced voltages and in single-phase induction motors.

8.4.2 Symmetrical components

The symmetrical component theory according to Fortescure recognizes that three-phase systems of phasors can be resolved into three balanced systems of phasors. The balanced sets of components are

1. The positive sequence system, consisting of three balanced phasors but displaced 120° to each other. The positive sequence phasors are equal in magnitude, and have the same sequence as the original unbalanced phasors.
2. The negative sequence system consists of three balanced phasors but displaced 120° to each other. The phasors are equal in magnitude and have the opposite sequence to the original unbalanced phasors.
3. The zero sequence system is made up of the three phasors equal in magnitude and phase angle. The subscripts of +, −, and 0 will represent the positive, negative, and zero sequences, respectively.

Figure 8.6 shows the three balanced phasors. From Fig. 8.6 we can see that the sequence phasors add up to give the original unbalanced phasors; thus,

$$V_a = V_a^+ + V_a^- + V_a^0 \tag{8.26}$$

$$V_b = V_b^+ + V_b^- + V_b^0 \tag{8.27}$$

$$V_c = V_c^+ + V_c^- + V_c^0 \tag{8.28}$$

Next we shall introduce an operator a that causes a rotation of 120° or $a = 1\underline{/120^\circ}$. This operator is a complex number of unit magnitude and angle 120°. Other ways of expressing this operator include

$$a = 1\underline{/120^\circ} = 1e^{j120} = -0.5 + j0.866$$

$$a^2 = 1\underline{/240^\circ} = 1e^{j240} = -0.5 - j0.866$$

$$a^3 = a^2 a = 1\underline{/360^\circ} = 1 + j0 = 1\underline{/0^\circ}$$

$$a^4 = a^2 a^2 = a = 1\underline{/0^\circ}\, 1\underline{/120^\circ} = 1\underline{/120^\circ} = -0.5 + j0.866$$

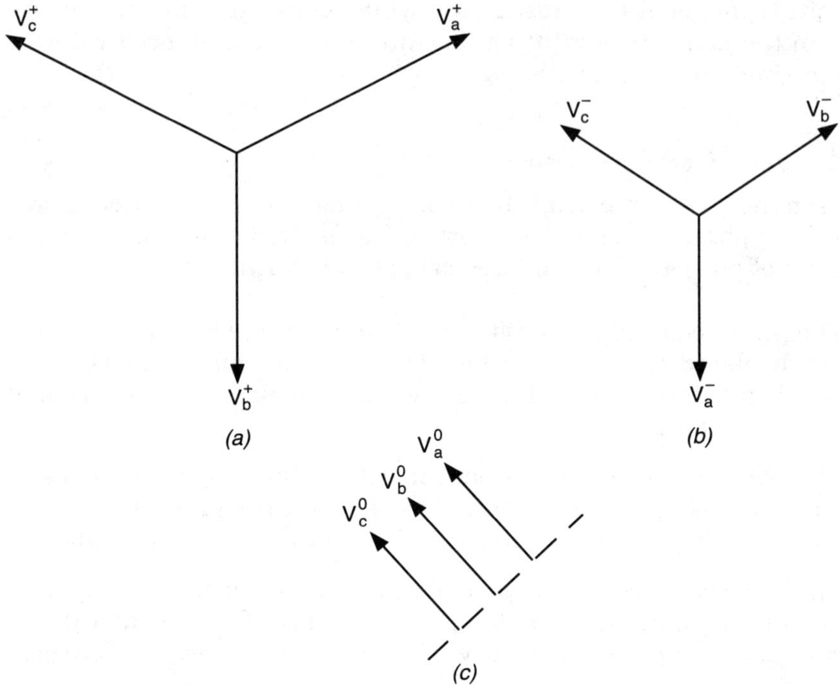

Figure 8.6 The three phases of balanced phasors: (*a*) positive sequence components, (*b*) negative sequence components, and (*c*) zero sequence components.

8.4.3 The determination of positive, negative, and zero sequence components of unsymmetrical phasors

Using Fig. 8.6*a*, *b*, and *c* we can say, for the balanced set, that

$$V_b^+ = a^2V_a^+ \qquad V_c^+ = aV_a^+ \tag{8.29}$$

$$V_b^- = aV_a^- \qquad V_c^- = aV_a^- \tag{8.30}$$

$$V_b^0 = V_a^0 = V_c^0 \tag{8.31}$$

Substituting Eqs. (8.29), (8.30), and (8.31) in Eqs. (8.27) and (8.28) we get

$$V_a = V_a^+ + V_a^- + V_a^0 \tag{8.32}$$

$$V_b = a^2V_a^+ + aV_a^- + V_a^0 \tag{8.33}$$

$$V_c = aV_a^+ + a^2V_a^- + V_a^0 \tag{8.34}$$

In matrix from Eqs. (8.32), (8.33), and (8.34) will be

$$\begin{bmatrix} V_a \\ V_b \\ V_c \end{bmatrix} = \begin{bmatrix} 1 & 1 & 1 \\ 1 & a^2 & a \\ 1 & a & a^2 \end{bmatrix} \begin{bmatrix} V_a^{0-} \\ V_a^{+} \\ V_a^{-} \end{bmatrix} \tag{8.35}$$

Let A be called the symmetrical components transformation matrix.

$$[A] = \begin{bmatrix} 1 & 1 & 1 \\ 1 & a^2 & a \\ 1 & a & a^2 \end{bmatrix}$$

Therefore, rewriting the above expression in terms of the sequence component voltage matrix we have

$$\begin{bmatrix} V_a^{0-} \\ V_a^{+} \\ V_a^{-} \end{bmatrix} = [A]^{-1} \begin{bmatrix} V_a \\ V_b \\ V_c \end{bmatrix} \tag{8.36}$$

The reader should be able to perform the matrix inversion of Eq. (8.36) to get

$$[A]^{-1} = \frac{1}{3} \begin{bmatrix} 1 & 1 & 1 \\ 1 & a & a^2 \\ 1 & a^2 & a \end{bmatrix} \tag{8.37}$$

Therefore, the desired result in matrix form is

$$\begin{bmatrix} V_a^{0} \\ V_a^{+} \\ V_a^{-} \end{bmatrix} = \frac{1}{3} \begin{bmatrix} 1 & 1 & 1 \\ 1 & a & a^2 \\ 1 & a^2 & a \end{bmatrix} \begin{bmatrix} V_a \\ V_b \\ V_c \end{bmatrix} \tag{8.38}$$

or written as separate equations

$$V_a^0 = \tfrac{1}{3}(V_a + V_b + V_c) \tag{8.39}$$

$$V_a^+ = \tfrac{1}{3}(V_a + aV_b + a^2 V_c) \tag{8.40}$$

$$V_a^- = \tfrac{1}{3}(V_a + a^2 V_b + aV_c) \tag{8.41}$$

The preceding equations could have been written for currents instead of voltage. Here we shall give the above equations in terms of currents, thus:

$$I_a = I_a^+ + I_a^- + I_a^0 \tag{8.42}$$

$$I_b = a^2 I_a^+ + a I_a^- + I_a^0 \tag{8.43}$$

$$I_c = a I_a^+ + a^2 I_a^- + I_a^0 \tag{8.44}$$

and

$$I_a^0 = \tfrac{1}{3}(I_a + I_b + I_c) \tag{8.45}$$

$$I_a^+ = \tfrac{1}{3}(I_a + aI_b + a^2 I_c) \tag{8.46}$$

$$I_a^- = \tfrac{1}{3}(I_a + a^2 I_b + aI_c) \tag{8.47}$$

In a given three-phase system the sum of the line currents is equal to the returning neutral current, that is,

$$I_a + I_b + I_c = I_n \tag{8.48}$$

Substituting I_n in Eq. (8.45) we get

$$I_a^0 = \tfrac{1}{3} I_n \tag{8.49}$$

$$I_n = 3I_a^0 \tag{8.50}$$

From Eqs. (8.49) and (8.50) we conclude that in a three-wire, three-phase system without neutral wire, the zero sequence component of the current is always zero. Only in a four-wire, three-phase system will the zero sequence exist!

We next consider few examples.

Example 8.2 Given the original unbalanced voltages $V_a = 6\underline{/0^\circ}$, $V_b = 2\underline{/-90^\circ}$, and $V_c = 9\underline{/135^\circ}$; find all of the voltage components for the positive, negative, and zero sequence systems.

solution Using Eq. (8.37) we have

$$\begin{bmatrix} V_a^0 \\ V_a^+ \\ V_a^- \end{bmatrix} = \frac{1}{3} \begin{bmatrix} 1 & 1 & 1 \\ 1 & a & a^2 \\ 1 & a^2 & a \end{bmatrix} \begin{bmatrix} 6\underline{/0^\circ} \\ 6\underline{/-90^\circ} \\ 9\underline{/135^\circ} \end{bmatrix}$$

$$V_a^0 = \tfrac{1}{3}(6\underline{/0^\circ} + 2\underline{/-90^\circ} + 9\underline{/135^\circ}) = \tfrac{1}{3}(6 - j2 - 6.364 + j6.364)$$

$$= \tfrac{1}{3}(-0.364 + j4.364) = -0.1213 + j1.4547 = 1.4597 - \underline{/94.77^\circ}$$

$$V_a^+ = \tfrac{1}{3}(6\underline{/0^\circ} + 2a\underline{/-90^\circ} + 9a^2\underline{/135^\circ})$$

$$= \tfrac{1}{3}[6 + (1\underline{/120^\circ})(2\underline{/-90^\circ}) + 1\underline{/-120^\circ}(9\underline{/135^\circ})]$$

$$= \tfrac{1}{3}(6 + 2\underline{/30^\circ} + 9\underline{/15^\circ})$$

$$= \tfrac{1}{3}(6 + 1.7321 + j1 + 8.6933 + j2.3294)$$

$$= \tfrac{1}{3}(16.4254 + j3.3294) = 5.4751 + j1.1098 = 5.5864\underline{/11.458^\circ}$$

$$V_a^- = \tfrac{1}{3}(6 + 2a^2\underline{/-90^\circ} + 9a\underline{/135^\circ}) = \tfrac{1}{3}(2\underline{/-210^\circ} + 6 + 9\underline{/255^\circ})$$

$$= \tfrac{1}{3}(6 - 1.7321 + j1 - 2.3294 - j8.6933)$$

$$= 0.6402 - j2.5644 = 2.6446\underline{/-75.86^\circ}$$

From Eqs. (8.14), (8.15), and (8.16) we have

$$V_b^0 = V_c^0 = V_a^0 = 1.4597\underline{/94.77^\circ}$$

$$V_b^+ = a^2V_a^+ = (1\underline{/-120^\circ})(5.5863\underline{/11.6^\circ}) = 5.58633\underline{/-108.4^\circ}$$

$$V_c^+ = aV_a^+(1\underline{/120^\circ})(5.5863\underline{/11.6^\circ}) = 5.5863\underline{/131.6^\circ}$$

$$V_b^- = aV_a^-(1\underline{/120^\circ})(2.6446\underline{/-75.86^\circ}) = 2.6446\underline{/44.14^\circ}$$

$$V_c^- = a^2V_a^-(1\underline{/120^\circ})(2.6446\underline{/-75.86^\circ}) = 2.6446\underline{/-195.86}$$

Example 8.3 Find the symmetrical components for the three currents $I_a = 100\underline{/0^\circ}$, $I_b = 160\underline{/-120^\circ}$, $I_c = 130\underline{/130^\circ}$.

solution Using Eqs. (8.45), (8.46), and (8.47) we have

$$\begin{bmatrix} I_a^0 \\ I_a^+ \\ I_a^- \end{bmatrix} = \frac{1}{3}\begin{bmatrix} 1 & 1 & 1 \\ 1 & a & a^2 \\ 1 & a^2 & a \end{bmatrix}\begin{bmatrix} I_a \\ I_b \\ I_c \end{bmatrix}$$

$$I_a^0 = \tfrac{1}{3}(I_a + I_b + I_c) = \tfrac{1}{3}(100\underline{/0^\circ} + 160\underline{/-120^\circ} + 130\underline{/130^\circ})$$

$$= \tfrac{1}{3}(100 - 80 - j138.56 - 83.576 + j99.59)$$

$$= \tfrac{1}{3}(-63.56 - j38.97) = -21.186 - j12.99 = 24.85\underline{/-148.49^\circ}$$

$$I_a^+ = \tfrac{1}{3}(I_a + aI_b + a^2I_c) = \tfrac{1}{3}[100\underline{/0^\circ} + (1\underline{/120^\circ})(160\underline{/-120^\circ}) + (1\underline{/-120^\circ})(130\underline{/130^\circ})]$$

$$= \tfrac{1}{3}(100 + 160 + 130\underline{/10^\circ}) = \tfrac{1}{3}(260 + 128.025 + j22.57)$$

$$= 129.56\underline{/3.3296^\circ}$$

$$I_a^- = \tfrac{1}{3}(I_a + a^2 I_b + aI_c) = \tfrac{1}{3}[100 + (1\underline{/-120^\circ})(160\underline{/-120^\circ}) + (1\underline{/120^\circ})(130\underline{/130^\circ})]$$

$$= \tfrac{1}{3}(100 + 160\underline{/-240^\circ} + 130\underline{/250^\circ})$$

$$= \tfrac{1}{3}(100 - 80 + j138.56 - 44.46 - j122.16)$$

$$= -8.1533 + j5.467 = 9.8165\underline{/146.157^\circ}$$

$$I_b^+ = a^2 I_a = (1\underline{/-120^\circ})(129.56\underline{/3.33^\circ}) = 129.56\underline{/-116.67^\circ}$$

$$I_c^+ = aI_a^+ = (1\underline{/120^\circ})(129.56\underline{/3.33^\circ}) = 129.56\underline{/123.3^\circ}$$

$$I_b^- = aI_a^- = (1\underline{/120^\circ})(9.8165\underline{/146.157^\circ}) = 9.8165\underline{/266.157^\circ}$$

$$I_c^- = a^2 I_a^- = (1\underline{/-120^\circ})(9.8165\underline{/146.157^\circ}) = 9.8165\underline{/26.16^\circ}$$

8.4.4 Δ-Δ Phase shift of symmetrical components in Y-Δ and Δ-Y transformer banks

If both primary and secondary windings are connected in the same way either in wye (Y-Y) or in delta (Δ-Δ) with the terminals joined, then there is no phase shift between the corresponding voltage of any phase to the neutral on either side, although a phase shift of 180° could occur if, for example, *A*, *B*, and *C* terminals were joined to form a wye point on the primary side and *a*, *b*, and *c* were joined on the secondary side and vice versa. The *A* and *a* phases on both sides are taken as reference and the corresponding phasor diagrams are shown in Fig. 8.7.

In Fig. 8.7 we see that $V_{AB} = V_{LV}\underline{/0^\circ}$ and $V_{ab} = V_{\mathrm{HV}}\underline{/30^\circ}$. Therefore, we show that all high-voltage quantities will lead the corresponding low-voltage counterparts by 30°. In order to standardize the delta-wye (wye-delta) connections so that the phase shift can be predicted with accuracy, the American National Standards Institute (ANSI) has written guidelines on terminal markings and connections for distribution and power transformers. The Standard states that:

> For either wye-delta or delta-wye connections, phases shall be labeled in such a way that positive-sequence quantities on the high-voltage side lead their corresponding positive-sequence quantities on the low-voltage side by 30°.

The effect on negative-sequence quantities is the reverse, that is, high-voltage values lag low-voltage values by 30°. The effect appears only for positive- and negative-sequence networks, the zero-sequence network quantities are not affected.

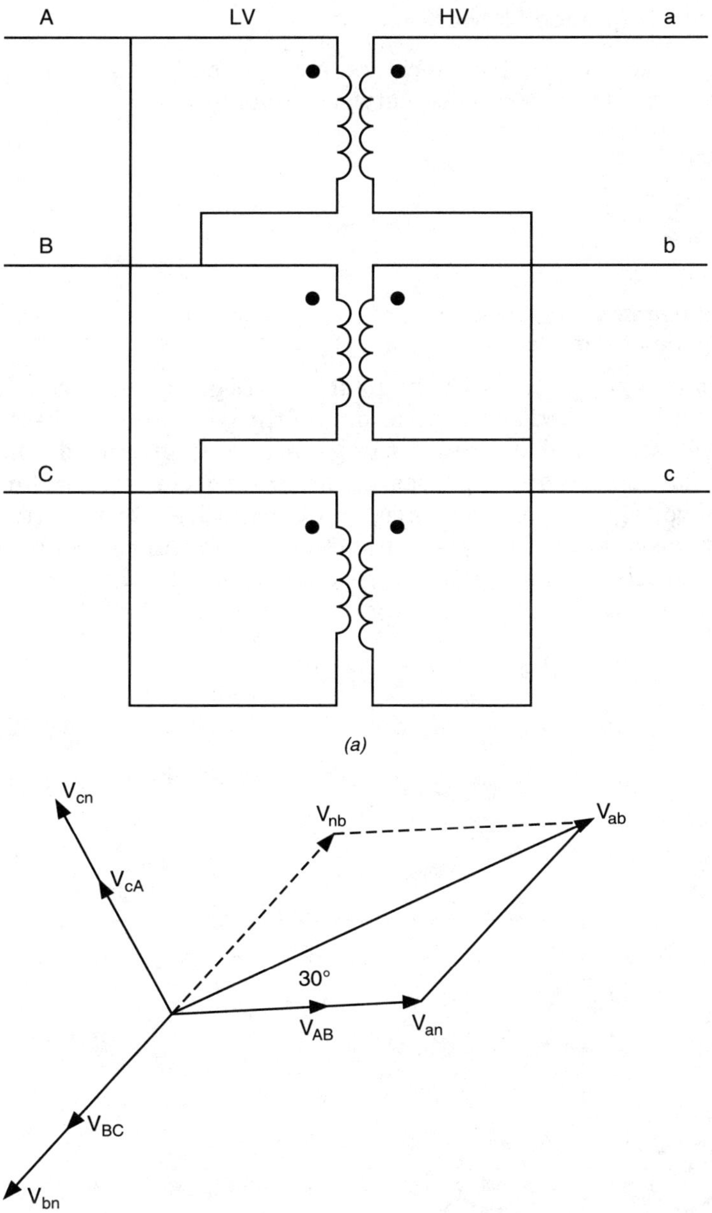

Figure 8.7 Wye-delta transformer: (*a*) ANSI standard LV-Δ; HV-Y connections; (*b*) Phasor diagram of the Δ-Y connection.

8.5 Models for Sequence Networks

The generators and motors, transformers, and transmission lines all can be modeled into three sequence networks, namely,

1. Positive sequence
2. Negative sequence
3. Zero sequence

8.5.1 Positive, negative, and zero sequences of generator model

Figure 8.8 shows a generator with the neutral part grounded through a reactor. If a fault is placed on the terminals of the generator, currents in all three phases will flow. Since the generator is grounded, the currents will flow through the phases to the ground via the reactor. The generated voltages are of positive sequence in nature. The negative and zero sequences do not possess voltage, only impedances. Figure 8.9 shows the positive, negative, and zero sequences of the generator.

From Fig. 8.9*a*, *b* and *c*, the following relations can be found:

$$V_a^+ = E_{af} - I_{af}^+ Z^+ \tag{8.51}$$

$$V_a^- = -I_{af}^- Z^- \tag{8.52}$$

$$V_a^0 = -I_{af}^0 Z^0 \tag{8.53}$$

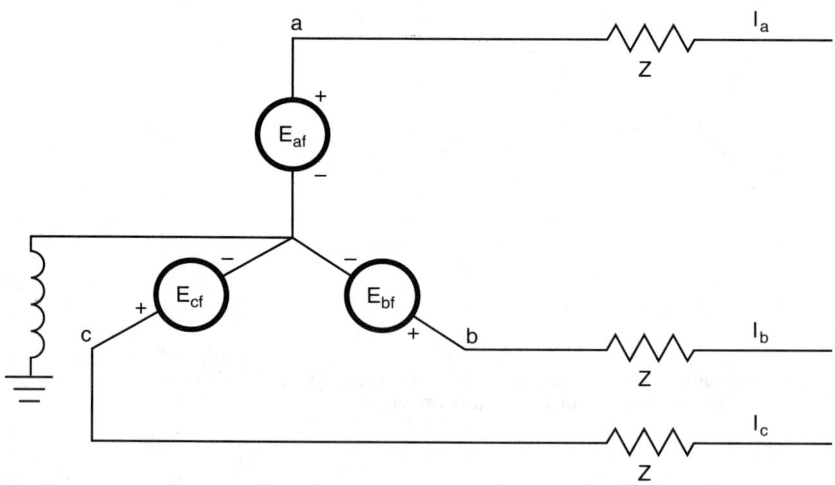

Figure 8.8 Generator with the neutral grounded through a reactor.

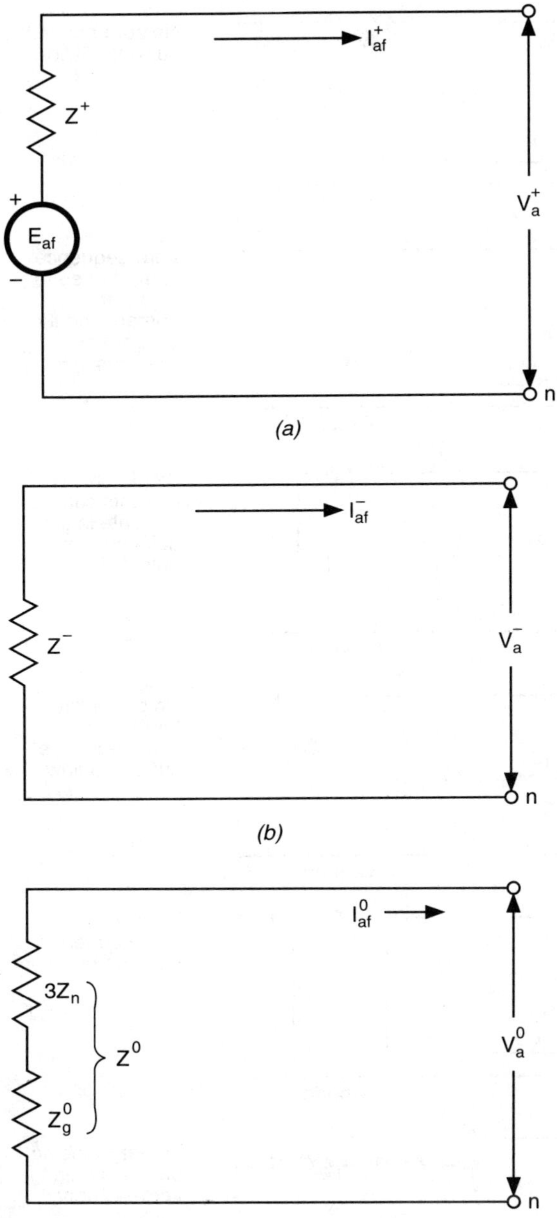

Figure 8.9 Generator sequence networks. (*a*) Positive-sequence network; (*b*) negative-sequence network; (*c*) zero-sequence network.

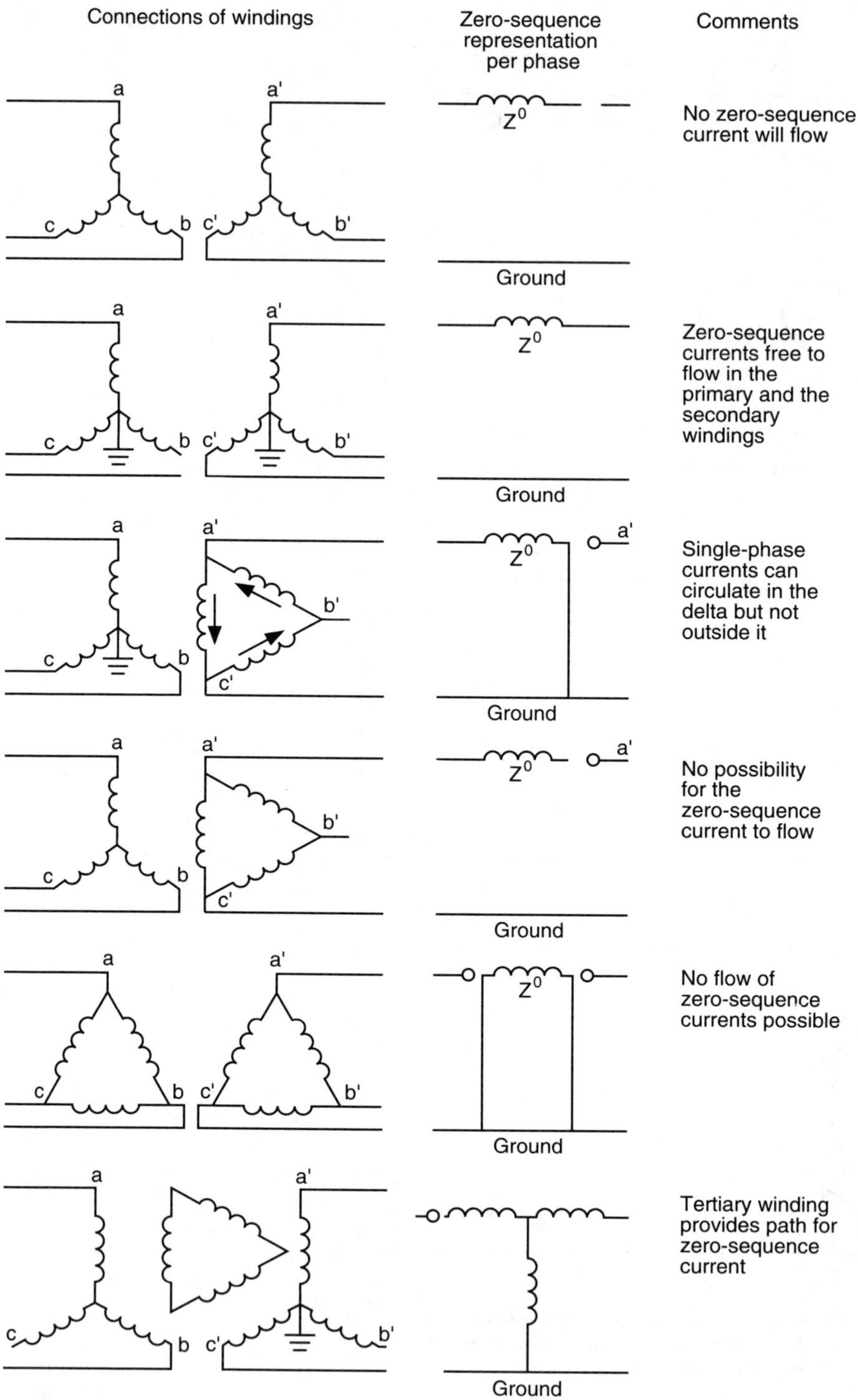

Figure 8.10 Zero sequence representation of transformers of various winding arrangements.

In matrix form

$$\begin{bmatrix} V_a^0 \\ V_a^+ \\ V_a^- \end{bmatrix} = \begin{bmatrix} 0 \\ E_{af} \\ 0 \end{bmatrix} - \begin{bmatrix} Z^0 & 0 & 0 \\ 0 & Z^+ & 0 \\ 0 & 0 & Z^- \end{bmatrix} \begin{bmatrix} I_{af}^0 \\ I_{af}^+ \\ I_{af}^- \end{bmatrix} \tag{8.54}$$

The total zero-sequence impedance (Z^0) is the sum of the impedance between the neutral and ground phase and the zero-sequence impedance. The current flowing in Z_n (shown in Fig. 8.9*c*) is $3I_a^0$. The voltage drop of zero sequence $-3I_a^0 Z_n - I_a^0 Z_g^0$ is equal to $-I_a^0(3Z_n^0 + Z_g^0)$. Therefore, the zero-sequence impedance is

$$Z^0 = 3Z_n + Z_g^0 \tag{8.55}$$

where Z_g^0 is the generator zero-sequence impedance.

8.5.2 Positive, negative, and zero sequences of the transformer model

The positive- and negative-sequence impedances are the normal balanced ones. The zero-sequence impedances of the transformer depend on the connection of the windings. A three-phase transformer may be made either from three identical single-phase transformers into a transformer bank or a three-phase transformer built on a single common core (shell-type or core-type). Figure 8.10 shows the zero-sequence representation of transformers for various winding arrangements. Zero-sequence currents in the windings on one side of a transformer must produce the corresponding ampere-turns in the other. But three-phase currents cannot flow in a wye connection without connecting to the ground. They can circulate around a delta winding, but not in the lines outside it.

8.5.3 Positive, negative, and zero sequences of transmission lines

The positive- and negative-sequence impedances are the normal balanced values. The zero-sequence impedance depends upon the nature of the return path through the ground if the fourth wire is provided. The per-phase transmission-line impedance is $Z_L = R_L + jX_L$. Normally the line resistance R_L is neglected.

Example 8.4 Consider the system shown in Fig. 8.11. The data associated with it is given in Table 8.1. The ratings of the generators and transformers are

Generator 1: 200 MVA, 18 kV, $X'' = 20\%$, $X^0 = 8\%$

Generator 2: 200 MVA, 18 kV, $X'' = 20\%$, $X^0 = 8\%$

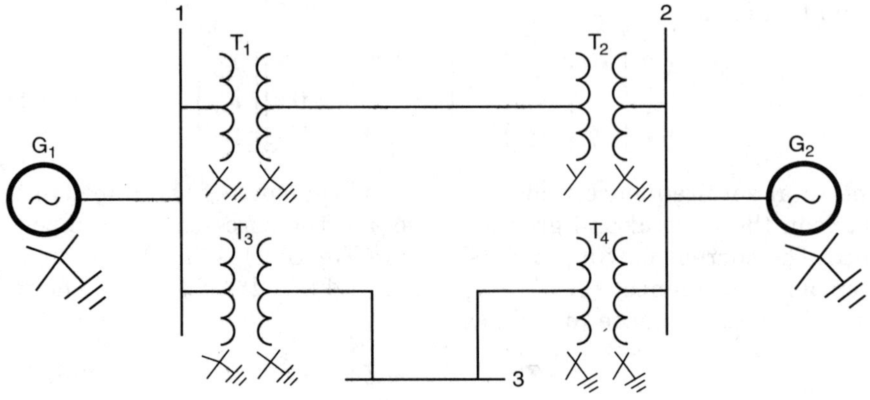

Figure 8.11 Transmission system for Example 8.4.

Transformers: 200 MVA, 138Y/18Y, $X = 10\%$, $X^0 = 8\%$

Transmission lines: $300\% \times 0.42 = 1.26$ pu

$300\% \times 0.210 = 0.63$ pu

Draw the positive-, negative-, and zero-sequence impedance of the system.

solution Simplifying Fig. 8.12*c* we get the network in Fig. 8.13.

Example 8.5 Two generators, each 60 MVA, 15 kV, 60 Hz, Y-connected, solidly grounded, are running at rated voltage (Fig. 314). Reactances are $X''_d = X^- = 0.12$ pu and $X^0 = 0.041$ pu, all on a 60-MVA base. The transformer's impedance is $Z^+_{tr} = Z^-_{tr} = Z^0_{tr} = j0.10$ pu, also on a 60-mA base. The line impedance from *e* to *g* (also at 60-MVA base) is $Z^+_L = Z^-_L = Z^0_{tr} = j0.03$; $Z^0_L = j0.08$ pu taking *d* as a point of fault on phases *b* and *c*. Find the subtransient fault current for

TABLE 8.1 For Example 8.4

Network component	MV rating	Voltage rating (kV)	X^+ (pu)	X^- (pu)	X^0 (pu)
G_1	200	18	0.2	0.2	0.08
G_2	200	18	0.2	0.2	0.08
T_1	200	18/138	0.1	0.1	0.08
T_2	200	18/138	0.1	0.1	0.08
T_3	200	18/138	0.1	0.1	0.08
T_4	200	18/138	0.1	0.1	0.08
T_{L12}	200	138	0.420	0.420	1.26
T_{L13}	200	138	0.210	0.210	0.63
T_{L23}	200	138	0.210	0.210	0.63

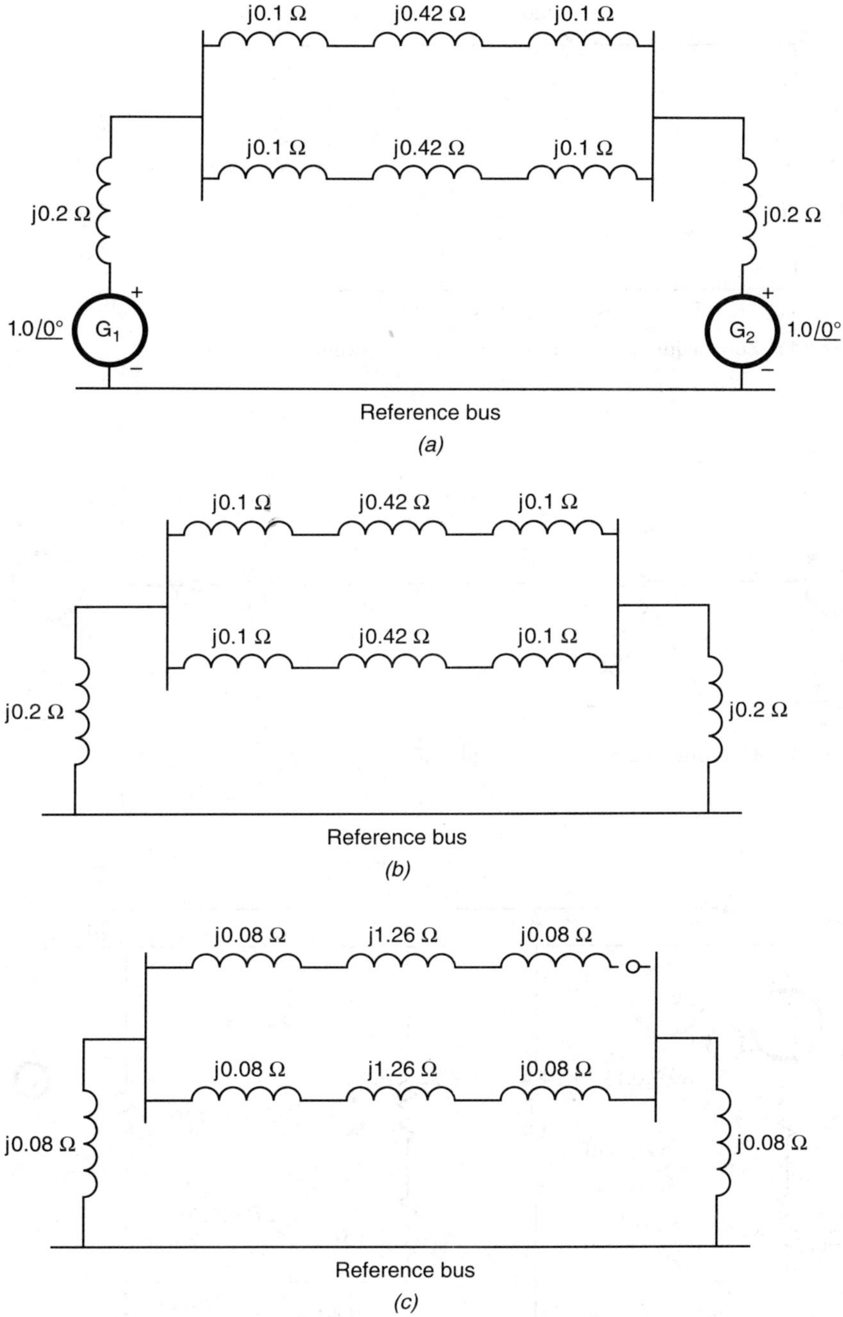

Figure 8.12 Sequence network for Example 8.4. (*a*) Positive-sequence network; (*b*) negative-sequence network; (*c*) zero-sequence network.

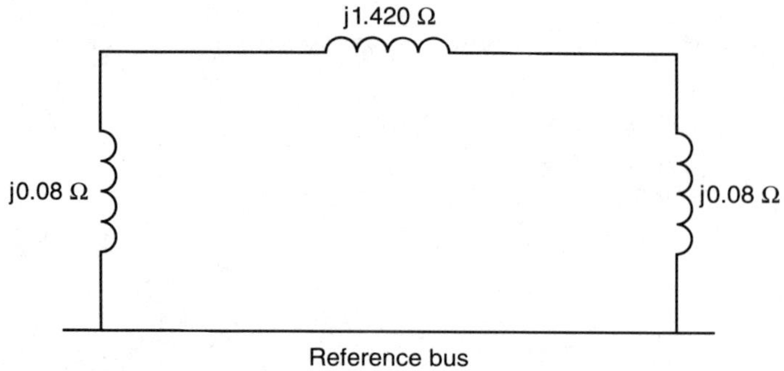

Figure 8.13 Zero-sequence resulting network for Example 8.4.

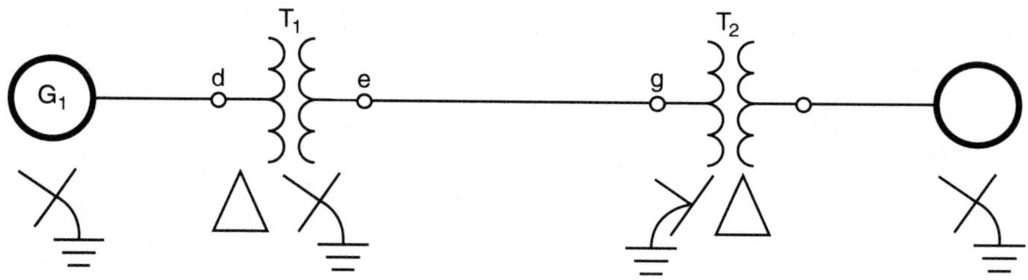

Figure 8.14 One-line diagram for Example 8.5.

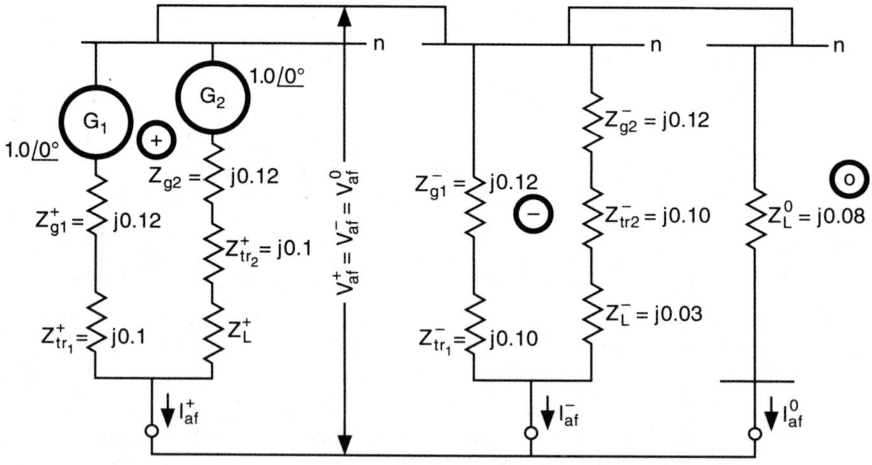

Figure 8.15 Sequence network for Example 8.5.

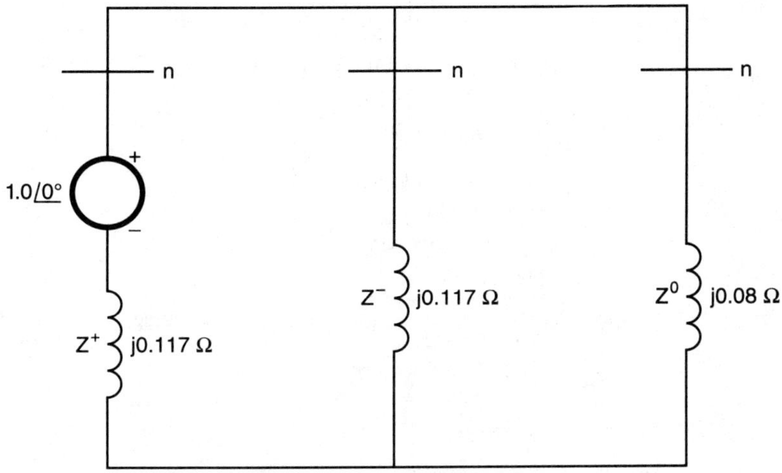

Figure 8.16 Reduced-sequence network for Example 8.5.

a double line to ground fault (DLG). Find the phase b and c fault currents as well as the line and phase voltages at the fault location d. The generators' voltages are $1\underline{/0^\circ}$ pu volts.

solution From Fig. 8.15

$$Z^+ = \frac{(Z_{g1}^+ + Z_{\text{tr1}}^+)(Z_{g2}^+ + Z_{\text{tr2}}^+ Z_L^+)}{Z_{g1}^+ + Z_{g2}^+ + Z_{\text{tr1}}^+ + Z_{\text{tr0}}^+ + Z_L^+}$$

$$= \frac{j(0.12 + 0.1)j(0.12 + 0.10 + 0.03)}{j(0.12 + 0.12 + 0.10 + 0.10 + 0.03)}$$

$$= +j\,\frac{(0.22)(0.25)}{0.47} = j0.117 \text{ pu}$$

$$Z^- = Z^+ = j0.117 \text{ pu}$$

$$Z^0 = j0.08 \text{ pu}$$

$$I_{af}^+ = \frac{E_a^+}{Z^+ + Z^- Z^0/(Z^- + Z^0)}$$

$$= \frac{1\underline{/10^\circ}}{j0.117 + (j0.117)(j0.08)/[j(0.117 + 0.08)]}$$

$$= \frac{1.0\underline{/0^\circ}}{j0.1645} = j - 6.078$$

From Eq. (8.49)

$$V_a^+ = E_a^+ - I_{af}^+ Z^+ = 1\underline{/0^\circ} - (-j6.0782)(j0.117) = 0.2888$$

$$I_{af}^- = \frac{V_a^+}{Z^-} = \frac{-0.2888}{j0.117} = j2.4688$$

$$I_{af}^0 = \frac{V_a^+}{Z^-} = \frac{-0.2888}{j0.08} = j3.61$$

$$I_{bf} = I_{af}^+\underline{/-120^\circ} + I_{af}^- + I_{af}^0$$

$$= 6.0782\underline{/-90^\circ}\ \underline{/-120^\circ} + 2.4688\underline{/90^\circ}\ \underline{/120^\circ} + j3.61$$

$$= 6.0782\underline{/-210^\circ} + 2.4688\underline{/210^\circ} + j3.61$$

$$= -7.4019 + j5.4147 = 9.171\underline{/143.8^\circ}$$

$$I_{cf} = I_{af}^+\underline{/120^\circ} + I_{af}^- + \underline{/-120^\circ} + I_{af}^0$$

$$= 6.0782\underline{/30^\circ} + 2.4688\underline{/-30^\circ} + j3.61$$

$$= 5.2639 + j3.0891 + 2.138 - j1.2884 + j0.361$$

$$= 7.4019 + j5.3647$$

The total fault current flowing into the fault is

$$I_n = I_{bf} + I_{cf} = -7.4019 + j5.4147 + 7.4019 + j5.3647 = j10.7794$$

Check: $I_n = 3I_{af}^0 = 3(j3.61) = j10.83$.

$$V_{af} = 3V_a^+ = 3(0.2888) = 0.8664 \text{ pu volt}$$

$$V_{ab} = -V_{af} = -0.8664 \text{ pu volt}$$

$$V_{ca} = V_{af} = 0.8664 \text{ pu volt}$$

8.6 Conditions for No Mutual Coupling between the Three Sequence Networks

The effect on impedance, in a three-phase system having balanced self-inductive reactance per phase X_L and mutual inductive reactance X_m between the pairs of phases, can be derived using Fig. 8.17.

Let

$$[V_{abc}] = [Z_{abc}][I_{abc}] \tag{8.56}$$

where $[Z_{abc}]$ is a 3×3 matrix of self- and mutual impedances.

$$[V_{abc}] = [A][V_{0+}] \tag{8.57}$$

and

$$[I_{abc}] = [A][I_{0+-}] \tag{8.58}$$

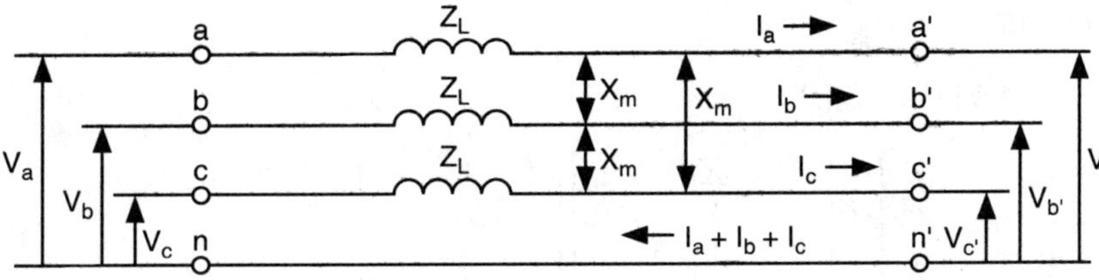

Figure 8.17 Three-phase system with balanced impedances.

Substituting Eqs. (8.57) and (8.58) in Eq. (8.56) we have

$$[A] = [V_{01+-}] = [Z_{ab}][A][V_{0+-}] \tag{8.59}$$

or

$$\begin{bmatrix} 1 & 1 & 1 \\ 1 & a^2 & a \\ 1 & a & a^2 \end{bmatrix} \begin{bmatrix} V_a^0 \\ V_a^+ \\ V_a^- \end{bmatrix} = [Z_{ab}] \begin{bmatrix} 1 & 1 & 1 \\ 1 & a^2 & a \\ 1 & a & a^2 \end{bmatrix} \begin{bmatrix} I_a^0 \\ I_a^+ \\ I_a^- \end{bmatrix} \tag{8.60}$$

Using Kirchhoff's voltage law in Fig. 8.17 we have

$$V_a - V_a^1 = jX_L I_a + jX_m I_b + jX_m I_c \tag{8.61}$$

$$V_b - V_b^1 = jX_m I_b + jX_L I_b + jX_m I_c \tag{8.62}$$

$$V_c - V_c^1 = jX_L I_a + jX_m I_b + jX_L I_c \tag{8.63}$$

or in matrix form:

$$\begin{bmatrix} V_a - V_a^1 \\ V_b - V_b^1 \\ V_c - V_c^1 \end{bmatrix} = j \begin{bmatrix} X_L & X_m & X_m \\ X_m & X_L & X_m \\ X_m & X_m & X_L \end{bmatrix} \begin{bmatrix} I_a \\ I_b \\ I_c \end{bmatrix} \tag{8.64}$$

From Eq. (8.59) we get

$$\begin{bmatrix} V_a^0 \\ V_a^+ \\ V_a^- \end{bmatrix} = [A]^{-1}[Z_{abc}][A] \begin{bmatrix} I_a^0 \\ I_a^+ \\ I_a^- \end{bmatrix} \tag{8.65}$$

Let

$$[Z_{0+-}] = [A]^{-1}[Z_{abc}][A] \tag{8.66}$$

Substituting Eq. (8.65) in Eq. (8.66) we get

$$\begin{bmatrix} V_a^0 \\ V_a^+ \\ V_a^- \end{bmatrix} = [Z_{0+-}] \begin{bmatrix} I_a^0 \\ I_a^+ \\ I_a^- \end{bmatrix} \tag{8.67}$$

Therefore,

$$[Z_{0+-}] = [A]^{-1}[Z_{abc}][A]$$

$$= \frac{1}{3}\begin{bmatrix} 1 & 1 & 1 \\ 1 & a & a^2 \\ 1 & a^2 & a \end{bmatrix} j \begin{bmatrix} X_L & X_m & X_m \\ X_m & X_L & X_m \\ X_m & X_m & X_L \end{bmatrix}\begin{bmatrix} 1 & 1 & 1 \\ 1 & a^2 & a \\ 1 & a & a^2 \end{bmatrix}$$

$$= j\,\frac{1}{3}\begin{bmatrix} 1 & 1 & 1 \\ 1 & a & a^2 \\ 1 & a^2 & a \end{bmatrix}\begin{bmatrix} X_L & X_m & X_m \\ X_m & X_L & X_m \\ X_m & X_m & X_L \end{bmatrix}\begin{bmatrix} 1 & 1 & 1 \\ 1 & a^2 & a \\ 1 & a & a^2 \end{bmatrix}$$

$$= j\,\frac{1}{3}\begin{bmatrix} X_L + X_m + X_m & X_m + X_L + X_m & X_m + X_m + X_L \\ X_L + aX_m + a^2X_m & X_m + aX_L + a^2X_m & X_m + aX_m + a^2X_L \\ X_L + a^2X_m + aX_m & X_m + a^2X_L + aX_m & X_m + a^2X_m + aX_L \end{bmatrix}$$

$$\times \begin{bmatrix} 1 & 1 & 1 \\ 1 & a^2 & a \\ 1 & a & a^2 \end{bmatrix}$$

$$= j\,\frac{1}{3}\begin{bmatrix} X_L + 2X_m & X_L + 2X_m & X_L + 2X_m \\ X_L - X_m & aX_L + (1 + a^2)X_m & a^2X_L + (1 + a)X_m \\ X_L - X_m & a^2X_L + (1 + a)X_m & aX_L + (1 + a^2)X_m \end{bmatrix}$$

$$\times \begin{bmatrix} 1 & 1 & 1 \\ 1 & a^2 & a \\ 1 & a & a^2 \end{bmatrix}$$

$$= j\begin{bmatrix} X_L + 2X_m & 0 & 0 \\ 0 & X_L - X_m & 0 \\ 0 & 0 & X_S - X_m \end{bmatrix} \tag{8.68}$$

8.7 Fault Calculations

8.7.1 Single line to ground fault (SLG)

Figure 8.18 shows a single line to ground fault with its neutral grounded. The terminal conditions are

$$I_b = 0 \tag{8.69}$$

$$I_c = 0 \tag{8.70}$$

$$V_a = 0 \tag{8.71}$$

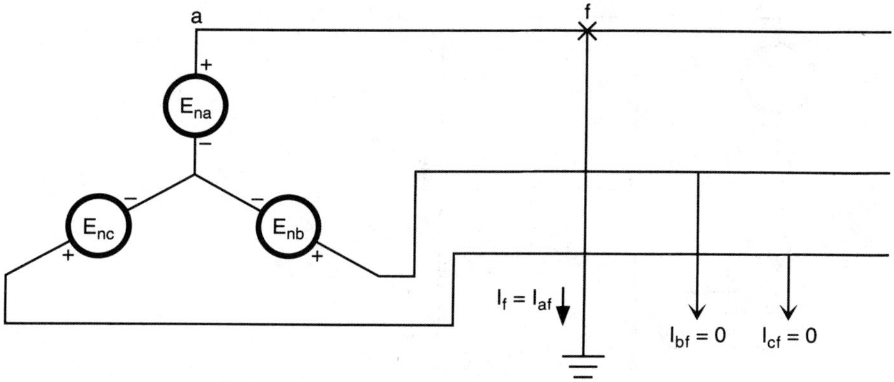

Figure 8.18 Single line to ground (SLG) fault in phase a.

The vector fault current is

$$\begin{bmatrix} I_{af} \\ I_{bf} \\ I_{cf} \end{bmatrix} = \begin{bmatrix} I_f \\ 0 \\ 0 \end{bmatrix} \tag{8.72}$$

where $I_{af} = I_f$. Finding symmetrical components of I_f we have

$$\begin{bmatrix} I_{af}^0 \\ I_{af}^+ \\ I_{af}^- \end{bmatrix} = \frac{1}{3}\begin{bmatrix} 1 & 1 & 1 \\ 1 & a & a^2 \\ 1 & a^2 & a \end{bmatrix}\begin{bmatrix} I_f \\ 0 \\ 0 \end{bmatrix} = \frac{I_f}{3}\begin{bmatrix} 1 \\ 1 \\ 1 \end{bmatrix} \tag{8.73}$$

Therefore,

$$I_{af}^0 = I_{af}^+ = I_{af}^- = \tfrac{1}{3} I_f \tag{8.74}$$

The voltage from ground to phase a is equal to zero, that is, $V_a = 0$, or

$$V_a = V_a^+ + V_a^- + V_a^0$$

or, from Eq. (8.54),

$$0 = E_{af}^+ - I_{af}^+ Z_a^- + (E_{af}^- - I_{af}^+ Z_a^-) + (E_{af}^0 - I_{af}^+ Z_a^0) \tag{8.75}$$

Since the generator produces only a balanced three-phase generated voltage of a positive sequence in nature, then the negative- and zero-sequence voltages are equal to zero. Therefore, Eq. (8.75) becomes

$$0 = E_a^+ - I_{af}^0 (Z_a^+ + Z_a^- + Z_a^0) \tag{8.76}$$

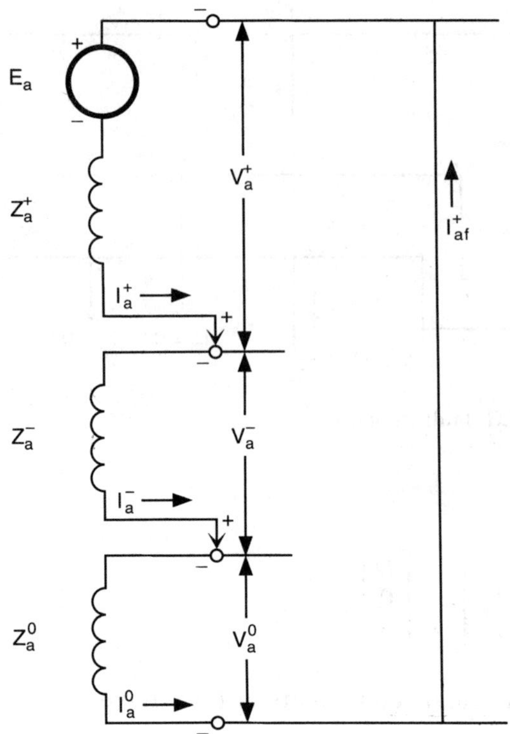

Figure 8.19 Single line to ground fault equivalent circuit of phase *a*.

Therefore,

$$I_{af}^{0} = \frac{E_a^+}{Z_a^+ + Z_a^- + Z_a^0} \tag{8.77}$$

From Eq. (8.77), Fig. 8.19 is drawn showing a single line to ground fault sequence network termination.

Example 8.6 A synchronous generator, 30 MVA, 15 kV, has the subtransient reactance $X_d'' = 0.11$ per unit. The negative and zero sequences are 0.12 pu and 0.04 pu, respectively. The generator is solidly grounded. Find the subtransient fault currents and voltages for a single line to ground, if per unit $E_{af} = 1\underline{/0^\circ}$.

solution From Eq. (8.72)

$$I_{af}^{0} = \frac{E_a^+}{Z_a^+ + Z_b^- + Z_c^-} = \frac{1\underline{/0^\circ}}{j0.11 + j0.12 + j0.04} = \frac{1\underline{/0^\circ}}{j0.27} = 3.7037\underline{/-90^\circ}$$

$$I_{af} = 3I_{af}^{0} = 3(3.7037\underline{/-90^\circ}) = 11.11\underline{/-90^\circ}$$

$$I_a^0 = I_{af}^0 = 3.7037\underline{/-90^\circ}$$

$$I_a^+ = I_{af}^0 = 3.7037\underline{/-90^\circ}$$

$$I_a^- = I_{af}^0 = 3.7037\underline{/-90^\circ}$$

The total phase a current is

$$I_a = I_a^+ + I_a^- + I_a^0 = 3.7037\underline{/-90^\circ} + 3.7037\underline{/-90^\circ}$$

$$= 3.7037\underline{/-90^\circ} = 11.11\underline{/-90^\circ}$$

$$V_{fa}^+ = E_a^- - I_a^+ Z_a^+ = 1\underline{/0^\circ} - (3.7037\underline{/-90^\circ})(0.111\underline{/90^\circ})$$

$$= 1 - 0.4074 = 0.5926 \text{ pu}$$

$$V_{fa}^- = -I_a^- Z_a^- = -(3.7037\underline{/-90^\circ})(0.12\underline{/90^\circ}) = +0.444 \text{ pu}$$

$$V_{fa}^0 = -I_a^0 Z_a^0 = -(3.7037\underline{/-90^\circ})(0.04\underline{/90^\circ}) = 0.1481 \text{ pu}$$

Line to ground voltages are

$$V_a = V_a^+ + V_a^- + V_a^0 = 0.5926 + (-0.444) + 0.1481 = 0.2967\underline{/0^\circ} \text{ pu}$$

$$V_b = a^2 V_a^+ + aV_{fa}^- + V_{fa}^0$$

$$= (-0.5 - j0.866)(0.5926) + (-0.5 + j0.866)(-0.444) + 0.1481$$

$$= -0.2963 - j0.5732 + 0.222 - j0.3845 + 0.1481 = 0.0738 - j0.8977$$

$$= 0.9007\underline{/-85.3^\circ}$$

$$V_0 = aV_{fa}^+ + a^2 V_{fa}^- + V_{fa}^0$$

$$= (-0.5 + j0.866)(0.5926) + (-0.5 - j0.866)(-0.444) + 0.1481$$

$$= -0.2963 + j0.5132 + 0.222 + j0.3845 + 0.1481 = 0.0738 + j0.8977$$

$$= 0.9007\underline{/85.3^\circ}$$

Line to line voltages are

$$V_{ab} = V_a - V_a = 0.29967 - (0.9007\underline{/-85.3^\circ}) = 0.2967 - 0.0738 + j0.8977$$

$$= 0.2229 + j0.8977 = 0.925\underline{/76.05^\circ}$$

$$V_{bc} = V_b - V_c = 0.0738 - j0.8977 - (0.0738 + j0.8977) = -j1.7954$$

$$= 1.7954\underline{/-90^\circ} = 1.7954\underline{/270^\circ}$$

$$V_{ca} = V_c - V_a = 0.0738 + j0.8977 - 0.2967$$

$$= -0.2229 + j0.8977 = 0.925\underline{/103.94^\circ}$$

Expressing V_{ab}, V_{bc}, and V_{ca} in volts we have

$$V_{ab} = (0.925\underline{/76.05^\circ})\frac{15}{\sqrt{3}} = 8.0107\underline{/76.05^\circ} \text{ kV}$$

$$V_{ac} = (1.7954\underline{/270°})\frac{15}{\sqrt{3}} = 15.55\underline{/270°}\text{ kV}$$

$$V_{ca} = (0.925\underline{/103.94°})\frac{15}{\sqrt{3}} = 8.0107\underline{/103.94°}\text{ kV}$$

8.7.2 Line to line fault

From Fig. 8.20 the terminal conditions are expressed by the following equations:

$$I_a = 0 \tag{8.78}$$

$$I_b = -I_c \tag{8.79}$$

$$V_b = V_c \tag{8.80}$$

From the fact that $V_b = V_c$ and using Eq. (8.38), we get

$$\begin{bmatrix} V_a^0 \\ V_a^+ \\ V_a^- \end{bmatrix} = \frac{1}{3}\begin{bmatrix} 1 & 1 & 1 \\ 1 & a & a^2 \\ 1 & a^2 & a \end{bmatrix}\begin{bmatrix} V_a \\ V_b \\ V_b \end{bmatrix} \tag{8.81}$$

$$V_a^0 = \tfrac{1}{3}(V_a + V_b + V_b) \tag{8.82}$$

$$V_a^+ = \tfrac{1}{3}(V_a + aV_b + a^2 V_b) \tag{8.83}$$

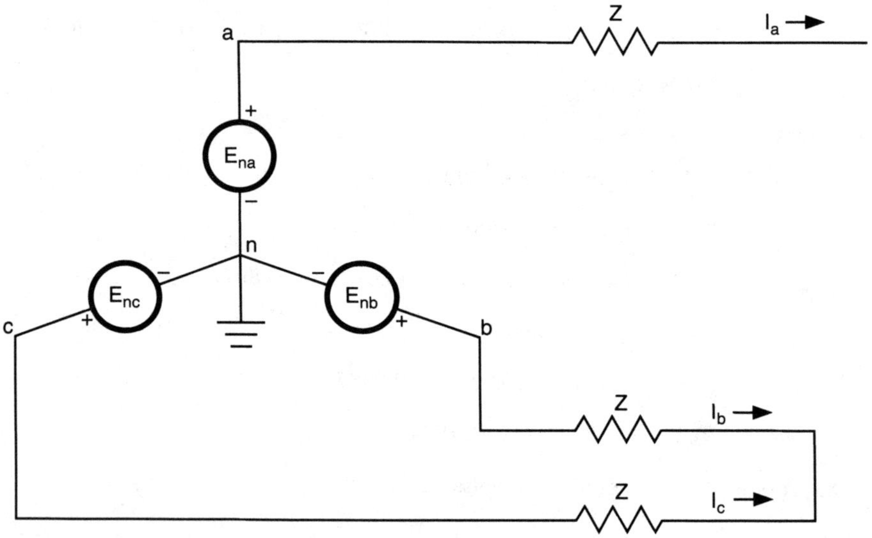

Figure 8.20 Line to line fault on phases b and c.

$$V_a^- = \tfrac{1}{3}(V_a + a^2 V_b + aV_c) \tag{8.84}$$

Therefore,

$$V_a^+ = V_a^- \tag{8.85}$$

Also using Eqs. (8.45), (8.46), and (8.47) we get

$$\begin{bmatrix} I_{af}^0 \\ I_{af}^+ \\ I_{af}^- \end{bmatrix} = \frac{1}{3}\begin{bmatrix} 1 & 1 & 1 \\ 1 & a & a^2 \\ 1 & a^2 & a \end{bmatrix} = \begin{bmatrix} 0 \\ I_b \\ -I_b \end{bmatrix} \tag{8.86}$$

Therefore,

$$I_{af}^0 = \tfrac{1}{3}(0 + I_b - I_b) \qquad \text{or} \qquad I_{af}^0 = 0 \tag{8.87}$$

$$I_{af}^+ = \frac{1}{3}(aI_b - a^2 I_b) = j\frac{I_b}{\sqrt{3}} \tag{8.88}$$

$$I_{af}^- = \frac{1}{3}(a^2 - a)I_b = -j\frac{I_b}{\sqrt{3}} \tag{8.89}$$

From Eqs. (8.88) and (8.89) we get

$$I_{af}^+ = -I_{af}^- \tag{8.90}$$

Equations (8.85), (8.87), and (8.90) give us a simple equivalent circuit for the solution of the line to line fault. Equation (8.87) tells us that there is no zero-sequence network and has no influence on any of the fault variables, since there is no ground connection at the fault.

From Fig. 8.21 we can obtain I_{af}^+ and I_{af}^- in the following procedure:

$$I_{af}^+ = \frac{E_{af}}{Z^+ + Z^-} \tag{8.91}$$

where $I_{af}^- = -I_{af}^+$.

Example 8.7 A line to line fault occurs at the terminals of the generator described in Example 8.6. Compute all fault currents and voltages.

solution

$$I_{af}^+ = -I_{af}^- = \frac{1\angle 0^\circ}{j0.11 + j0.12} = \frac{1\angle 0^\circ}{0.23\angle 90^\circ} = 4.348\angle -90^\circ$$

$$I_{af}^0 = 0$$

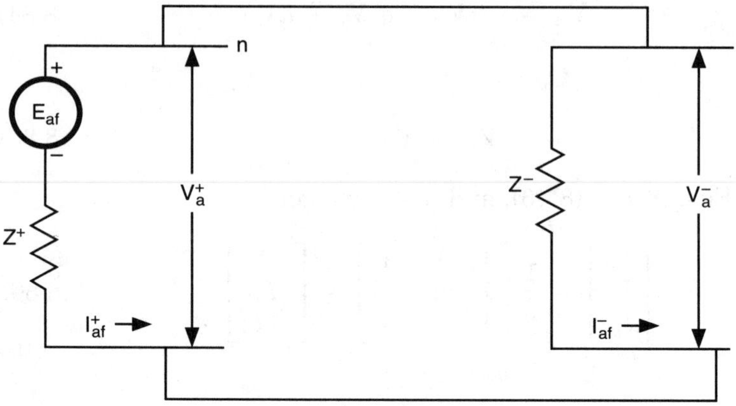

Figure 8.21 Sequence network termination for a line to line fault between phase banks.

The phase currents are found from Eqs. (8.42), (8.43), and (8.44).

$$\begin{bmatrix} I_a \\ I_b \\ I_c \end{bmatrix} = \begin{bmatrix} 1 & 1 & 1 \\ 1 & a^2 & a \\ 1 & a & a^2 \end{bmatrix} \begin{bmatrix} 1 \\ -j4.348 \\ j4.348 \end{bmatrix}$$

$$I_a = -j4.348 + j4.348 = 0$$

$$I_b = a^2(-j4.348) + a(j4.348)$$

$$= (-0.5 - j0.866)(-j4.348) + (-0.5 + j0.866)(j4.348)$$

$$= j2.1740 - 3.7654 - j2.174 - 3.7654 = -7.5308 \text{ pu}$$

$$I_c = (-0.5 - j0.866)(-j4.348) + (-0.5 - j0.866)(j4.348)$$

$$= j2.174 + 3.7654 - j2.174 + 3.7654 = 7.5308 \text{ pu}$$

$$\text{Base current} = \frac{30{,}000}{\sqrt{3}(15)} = 1154.7 \text{ A}$$

$$I_b = -7.5308(1154.7) = -8695.82 = 8695.82\underline{/180^\circ}$$

$$I_c = 7.5308(1154.7) = 8695.82\underline{/0^\circ} \text{ A}$$

$$V_{af}^+ = V_{af}^- = I_a^+ Z_a^+ = (-j4.348)(j0.11) = 0.4783 \text{ pu}$$

$$V_{af}^0 = 0$$

The phase voltages are

$$\begin{bmatrix} V_a \\ V_b \\ V_c \end{bmatrix} = \begin{bmatrix} 1 & 1 & 1 \\ 1 & a^2 & a \\ 1 & a & a^2 \end{bmatrix} \begin{bmatrix} 0 \\ 0.478 \\ 0.478 \end{bmatrix}$$

$$V_a = 0.478 + 0.476 = 0.9566 \text{ pu}$$

$$V_b = (-0.5 - j0.866)(0.478) + (0.5 + j0.866)(0.476)$$

$$= -0.239 - j0.4139 - 0.239 + j0.4139 = -0.478 \text{ pu}$$

$$V_c = (-0.5 + j0.866)(0.478) + (-0.5 - j0.866)(0.478)$$

$$= -0.239 + j0.4139 - 0.239 - j0.4139 = -0.478 \text{ pu}$$

Line to line voltages are

$$V_{ab} = V_b - V_c = 0.9566 - (-0.478) = 1.4346\underline{/0^\circ} \text{ pu}$$

$$V_{bc} = V_b - V_c = -0.478 - (0.478) = 0 \text{ pu}$$

$$V_{ca} = V_c - V_a = -0.478 - (0.9566) = 1.4346\underline{/180^\circ} \text{ pu}$$

Expressing the volts in actual values we get

$$V_{ab} = 1.4346\left(\frac{15}{\sqrt{3}}\right) = 12.474\underline{/0^\circ} \text{ kV}$$

$$V_{bc} = 0 \text{ kV}$$

$$V_{ca} = -1.4346\left(\frac{15}{\sqrt{3}}\right) = 12.424\underline{/180^\circ} \text{ kV}$$

8.7.3 Double line to ground fault (DLG)

Figure 8.22 shows a three-phase system with phases b and c both shorted to ground. The terminal conditions are

$$I_{af} = 0 \tag{8.92}$$

$$V_b = V_c = 0 \tag{8.93}$$

Also,

$$I_{af} = I_{af}^0 + I_{af}^+ + I_{af}^- = 0 \tag{8.94}$$

Since $V_b = 0$ and $V_c = 0$, from Eq. (8.38) we get

$$\begin{bmatrix} V_{af}^0 \\ V_{af}^+ \\ V_{af}^- \end{bmatrix} = \frac{1}{3}\begin{bmatrix} 1 & 1 & 1 \\ 1 & a & a^2 \\ 1 & a^2 & a \end{bmatrix}\begin{bmatrix} V_{af} \\ 0 \\ 0 \end{bmatrix} \tag{8.95}$$

$$V_{af}^0 = ⅓V_{af} \qquad V_{af}^+ = ⅓V_{af} \qquad V_{af}^- = ⅓V_{af}$$

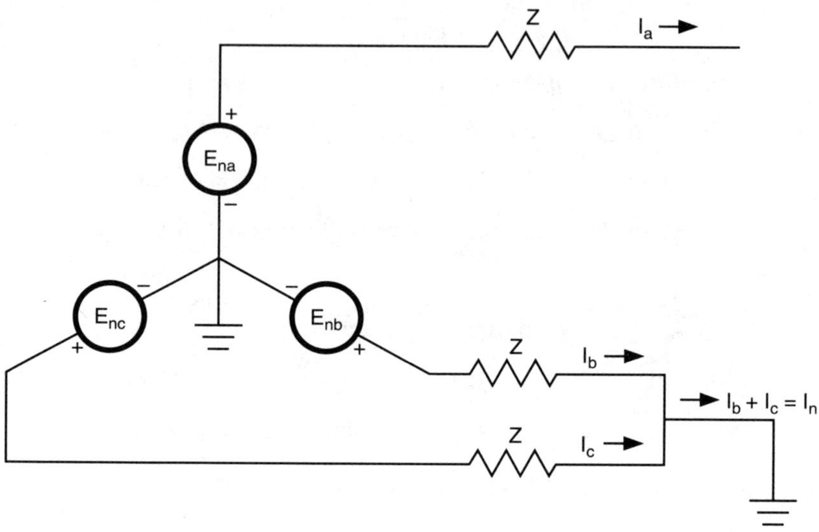

Figure 8.22 Double line to ground (DGL) fault.

It follows that

$$V^0_{af} = V^+_{af} = V^-_{af}$$

From Eqs. (8.94) and (8.95) the double to ground sequence network is obtained, as shown in Fig. 8.23. Solving for the sequence fault currents I^+_{af}, I^-_{af}, and I^0_{af} we get

$$I_{bf} = I^+_{bf} + I^-_{bf} + I^0_{bf} = a^2 I^+_{af} + aI^-_{af} + I^0_{af} \quad (8.96)$$

$$I_{cf} = aI^+_{af} + a^2 I^-_{af} + I^0_{af} \quad (8.97)$$

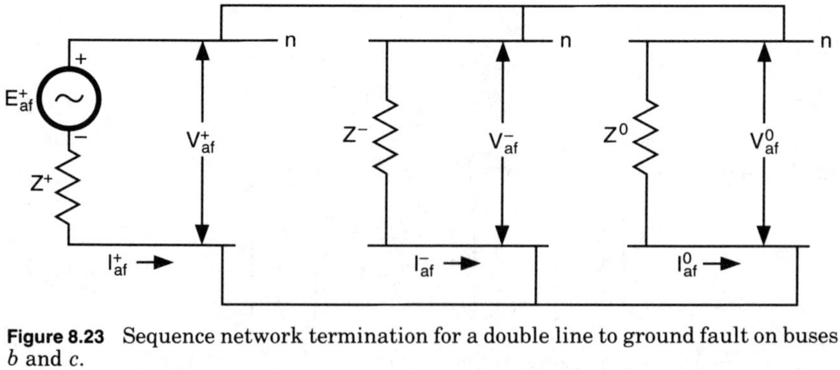

Figure 8.23 Sequence network termination for a double line to ground fault on buses *b* and *c*.

From Fig. 8.23 the currents I_{af}^{+}, I_{af}^{-}, and I_{af}^{0} can be found as follows:

$$I_{af}^{+} = \frac{E_{af}}{Z^{+} + Z^{-}Z^{0}/(Z^{-} + Z^{0})} \tag{8.98}$$

$$V_{af}^{+} = E_{af}^{+} - I_{af}^{+}Z^{+} \tag{8.99}$$

$$I_{af}^{-} = -\frac{V_{af}^{+}}{Z^{-}} \tag{8.100}$$

$$I_{af}^{0} = -\frac{V_{af}^{+}}{Z^{0}} \tag{8.101}$$

Example 8.8 A double line to ground occurs at the terminals of the generator described in Example 8.6. Compute all fault currents and voltages.

solution

$$I_{af}^{+} = \frac{1\angle 0^\circ}{j0.11 + (j0.12)(j0.04)/(j0.12 + j0.04)} = \frac{1\angle 0^\circ}{j0.11 + -0.0048/j0.16}$$

$$= \frac{1\angle 0^\circ}{j0.11 + j0.03} = \frac{1\angle 0^\circ}{j014} = 7.1429\angle -90^\circ = -j7.14296$$

From Eq. (8.96) we get

$$V_{af}^{+} = V_{af}^{-} = V_{af} = E_{af}^{+} - I_{af}^{+}Z^{+} = 1\angle 0^\circ - (-j7.1429)(j0.11)$$

$$= 1 - 0.7857 = 0.2143 \text{ pu}$$

From Eqs. (8.99) and (8.101) we get

$$I_{af}^{-} = \frac{-V_{af}^{+}}{Z^{-}} = \frac{-0.2143}{0.12\angle 90^\circ} = -1.7858\angle 90^\circ = -(-j1.7858)$$

$$= j1.7858 \text{ pu}$$

$$I_{af}^{0} = \frac{-V_{af}^{+}}{Z^{0}} = \frac{-0.2143}{j0.04} = j5.3575 \text{ pu}$$

From Eqs. (8.42), (8.43), and (8.44) we get

$$\begin{bmatrix} I_a \\ I_b \\ I_c \end{bmatrix} = \begin{bmatrix} 1 & 1 & 1 \\ 1 & a^2 & a \\ 1 & a & a^2 \end{bmatrix} \begin{bmatrix} j5.36 \\ -j7.143 \\ j1.786 \end{bmatrix}$$

Therefore,

$$I_a = j5.36 - j7.143 + j1.786 = 0$$

$$I_b = j5.36 - (-0.5 - j0.866)(-j7.143) + (-0.5 + j0.866)(j1.786)$$

$$= j5.36 + j3.57515 - 6.1858 - j0.893 - 1.5467 = -7.7325 + j8.0385$$

$$= 11.154\angle 133.89^\circ \text{ pu}$$

$$I_c = j5.36 + (-0.5 + j0.866)(-j7.143) + (-0.5 - j0.866)(j1.786)$$

$$= j5.36 + j3.5715 + 6.1858 - j0.8493 + 1.5467 = 7.7325 + j8.0385$$

$$= 11.154\underline{/46.11°}$$

From Fig. 8.21:

$$I_n = I_b + I_c = -7.7325 + j8.0385 + 7.7325 + j8.0385$$

$$= j16.077 = 16.077\underline{/-90°} \text{ pu}$$

$$V_a = V_a^+ + V_a^- + V_a^0 = 3V_a^+ = 3(0.2143) = 0.6429 \text{ pu}$$

$$V_b = V_c = 0$$

$$V_{ab} = V_a - V_b = 0.6429 \text{ pu}$$

$$V_{bc} = 0$$

$$V_{ca} = V_c - V_a = -0.6429 \text{ pu}$$

Expressed in actual values of amperes and volts we get

$$I_{af} = 0$$

$$I_{bf} = 1154.7(11.154\underline{/133.89°}) = 12{,}879.5\underline{/133.89°} \text{ A}$$

$$I_{cf} = 1154.7(11.154\underline{/46.11°}) = 12{,}879.5\underline{/46.11°} \text{ A}$$

$$V_{ab} = 0.6429\left(\frac{15}{\sqrt{3}}\right) = 5.5677 \text{ kV}$$

$$V_{bc} = 0$$

$$V_{ca} = 0.6429\left(\frac{15}{\sqrt{3}}\right) = -5.5677 \text{ kV}$$

Problems

1 Calculate the symmetrical components for phase currents of $I_a = 120\underline{/0°}$, $I_b = 190\underline{/-76°}$, and $I_c = 85\underline{/1558.2°}$.

2 Given unbalanced phase currents of $I_a = 110\underline{/90°}$, $I_b = 120\underline{/0°}$, and $I_c = 50\underline{/3}$ A,

a find the symmetrical components

b draw a phasor diagram showing I_a^0, I_a^+, I_a^-, I_b^+, I_b^-, I_b^0, I_c^+, I_c^-, I_c^0.

3 Determine the symmetrical components for phase currents of $I_a = 120\underline{/20°}$, $I_b = 90\underline{/-60°}$, and $I_c = 133\underline{/0°}$. Draw a phasor diagram showing all the nine symmetrical components.

4 A 13.8-kV bus bar is fed from three synchronous generators having the following ratings and reactances: 30 MVA, 60 Hz, $X'_d = j0.08$ pu; 40 MVA, 60 Hz, $X'_d = j0.1$ pu; 20 MVA, 60 Hz, $X'_d = j0.09$ pu. Find the fault current and MVA if a three-phase symmetrical fault occurs on the bus bars. Neglecting the resistance, take the base values to be 13.8 kV and 40 MVA.

5 From the system shown in Fig. 8.11 and the data given in Table 8.1 a single line to ground fault (SLG) has been placed at bus 3. Find the following:

a Thevenin equivalent positive-sequence impedance
b Thevenin equivalent negative-sequence impedance
c Thevenin equivalent zero-sequence impedance
d Positive-, negative-, and zero-sequence currents and voltages

6 A synchronous generator, which generates 1 pu voltage, is connected through a wye-wye transformer, reactance 0.20 pu, to two lines in parallel. At the other end the lines are connected through a wye-wye transformer of reactance 0.1 pu to a second synchronous generator, which generates 1.0 pu voltage (Fig. 8.24). For both transformers $X^+_{\text{tr}_1} = X^-_{\text{tr}_1} = X^0_{\text{tr}_1} = 0.2$ and $X^+_{\text{tr}_2} = X^-_{\text{tr}_2} = X^0_{\text{tr}_2} = 0.1$. The following per-unit reactances, all referred to the same base, are

Generator 1	$X^+ = j0.3$	$X^- = j0.3$	$X^0 = j0.05$
Generator 2	$X^+ = j0.25$	$X^- = j0.15$	$X^0 = j0.03$

For each line $X^+_{L1} = X^-_{L1} = 0.3$; $X^0_{L1} = 0.07$. Determine the current fed into a double line to ground fault on the line side terminals of the transformer fed from 1. All generators and transformers are solidly grounded.

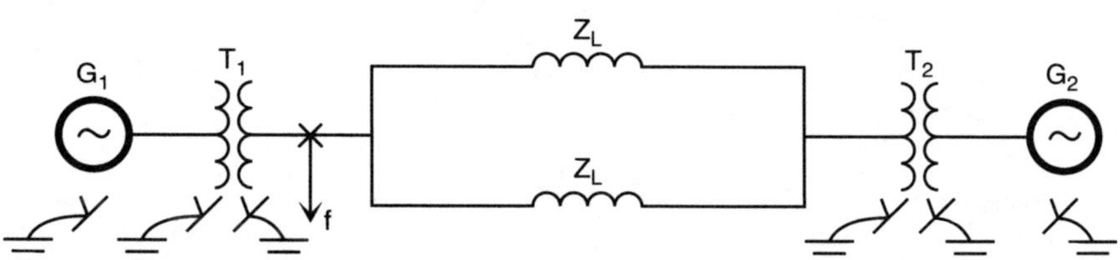

Figure 8.24 Transmission system for Prob. 8.6.

7 Repeat Prob. 8.5 for a double line to ground fault.

8 Repeat Prob. 8.5 for a line to line fault.

9 A generator is subjected to a single line to ground terminal fault. Assume that per unit reactances based on the generator rating are $X^+ = X^-$ 0.11 and $X^0 = 0.04$. If a ground reactor is added to the generator to limit the fault current to 6 pu, calculate the fault voltages of the two remaining healthy phases for $E_a = 1\angle 0^\circ$ V.

10 A single line to ground (SLG) fault occurs in a radial transmission system. Between the source of supply (an infinite bus) of voltage $1\angle 0^\circ$ pu to a point of fault the sequence impedances are $Z_1 = 0.3 + j0.6$ pu, $Z_2 = 0.3 + j0.55$ pu, $Z_0 = 1 + j0.78$ pu. The fault path to ground has a resistance of 0.66 pu. Find the fault current and the voltage at the point of fault.

Bibliography

Bergen, Arthur R.: *Power System Analysis,* Prentice-Hall, Englewood Cliffs, N.J., 1986.

Gross, Charles A.: *Power System Analysis,* 2d ed., John Wiley, New York, 1986.

Guide, A. E., and W. Paterson: *Electrical Power Systems,* Oliver and Boyd, Edinburgh, 1969.

Gunen, Turran: *Modern Power System Analysis,* John Wiley, New York, 1988.

Neuenswander, John R.: *Modern Power Systems,* International Textbook, Scranton, Penn., 1971.

Stevenson, William D. Jr.: *Elements of Power Systems Analysis,* 4th ed., McGraw-Hill, New York, 1982.

Weedy, B. M.: *Electric Power Systems,* John Wiley, New York, 1979.

Appendix

Complex Numbers, Phasors, Impedances, and Polyphase Systems

A.1 Complex Numbers

A knowledge of complex algebra is the most important tool for the study of phasors. A phasor is defined as a complex number that carries the amplitude and phase angle information of a sinusoidal function. There are two ways to designate a complex number, namely, the cartesian or rectangular form and the polar or trigonometric (exponential) form. Figure A.1 shows a complex number plane. A complex number

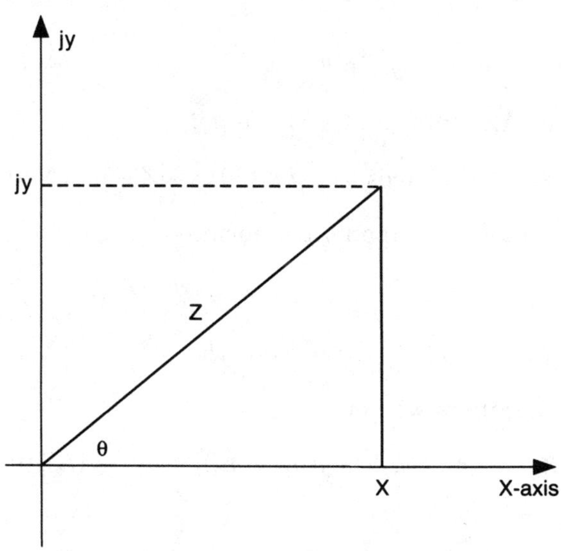

Figure A.1 Graphical representation of $x + jy$.

is defined by its coordinates x and jy along the real and imaginary axes, respectively.

Let the complex number Z in its cartesian or rectangular form be

$$Z = x + jy \tag{A.1}$$

where

$$j = \sqrt{-1} \qquad j^2 = -1 \tag{A.2}$$

To change Eq. (A.1) which is expressed in rectangular form, to polar or trigonometric form, the magnitude of Z is given by

$$Z = \sqrt{x^2 + y^2} \tag{A.3}$$

and its angular form is

$$\underline{/Z} = \theta = \tan^{-1} \frac{y}{x} \tag{A.4}$$

Expressing this in polar form, we get

$$Z = |Z|\underline{/\theta} \tag{A.5}$$

From Fig. A.1 we have

$$x = Z \cos \theta \qquad \text{and} \qquad y = Z \sin \theta \tag{A.6}$$

From Euler's formula

$$e^{j\theta} = \cos \theta + j \sin \theta \tag{A.7}$$

Substituting Eqs. (A.6) and (A.7) into Eq. (A.1), we get

$$Z = |Z| \cos \theta + j|Z| \sin \theta = |Z|(\cos \theta + j \sin \theta) = |Z|e^{j\theta} \tag{A.8}$$

Z in Eq. (A.8) has been expressed in exponential form.

Addition (subtraction). Let

$$Z_1 = a_1 + jb_1 \qquad \text{and} \qquad Z_2 = a_2 + jb_2 \tag{A.9}$$

Adding these two complex numbers we get

$$Z = Z_1 + Z_2 = (a_1 + a_2) + j(b_1 + b_2) \tag{A.10}$$

Subtracting

$$Z = Z_1 - Z_2 = (a_1 - a_2) + j(b_1 - b_2) \tag{A.11}$$

Writing Eqs. (A.10) and (A.11) in polar form, we get

$$Z = (a_1 + a_2) + j(b_1 + b_2) = |Z|\angle\phi_1 = |Z|e^{j\phi_1} \qquad \text{(A.12)}$$

where $Z = \sqrt{(a_1 + a_2)^2 + (b_1 + b_2)^2}$ and $\phi = \tan^{-1}[(b_1 + b_2)/(a_1 + a_2)]$.

From Eq. (A.11), we have

$$Z = (a_1 - a_2) + j(b_1 - b_2) = |Z|\angle\phi_2 = |Z|e^{j\phi_2} \qquad \text{(A.13)}$$

where Z and ϕ can be calculated in the same way as in Eq. (A.12).

Example A.1 Given $Z_1 = 8 + j19$; $Z_2 = 15 - j5$. Find

1. The sum of the two complex numbers
2. The difference of the two complex numbers.

Put the final results into polar form.

solution

1. $Z = Z_1 + Z_2 = (8 + 15) + j(19 - 5) = 26.96\angle 31.32°$
2. $Z = Z_1 - Z_2 = (8 - 15) + j(19 + 5) = 25.00\angle 106.3°$

Multiplication (division). Multiplication or division of complex numbers can be performed in rectangular or polar forms.

Using rectangular form:

$$\begin{aligned} Z &= Z_1Z_2 = (a_1 + jb_1)(a_2 + jb_2) \\ &= a_1a_2 - b_1b_2 + j(a_2b_1 + a_1b_2) \end{aligned} \qquad \text{(A.14)}$$

$$\begin{aligned} Z &= \frac{Z_1}{Z_2} = \frac{a_1 + jb_1}{a_2 + jb_2}\left(\frac{a_2 - jb_2}{a_2 - jb_2}\right) \\ &= \frac{a_1a_2 - ja_1b_2 + ja_2b_1 + b_1b_2}{a_2^2 + b_2^2} \\ &= \frac{(a_1a_2 + b_1b_2) + j(a_2b_1 - a_1b_2)}{a_2^2 + b_2^2} \end{aligned} \qquad \text{(A.15)}$$

A better way to multiply or divide complex numbers is first to convert all rectangular forms into polar forms. Putting Eqs. (A.14) and (A.15) in polar form, we get

$$Z = |Z_1|\underline{/\phi_1}(|Z_2|\underline{/\phi_2} \tag{A.16}$$

where
$$|Z_1| = \sqrt{a_1^2 + b_1^2}$$
$$|Z_2| = \sqrt{a_2^2 + b_2^2}$$
$$\phi_1 = \tan^{-1}\left(\frac{b_1}{a_1}\right)$$
$$\phi_2 = \tan^{-1}\left(\frac{b_2}{a_2}\right)$$

and

$$Z = \frac{Z_1}{Z_2} = \frac{a_1 + jb_1}{a_2 + jb_2} = \frac{|Z_1|\underline{/\phi_1}}{|Z_2|\underline{/\phi_2}} = \frac{|Z_1|}{|Z_2|}\underline{/\phi_1 - \phi_2} \tag{A.17}$$

Example A.2 Given $Z_1 = 8 + j16\ \Omega$; $Z_2 = 10 - j2\ \Omega$. Find

1. The product of Z_1Z_2
2. The value of Z_1/Z_2

solution

1. $Z = Z_1Z_2 = (8 + j16)(10 - j2) = 182.29\underline{/-52.12°}$
2. $Z = \dfrac{Z_1}{Z_2} = \dfrac{8 + j16}{10 - j12} = 1.756\underline{/-74.74°}$

Power of complex numbers. Applying the law of exponents to a complex number in polar form:

$$Z^k = (a + jb)^k = (|Z|e^{j\theta})^k = |Z|^k e^{jk\theta} = |Z|^k\underline{/-k\theta} \tag{A.18}$$

Roots of a complex number. To extract the kth root of a complex number is equivalent to raising the complex number to the $(1/k)$th power.

$$Z^{1/k} = (a + jb)^{1/k} = (|Z|e^{j\theta})^{1/k} = |Z|^{1/k}\underline{/\theta/k} \tag{A.19}$$

Example A.3 Given $Z = (7.5 - j2)^{1/3}$, find the value of Z.

solution

$$Z = (7.5 - j2)^{1/3} = (7.762e^{-j14.93})^{1/3} = 1.9715 - j0.1718$$

A.2 Phasors

A phasor is a complex number that carries the amplitude and phase angle information of a sinusoidal function. Consider the general sinusoidal function $f(t)$ as

$$f(t) = F_{\max} \cos(\omega t - \phi) \tag{A.20}$$

where $F_{\max}$ = maximum value
ω = angular velocity in rad/s
ϕ = phase angle behind the reference phasor

The phasor concept is rooted in Euler's identity which relates the exponential function to the trigonometric function.

$$e^{\pm j\theta} = \cos\theta \pm j\sin\theta \tag{A.21}$$

A.3 Impedances of Passive Circuit Elements in Phasor Domain

In this section the response of passive circuit elements to the steady-state sinusoidal source function is found for two main reasons: First, to establish the relationship between the phasor current and phasor voltage at the terminals of the elements. Second, to produce an opportunity to show the manner in which the response can be found easily and directly by the use of phasor representations of sinusoids. Table A.1 shows the three passive circuit elements carrying a sinusoidal current and their phasor equivalent circuits.

From Eq. (A.20) we have:

$$\begin{aligned} f(t) &= F_{\max}\cos(\omega t - \phi) \\ &= \mathrm{Re}[F_{\max}\cos(\omega t - \phi) + jF_{\max}\sin(\omega t - \phi)] \\ &= \mathrm{Re}[F_{\max}e^{j(\omega t-\phi)}] = \sqrt{2}\,\mathrm{Re}\left[\left(\frac{F_{\max}}{\sqrt{2}}\right)e^{j\phi}e^{j\omega t}\right] \end{aligned} \tag{A.22}$$

The coefficient of the exponential $e^{j\omega t}$ is a complex number that carries the amplitude and phase angle of a given sinusoidal function. This complex number is the phasor representation or phasor transform of the given sinusoidal function. Therefore, the phasor is defined by

$$\mathbf{F} = \left(\frac{F_{\max}}{\sqrt{2}}\right)e^{j\phi} = F_{\mathrm{rms}}e^{j\phi} = F\angle\phi \tag{A.23}$$

TABLE A.1 Three Passive Circuit Elements

Elements carrying a sinusoidal current	The phasor domain equivalent circuits

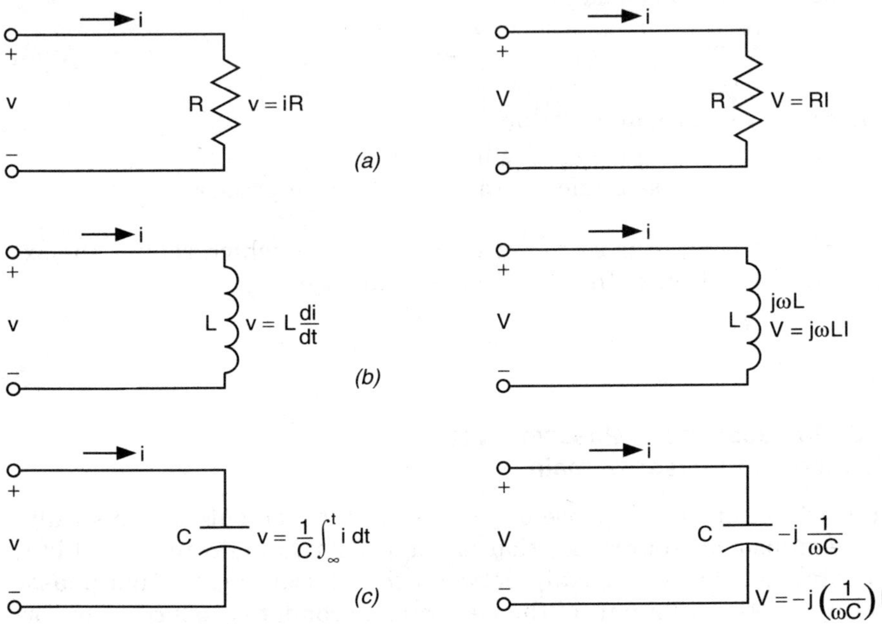

Therefore, Eq. (A.20) becomes

$$f(t) = \sqrt{2}\text{Re}(Fe^{j\omega t}) \tag{A.24}$$

The quantity F is defined as a phasor representation of $f(t)$.

Example A.4 Find the phasor transform of the following trigonometric function: $V = 300 \cos(20{,}000\pi t + 45°) - 100 \sin(20{,}000\pi t + 30°)$.

solution

$$V = [300 \cos(20{,}000\pi t + 45°) - 100 \sin(20{,}000\pi t + 30°)]$$

$$= [300 \cos(20{,}000\ \pi t + 45°) - 100 \cos(20{,}000\pi t + 30° - 90°]^*$$

$$= \left(\frac{300}{\sqrt{2}}\right) \underline{/45°} - \left(\frac{100}{\sqrt{2}}\right) \underline{/60°} = 240.2\underline{/61.52°}$$

*$\sin x = \cos(x - 90°)$

From Table A.1, we have the following:

1. *Resistor.* From Ohm's law it follows that if the current in a resistor varies sinusoidally with time as

$$i = I_m \cos(\omega t - \phi) \tag{A.25}$$

then the voltage at the terminals of the resistor will be

$$v = R[I_m \cos(\omega t - \phi)] = RI_m \cos(\omega t - \phi) \tag{A.26}$$

where I_m is the maximum current.

In the phasor domain, we have

$$\mathbf{V} = R\left(\frac{I_m}{\sqrt{2}}\right) e^{-j\phi} = R\left(\frac{I_m}{\sqrt{2}}\right) \underline{/\phi} = R\mathbf{I} \tag{A.27}$$

where

$$\mathbf{I} = \frac{I_m}{\sqrt{2}} e^{-j\phi} \tag{A.28}$$

2. *Inductor.* The relationship between phasor current and phasor voltage at the terminals of the inductor is given by assuming a sinusoidal current and using the expression for the voltage:

$$V = L\frac{di}{dt} = -\omega LI_m \sin(\omega t - \phi) \tag{A.29}$$

Using the cosine formula, Eq. (A.29) becomes

$$V = L\frac{di}{dt} = -\omega LI_m \cos(\omega t - \phi - 90°) \tag{A.30}$$

Therefore, representing Eq. (A.30) in phasor domain, we have

$$V = -\omega LI_m e^{-j(\phi + 90°)} = -\omega LI_m e^{-j\phi} e^{-j90°}$$

$$= j\omega LI_m e^{-j\phi} = j\omega LI \tag{A.31}$$

We define the impedance as a ratio of phasor voltage to phasor current, for an inductor:

$$Z_L = j\frac{\omega LI}{I} = j\omega L \tag{A.32}$$

where

$$X_L = j\omega L \qquad \text{(inductive reactance)} \tag{A.33}$$

3. *Capacitor.* The relationship between the phasor current and phasor voltage at the terminals of a capacitor is found by integrating the voltage across the terminals of the capacitor:

$$v = \frac{I}{C}\int_0^t i\,dt = \frac{I}{C}\int_0^t I_m \cos(\omega t - \phi)\,dt$$

$$= \frac{1}{\omega C} I_m \sin(\omega t - \phi)$$

$$= \frac{1}{C} I_m \cos(\omega t - \phi - 90°) = \frac{1}{C} I_m e^{-j\phi} e^{-j90°} e^{j\omega t} \qquad \text{(A.34)}$$

The phasor representation of Eq. (A.34) is

$$\mathbf{v} = \frac{1}{\omega C} I_m e^{-j(\phi + 90°)} = -\frac{1}{\omega C} I_m e^{-j\phi} = -j\frac{1}{\omega C}\mathbf{I} \qquad \text{(A.35)}$$

From Eq. (A.35), the capacitive impedance is given by

$$Z_c = -j\frac{I}{\omega c} = -jX_c \qquad \text{(A.36)}$$

where

$$X_c = -\frac{1}{j\omega c} \quad \text{(capacitive reactance)} \qquad \text{(A.37)}$$

Example A.5 A 60-Ω resistor, a 30-mH inductor, and 5-μF capacitor are connected in series across the terminals of a sinusoidal voltage source as shown in Fig. A.2. The steady-state expression for the voltage is $v = 650 \cos(5000t + 30°)$.

1. Draw the phasor equivalent circuit.
2. Determine the steady-state current i by the phasor method.

solution

1. From the expression $v = 650 \cos(5000t + 30°)$, $\omega = 5000$ rad/s. Therefore,

$$Z_L = j\omega L = j5000 \times 10^{-3} = j150\ \Omega$$

$$Z_c = -j\frac{1}{\omega C} = -j\frac{1}{5000(5 \times 10^{-6})} = -j40\ \Omega$$

The phasor representation of the voltage is

$$V = \frac{650}{\sqrt{2}}\underline{/30°} = 459.6\underline{/30°}$$

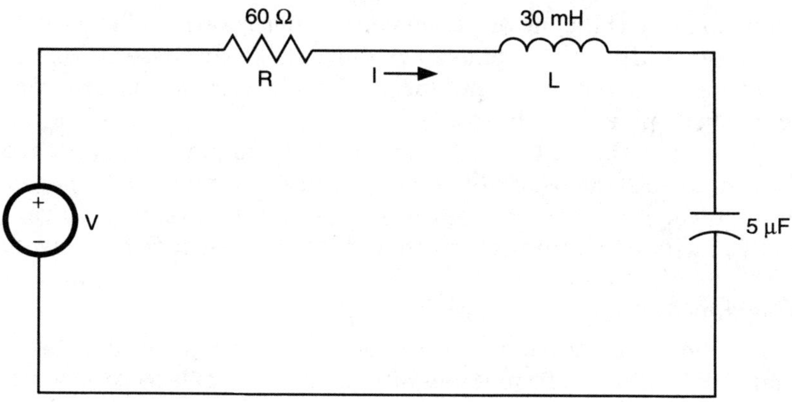

Figure A.2 Example A.5.

The voltage at the source must be changed to rms value. The phasor equivalent circuit is shown in Fig. A.3.

2. Steady-state current is

$$I = \frac{\mathbf{V}}{Z_T} = \frac{459.6\underline{/30^\circ}}{60 + j(150 - 40)} = 3.67\underline{/-31.38^\circ}\ \text{A}$$

Therefore the steady-state expression for current i is

$$i = 3.67 \cos (5000t - 31.38^\circ)\ \text{A}$$

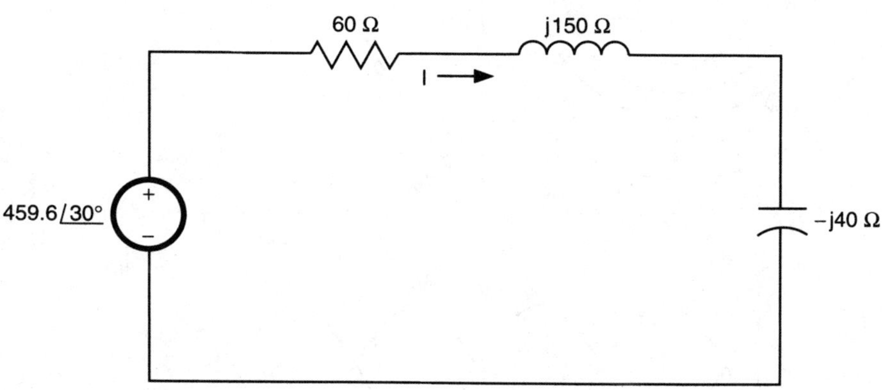

Figure A.3 The phasor equivalent circuit.

A.4 Polyphase Power Circuits

A.4.1 Introduction

The generation, transmission, distribution, and utilization of a large block of power is accomplished by means of polyphase generators, which develop more than one ac phase voltage per rotation of the rotor. In this chapter, only a three-phase system will be discussed. Almost all generators used today for production of electric power are constructed as three-phase machines.

The waveform of the voltages at the buses of the power system are sinusoidal in nature and oscillate at a constant frequency. Since power systems have constant frequency, voltages and currents may be represented by phasors, and throughout this book, phasors have been used.

A.4.2 Three-phase circuits

Consider an elementary three-phase ac generator, which generates three-phase sinusoidal voltages of equal magnitude but are displaced from each other by 120°. The elementary ac generator shown in Fig. A.4 has three

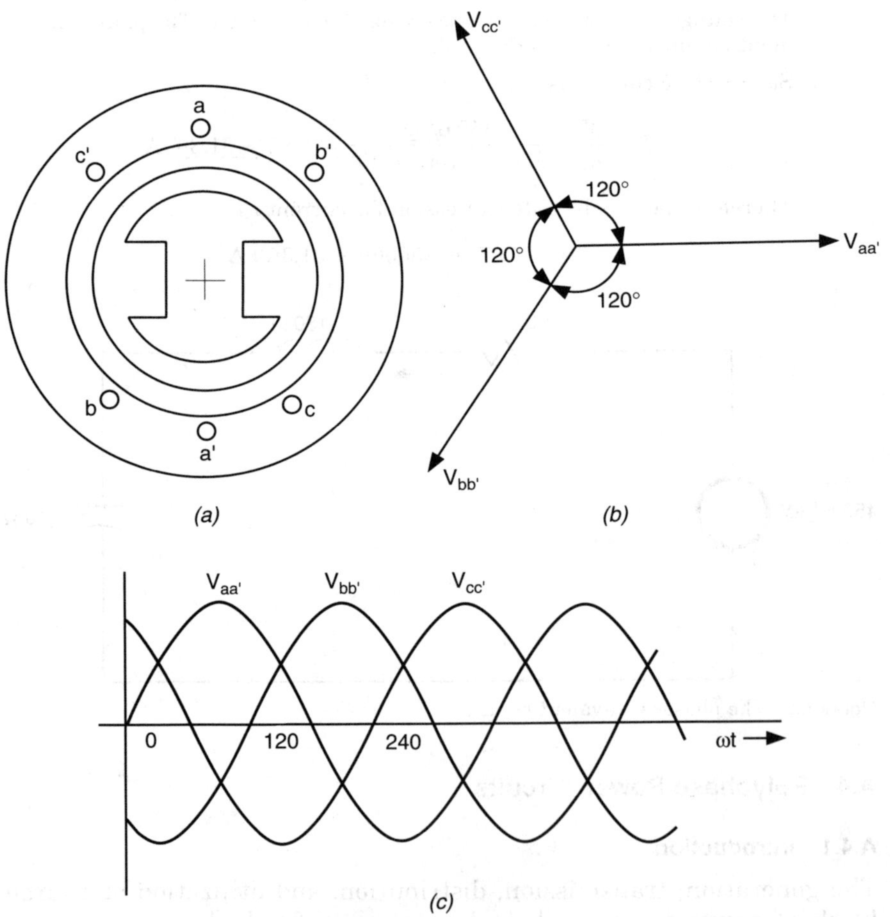

Figure A.4 Balanced three-phase voltages. (*a*) Elementary three-phase generator. (*b*) Phasor representation. (*c*) Instantaneous voltages.

identical coils of equal turns. The two sides of each coil and their terminals are shown in Fig. A.4*a* by *a-a'*, *b-b'*, and *c-c'*.

Defining $V_{aa'}$ as the potential of terminal a' with respect to a, can be described by the voltages

$$
\begin{aligned}
V_{aa'} &= V_m \cos(\omega t - \phi) = \sqrt{2}V \cos(\omega t - \phi) \\
V_{bb'} &= \sqrt{2}V \cos(\omega t - \phi - 120^\circ) \\
V_{cc'} &= \sqrt{2}V \cos(\omega t - \phi + 120^\circ)
\end{aligned}
\tag{A.38}
$$

Putting Eq. (A.38) into phasors we get

$$
\begin{aligned}
V_{aa'} &= V\underline{/-\phi} \\
V_{bb'} &= V\underline{/-\phi - 120^\circ} \\
V_{cc'} &= V\underline{/-\phi + 120^\circ}
\end{aligned}
\tag{A.39}
$$

These three-phase voltage sources and loads can be connected either into wye (star) or into delta.

Wye (star)-connected load. If the three primed coils a', b', and c' of the generator in Fig. A.4*a* are connected together, then the unprimed terminals a, b, and c constitute the wye(Y)-connected generator terminals. If a Y-connected load is connected to the generator terminals, then phase and line voltages of the load can be computed.

Figure A.5*a* shows a schematic diagram of a Y-connected load, connected to a Y-connected three-phase generator and Fig. A.5*b* shows its phasor diagram. The line to line voltage in Fig. A.5*a* and *b* is given by

$$V_{ab} = V_{a'n} + V_{nb'} = V_{a'n} - V_{b'n} \tag{A.40}$$

and from the phasor diagram

$$X = V_{a'n} \cos\theta = V_{a'n} \cos 30^\circ$$

But

$$V_{ab} = 2X = 2V_{a'n} \cos 30^\circ = \sqrt{3}V_{a'n} \tag{A.41}$$

From Eq. (A.41) the line to line voltage in a wye-connected load is given as the square root of three times the phase voltage. The line

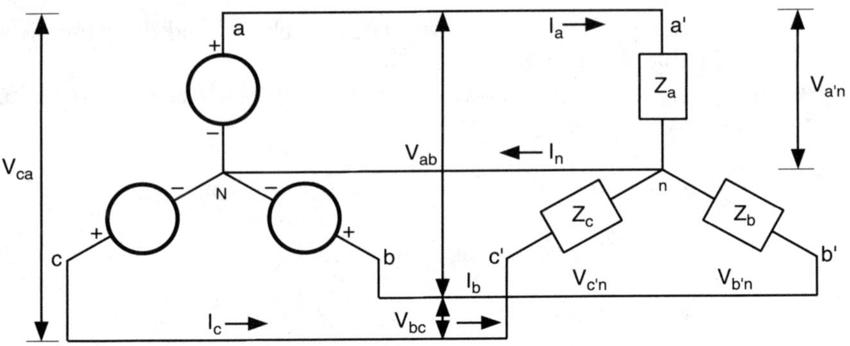

(a)

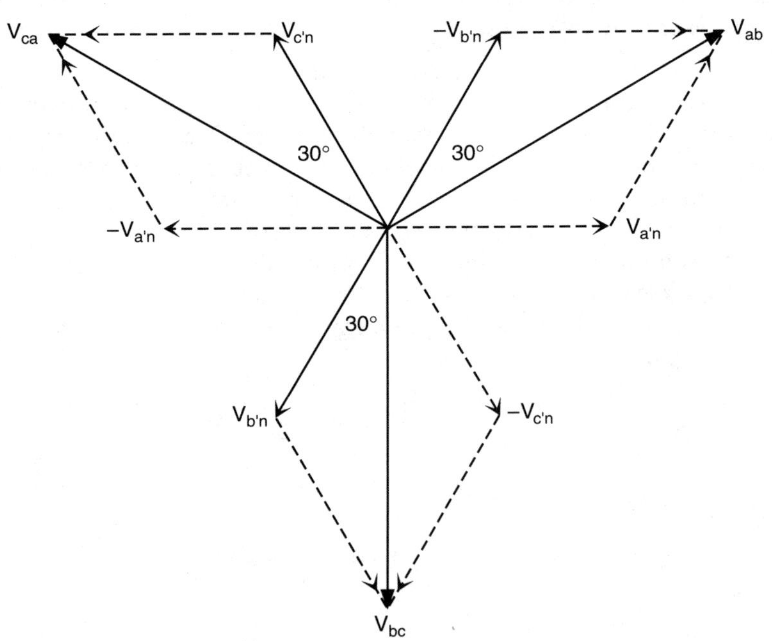

(b)

Figure A.5 (*a*) A wye-connected load with wye-connected source. (*b*) Phasor diagram of wye-connected load.

voltage V_{ab} leads the phase voltage $V_{a'n}$ by 30°. From Fig. A.5*b* the three line to line voltages are given as

$$\begin{aligned} V_{ab} &= \sqrt{3}V_{a'n}e^{j30^\circ} \\ V_{bc} &= \sqrt{3}V_{a'n}e^{-j90^\circ} \\ V_{ca} &= \sqrt{3}V_{a'n}e^{-j210^\circ} \end{aligned} \tag{A.42}$$

In general it can be stated that in a Y-connection, the line to line voltages are equal to the cube root times the phasor voltages ($V_L = \sqrt{3}V_\phi$). The line currents, on the other hand, are equal to the phase currents ($I_L = I_\phi$).

Example A.6 A 400-V three-phase supply feeds a Δ-connected load with the following branch impedances: $Z_a = j100\ \Omega$, $Z_b = j100\ \Omega$, and $Z_c = j100\ \Omega$. Calculate

1. All the line and phase voltages and currents
2. Draw a phasor diagram

solution Using the line voltages derived in this section, the phase and line currents can be computed. The phase currents are

$$\begin{aligned} I_{ab} &= \frac{V_{ab}}{Z} = \frac{\sqrt{3}V_{an}e^{j30}}{|Z|\underline{/\theta_z}} = \frac{\sqrt{3}V_{an}e^{j(30^\circ-\theta_z)}}{|Z|} \\ I_{bc} &= \frac{V_{bc}}{Z} = \frac{\sqrt{3}V_{an}e^{-j90}}{|Z|\underline{/\theta_z}} = \frac{\sqrt{3}V_{an}e^{-j(90^\circ+\theta_z)}}{|Z|} \\ I_{ca} &= \frac{V_{ca}}{Z} = \frac{\sqrt{3}V_{an}e^{-j120}}{|Z|\underline{/\theta_z}} = \frac{\sqrt{3}V_{an}e^{-j(210^\circ+\theta_z)}}{|Z|} \end{aligned} \tag{A.43}$$

The line currents are then given by

$$\begin{aligned} I_a &= I_{ab} - I_{ca} = \frac{\sqrt{3}V_{an}e^{-j\theta_z}(e^{j30^\circ} - e^{-j210^\circ})}{|Z|} = \frac{3V_{an}e^{-j\theta}}{|Z|} \\ I_b &= I_{bc} - I_{ab} = \frac{\sqrt{3}V_{an}e^{-j\theta_z}(e^{-j90^\circ} - e^{j30^\circ})}{|Z|} = \frac{3V_{an}e^{-j(120^\circ+\theta_z)}}{|Z|} \\ I_c &= I_{ca} - I_{bc} = \frac{\sqrt{3}V_{an}e^{-j\theta_z}(e^{-j120^\circ} - e^{-j90^\circ})}{|Z|} = \frac{3V_{an}e^{-j(240^\circ+\theta_z)}}{|Z|} \end{aligned} \tag{A.44}$$

1. Phase voltages

$$\begin{aligned} V_{an} &= \frac{400}{\sqrt{3}}\underline{/0^\circ}\ \text{V} \\ V_{bn} &= 230.9\underline{/-120^\circ}\ \text{V} \\ V_{cn} &= 230.9\underline{/120^\circ}\ \text{V} \end{aligned}$$

2. Line to line voltages

$$V_{ab} = \sqrt{3}V_{an}\underline{/30^\circ} = \sqrt{3}(230.9)\underline{/30^\circ} = 400\underline{/30^\circ}\text{ V}$$

$$V_{bc} = \sqrt{3}V_{an}\underline{/-90^\circ} = \sqrt{3}(230.9)\underline{/-120^\circ - 90^\circ} = 400\underline{/-210^\circ}\text{ V}$$

$$V_{ac} = \sqrt{3}V_{an}\underline{/-210^\circ} = \sqrt{3}(230.9)\underline{/-210^\circ + 120^\circ} = 400\underline{/-90^\circ}\text{ V}$$

Line and phase currents:

$$I_{an} = \frac{V_{an}}{Z} = \frac{230.9\underline{/0^\circ}}{100\underline{/90^\circ}} = 2.309\underline{/-90^\circ}\text{ A}$$

$$I_{bn} = \frac{V_{bn}}{Z} = \frac{230.9\underline{/-120^\circ}}{100\underline{/90^\circ}} = 2.309\underline{/-210^\circ}\text{ A}$$

$$I_{cn} = \frac{V_{cn}}{Z} = \frac{230.9\underline{/120^\circ}}{100\underline{/90^\circ}} = 2.309\underline{/30^\circ}\text{ A}$$

Delta-connected load. If delta-connected load terminals are connected to the terminals of a Y-connected generator, the phase and line voltages and currents can be computed using Fig. A.6*a* and following the same procedure used in the previous section. Figure A.6*b* is its phasor diagram. Students should try to find the expressions for an exercise.

The three-phase unbalanced loads. In more complex unbalanced networks, the method of symmetrical components is used, but in simple situations conventional network theory can be applied. The symmetrical component theory recognizes that any unbalanced three-phase system of phasors may be resolved into three balanced systems of phasors, namely, the positive-sequence, the negative-sequence, and the zero-sequence systems. In this section the network method will be developed. Let three unbalanced loads Z_a, Z_b, and Z_c be connected in Y, as shown in Fig. A.7.

Using the mesh (loop) and matrix inversion methods, the unknown loop currents can be found. Finally the line and phase currents and voltages can be computed. Using the mesh method, we obtain

$$\begin{bmatrix} V_{ab} \\ V_{bc} \\ V_{ca} \end{bmatrix} = \begin{bmatrix} Z_a + Z_b & -Z_a & -Z_b \\ -Z_a & Z_a + Z_c & -Z_c \\ -Z_b & -Z_c & Z_b + Z_c \end{bmatrix} \begin{bmatrix} I_1 \\ I_2 \\ I_3 \end{bmatrix} \tag{A.45}$$

Taking the inverse of matrix (A.45), expression (A.46) is found:

$$\begin{bmatrix} I_1 \\ I_2 \\ I_3 \end{bmatrix} = \begin{bmatrix} Z_a + Z_b & -Z_a & -Z_b \\ -Z_a & Z_a + Z_c & -Z_c \\ -Z_b & -Z_c & Z_b + Z_c \end{bmatrix}^{-1} \begin{bmatrix} V_{ab} \\ V_{bc} \\ V_{ca} \end{bmatrix}$$

$$= [Z]^{-1} \begin{bmatrix} V_{ab} \\ V_{bc} \\ V_{ca} \end{bmatrix} \tag{A.46}$$

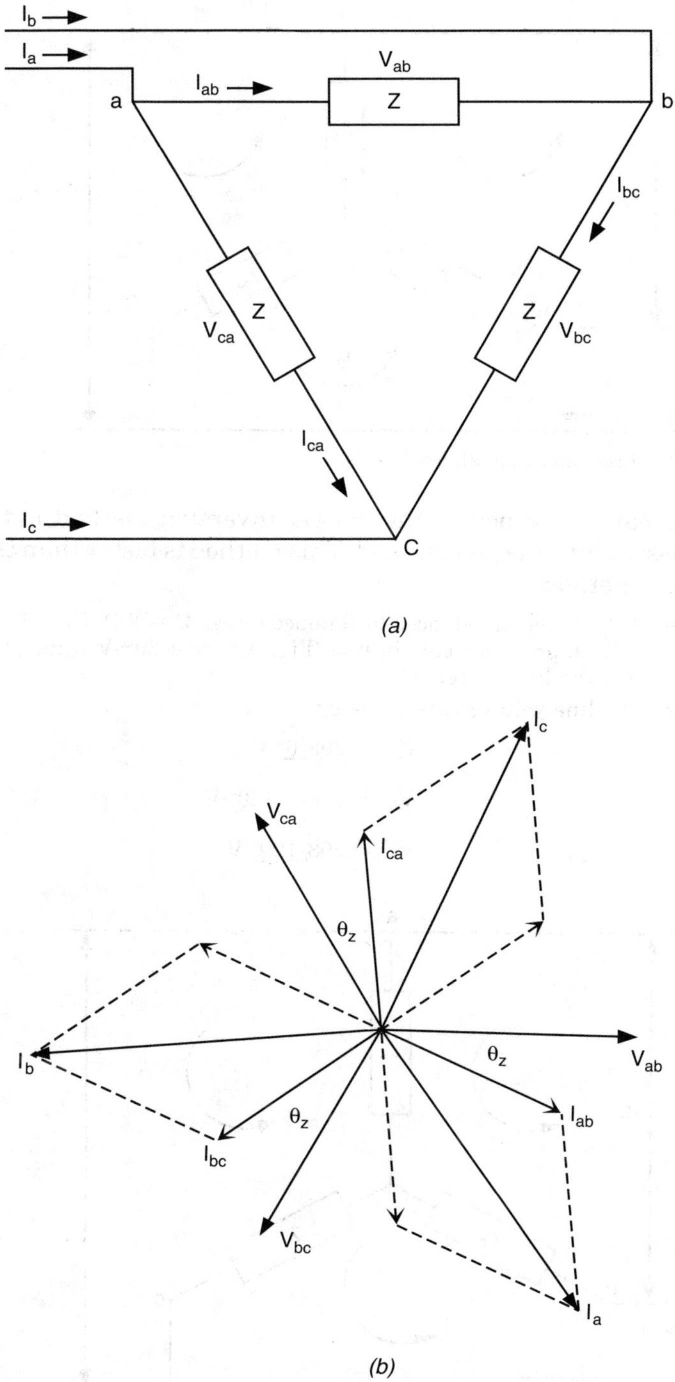

Figure A.6 (*a*) Delta-connected load. (*b*) Phasor diagram of a delta-connected load.

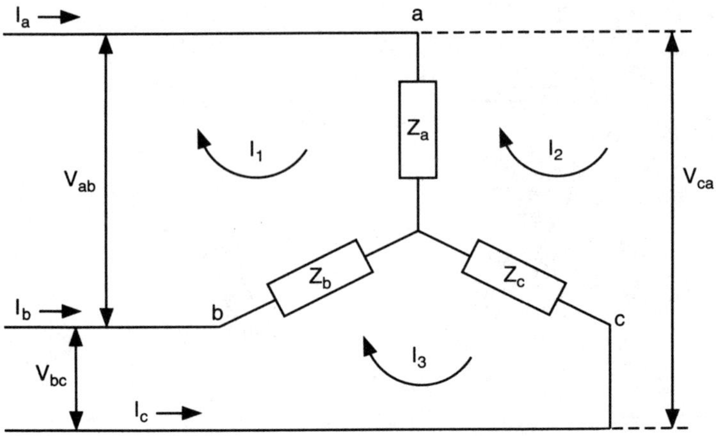

Figure A.7 Three-phase unbalanced loads.

An alternative method to the matrix inversion method is the use of the gaussian elimination method. This method is faster than the matrix inversion method.

Example A.7 Three unbalanced load impedances, $Z_a = 2\ \Omega$, $Z_b = 1 + j2\ \Omega$, and $Z_c = 3 - j2\ \Omega$, are connected in wye (Fig. A.8) to a 208-V three-phase, 60-Hz supply. Find the line currents.

solution The line voltages are given by

$$V_{an} = 208\underline{/0^\circ}\text{ V}$$

$$V_{bc} = 208\underline{/-120^\circ}\text{ V}$$

$$V_{ca} = 208\underline{/120^\circ}\text{ V}$$

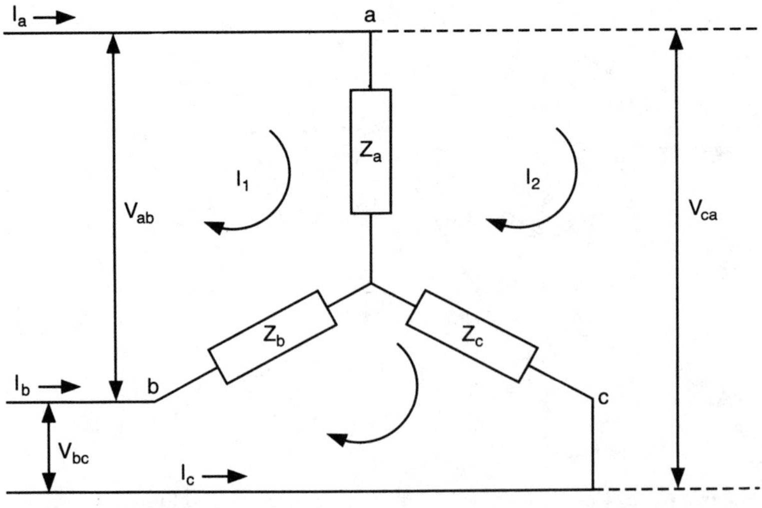

Figure A.8 Example A.7.

Using the method of matrix inversion, we get

$$\begin{bmatrix} I_1 \\ I_2 \\ I_3 \end{bmatrix} = \begin{bmatrix} 3 + j2 & -2 & -1 - j2 \\ -2 & 5 - j2 & -3 + j2 \\ -1 - j2 & -3 + j^2 & 4 \end{bmatrix}^{-1} \begin{bmatrix} V_{ab} \\ V_{bc} \\ V_{ca} \end{bmatrix}$$

$$\begin{bmatrix} I_1 \\ I_2 \\ I_3 \end{bmatrix} = \begin{bmatrix} 155\underline{/105^\circ} & 155\underline{/105^\circ} & 50\underline{/143^\circ} \\ 155\underline{/105^\circ} & 154.9\underline{/104^\circ} & 116\underline{/70^\circ} \\ 50\underline{/143^\circ} & 116\underline{/70^\circ} & 154.2\underline{/1049^\circ} \end{bmatrix} \begin{bmatrix} 208\underline{/0^\circ} \\ 208\underline{/-120^\circ} \\ 208\underline{/120^\circ} \end{bmatrix}$$

Therefore,

$$I_1 = [155\underline{/105^\circ}(208)] + [155\underline{/105^\circ}(208\underline{/-120^\circ})] + [50\underline{/-143^\circ}(208\underline{/-120^\circ})]$$

$$= 21{,}529.6 + j12{,}474.6 = 24.88\underline{/30^\circ} \text{ kA}$$

$$I_2 = [155\underline{/105^\circ}(208)] + [154.9\underline{/104^\circ}(208\underline{/-120^\circ})] + [116\underline{/70^\circ}(208\underline{/120^\circ})]$$

$$= 1134.6 + j18{,}070.8 = 18.106\underline{/86.4^\circ} \text{ kA}$$

$$I_3 = -873.92 + j19{,}299.7 = 89.49\underline{/167.5^\circ} \text{ kA}$$

Therefore, the line currents are

$$I_a = I_1 - I_2 = 21{,}529.6 + j12{,}474.6 - (1134.6 + j18{,}070.8)$$

$$= 20{,}395 - j5596.2 = 21.14\underline{/-15.3^\circ} \text{ kA}$$

$$I_b = I_3 - I_1 = -87{,}392 + j19{,}299.7 - 21{,}529.6 - j12{,}474.6$$

$$= -108{,}921.6 + j6825.1 = 109.13\underline{/176^\circ} \text{ kA}$$

$$I_c = I_2 - I_3 = 1134.6 + j18{,}070.8 + 87{,}392 - j19{,}299.7$$

$$= 88{,}526.6 - j1228.9 = 88.53\underline{/-0.79^\circ} \text{ kA}$$

A.4.3 AC power calculations

Real and reactive power. Taking any general type of load, the current and voltage will not be in phase, but will be out of phase by an angle Ψ. When this angle is positive, the voltage leads the current, and the load is inductive. Let

$$v = V_m \cos \omega \qquad i = I_m \cos (\omega t + \Psi) \tag{A.47}$$

Power transmitted is

$$p = vi = V_m I_m \cos \omega t \cos (\omega t + \Psi) \tag{A.48}$$

Using trigonometric identity,

$$\cos \omega t \cos (\omega t + \Psi) = \tfrac{1}{2}[\cos \Psi + \cos (2\omega t + \Psi)] \tag{A.49}$$

Substituting Eq. (A.49) into Eq. (A.48), the desired expression for instantaneous power is given by

$$p = vi = \frac{V_m I_m}{2} [\cos \psi + \cos (2\omega t + \psi)] \qquad (A.50)$$

But

$$\cos (2\omega t + \Psi) = \cos \Psi \cos 2\omega t - \sin \Psi \sin 2\omega t \qquad (A.51)$$

Substituting Eq. (A.51) into (A.50) the power equation becomes

$$p = \frac{V_m I_m}{2} (\cos \psi + \cos \psi \cos 2\omega t - \sin \psi \sin 2\omega t)$$

$$= VI \cos \Psi (1 + \cos 2\omega t) - VI \sin \Psi \sin 2\omega t \qquad (A.52)$$

The first component of Eq. (A.52) is the average power. The second component has a zero average because there is a periodic energy storage at half a cycle and return of energy at the second half of a cycle. Graphically the average power is shown in Fig. A.9.

Introducing the following quantities:

$$P = |V||I| \cos \Psi \qquad \text{(real power, W)} \qquad (A.53)$$

$$Q = |V||I| \sin \Psi \qquad \text{(reactive power, VAR)} \qquad (A.54)$$

Substituting Eqs. (A.53) and (A.54) into Eq. (A.52) we have

$$p = P(1 + \cos 2\omega t) - Q \sin 2\omega t \qquad \text{W} \qquad (A.52a)$$

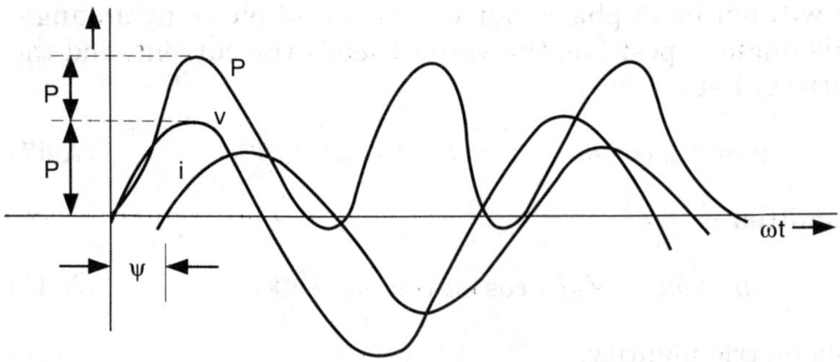

Figure A.9 Instantaneous voltage, current, and power.

Complex power. If V and I are the phasor voltage and phasor current per phase in a balanced three-phase, the complex power is given by

$$S = P + jQ \qquad \text{VA} \tag{A.55}$$

where P and Q are the real and reactive power, respectively. Let

$$V = |V|e^{j\alpha} \qquad \text{V} \tag{A.56}$$

$$I = |I|e^{j\beta} \qquad \text{A} \tag{A.57}$$

Introducing the conjugate of the current,

$$I^* = |I|e^{-j\beta} \tag{A.58}$$

Equation (A.55) can be written as

$$S = VI^* = ZII^* = Z|I|^2 = j\omega L|I|^2 \tag{A.59}$$

Substituting Eqs. (A.56) and (A.58) into Eq. (A.59) we have

$$\begin{aligned} S &= |V||I|e^{j(\alpha-\beta)} = |V||I|e^{j\Psi} \\ &= |V||I| \cos \Psi + j|V||I| \sin \Psi \end{aligned} \tag{A.60}$$

Combining Eqs. (A.55) and (A.60) yields

$$S = VI^* = P + jQ = S \cos \Psi + jS \sin \Psi \tag{A.61}$$

Complex power in a three-phase wye- or delta-connected generator can be written as

$$S_a = V_a I_a^* = P_a + jQ_a \tag{A.62}$$

$$S_b = V_b I_b^* = P_b + jQ_b \tag{A.63}$$

$$S_c = V_c I_c^* = P_c + jQ_c \tag{A.64}$$

where a, b, and c are the three phases of the generator. Since complex power is a complex quantity, and if Eqs. (A.62), (A.63), and (A.64) are added up, we get

$$S_T = P_T + jQ_T = (P_a + P_b + P_c) + j(Q_a + Q_b + Q_c) \qquad \text{VA} \tag{A.65}$$

where S_T is the total apparent power generated by a balanced three-phase wye- or delta-connected generator. But

$$P_a = |V_a||I_a| \cos \psi = |V_a||I_a|e^{j\psi} \quad \text{(A.66)}$$

$$P_b = |V_b||I_b| \cos \psi = |V_b||I_b|e^{j\psi} \quad \text{(A.67)}$$

$$P_c = |V_c||I_c| \cos \psi = |V_c||I_c|e^{j\psi} \quad \text{(A.68)}$$

and since

$$|V_a| = |V_b| = |V_c| = |V| \quad \text{and} \quad |I_a| = |I_b| = |I_c| = |I| \quad \text{(A.69)}$$

then substituting Eqs. (A.66), (A.67), and (A.68) into (A.65), we get

$$S_T = P_T + jQ_T = 3S = (P_a + P_b + P_c) + j(Q_a + Q_b + Q_c)$$

$$= 3(P + jQ) = 3(VI^*) \quad \text{(A.70)}$$

Equation (A.70) gives the total complex power generated by a balanced three-phase wye- or delta-connected generator. The total power can be obtained by multiplying the phase power by three. Figure A.10 shows the voltage and current conjugate phasor diagram.

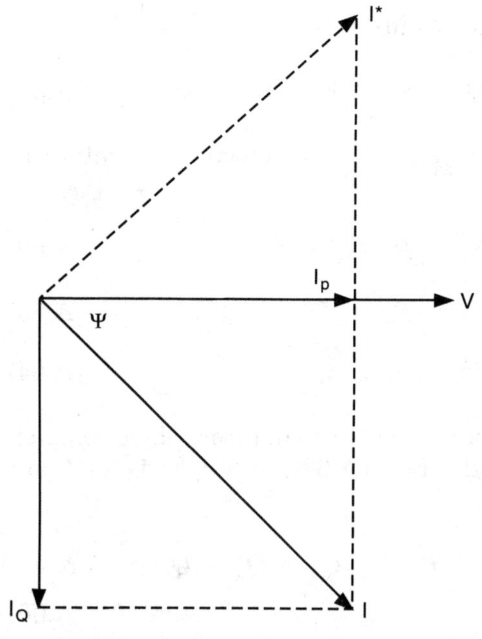

Figure A.10 The phasor voltage and current and its conjugate.

From Fig. A.10 we get

$$I_p = I \cos \Psi \qquad I_Q = I \sin \Psi \tag{A.71}$$

Substituting Eq. (A.69) into Eq. (A.58) we get

$$S = |V||I_p| + j|V||I_Q| \tag{A.72}$$

Power in three-phase systems. From page 282 and Eq. (A.52) it was obvious that the instantaneous power for a single-phase sinusoidal source varies sinusoidally at twice the frequency of the source. By Eq. (A.50) the expression for the instantaneous power is

$$P = \frac{V_m I_m}{2} \cos \omega t + \frac{V_m I_m}{2} \cos (2\omega t + \psi) \tag{A.73}$$

The effective value of the instantaneous power is given by

$$p = VI \cos \Psi + VI \cos (2\omega t + \Psi) \tag{A.74}$$

Applying Eq. (A.74) to each phase of the three-phase load we have

$$P_a = V_p I_p \cos \Psi + V_p I_p \cos (2\omega t + \Psi) \tag{A.75}$$

$$P_b = V_p I_p \cos \Psi + V_p I_p \cos (2\omega t + \Psi - 120°) \tag{A.76}$$

$$P_c = V_p I_p \cos \Psi + V_p I_p \cos (2\omega t + \Psi + 120°) \tag{A.77}$$

where a, b, and c are the three phases. Phase a has been taken as a reference phase. The total instantaneous power in the three-phase system is equal to three times the average power per phase. Mathematically, the total average power can be expressed as

$$P = \frac{I}{T} \int_0^T (P_a + P_b + P_c)\, dt \tag{A.78}$$

or

$$P = 3V_p I_p \cos \Psi \tag{A.79}$$

For a wye-connected load the three-phase power is

$$P = \frac{3V_L I_L}{\sqrt{3}} \cos \psi = \sqrt{3} V_L I_L \cos \psi \tag{A.80}$$

For a delta-connected load the three-phase power is

$$P = \frac{3V_L I_L}{\sqrt{3}} \cos \psi = \sqrt{3} V_L I_L \cos \psi \tag{A.81}$$

Power factor is the relation of the average power to that of the apparent power. Thus

$$\text{Power factor (pf)} = \cos \Psi = \frac{P}{\sqrt{P^2 + Q^2}} \tag{A.82}$$

and

$$\psi = \cos^{-1}\left(\frac{P}{\sqrt{P^2 + Q^2}}\right) = \tan^{-1}\left(\frac{Q}{\sqrt{P^2 + Q^2}}\right) \tag{A.83}$$

Example A.8 Two machines are connected as shown in Fig. A.11. Machine 1 has $V_1 = 120\underline{/0^\circ}$ V and machine 2 has $V_2 = 120\underline{/30^\circ}$ V. An impedance of $Z = 2 - j5\ \Omega$ is between the two machines. Calculate

1. Power delivered by each machine
2. The values of real power (P) and reactive power (Q) absorbed by the impedance.

solution

1. Power for each machine:

$$I = \frac{120\underline{/0^\circ} - 120\underline{/30^\circ}}{2 - j5} = \frac{62.11\underline{/-74.9^\circ}}{5.385\underline{/-68.19^\circ}} = 11.53\underline{/-6.71^\circ}\text{ A}$$

From Eq. (A.59), we have power for each machine:

$$S_1 = V_1 I^* = (120\underline{/0^\circ})(11.45 + j1.3472) = 1383.6\underline{/6.71^\circ}\text{ VA}$$

$$S_2 = V_2 I^* = (120\underline{/0^\circ})(11.45 + j1.3472) = 1383.6\underline{/36.71^\circ}\text{ VA}$$

2. The values of real power P and reactive power Q absorbed by the impedance:

$$S = I^2 Z = (11.53\underline{/-6.81^\circ})^2(5.385\underline{/-68.19^\circ})$$

$$= 715.9\underline{/-81.81^\circ} = 101.98 - j708.58\text{ VA}$$

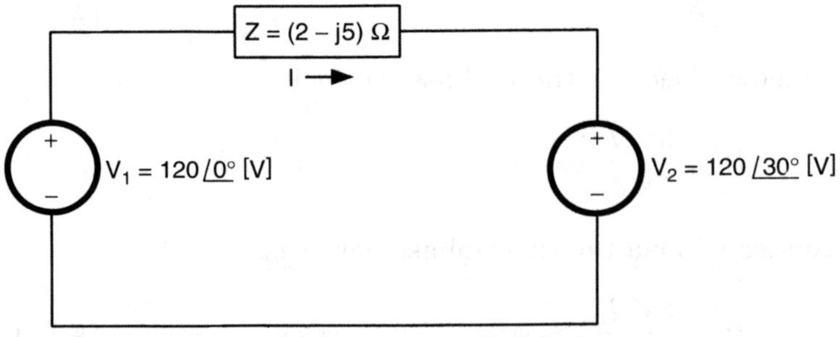

Figure A.11 Example A.8.

Therefore,

$$P = 101.98 \text{ W} \qquad \text{(real power)}$$

$$Q = -708.5 \text{ VAR} \qquad \text{(capacitive reactive power)}$$

Problems

1 The line voltage of a balanced three-phase delta-connected load of $Z = 5\angle 60^\circ$ is given as 120 V. Find
a The phase and line currents.
b Draw a well-labeled phasor diagram.

2 A three-phase, 20-kVA, 440-V, 60-Hz generator is operating at full load under balanced steady-state conditions. Calculate the magnitudes of the generator line voltages and currents when
a The generator windings are wye-connected.
b The generator windings are delta-connected.

3 A wye-connected balanced load with a per-phase impedance of $(10 + j9)$ Ω is supplied from a 220-V, 60-Hz, three-phase source.
a Calculate the line currents, the total voltamperes, the power factor, the real power, and the reactive power absorbed by the load.
b Draw the phasor diagram showing all the voltages and currents with V_{na} as the reference.
c If the star point of the load is connected to the system neutral through an ammeter, what would be the meter reading?

4 In Fig. A.5 assume that $E_a = 110\angle 0^\circ$ V, $E_b = 100\angle -120^\circ$ V, $E_c = 100\angle 240^\circ$ V, and $Z_a = 5 + j0\ \Omega$, $Z_b = 0 - j5\ \Omega$ and $Z_c = 3 + j4\ \Omega$. Calculate
a The generated currents I_a, I_b, and I_c.
b Draw a phasor diagram.

5 Three impedances Z each consist of a 9-Ω resistor, a 60-mH inductor, and a 200-mF capacitor, all connected in series. The impedances are connected in delta across the phase terminals of a three-phase generator that delivers a line voltage of 3 kV, 60 Hz. Calculate
a Line currents
b Phase currents
c Delivered power, three-phase, real and reactive

6 A 415-V, 60-Hz, three-phase supply is connected to a balanced wye load of impedance per phase $(6 + j7)$ Ω by an interconnector of impedance per phase $(0.6 + j0)$ Ω. Find the power dissipated in the interconnector. The load power factor is brought to unity by connecting a capacitor of magnitude C in delta across the load. Find C and the loss in the interconnector with the capacitors in circuit.

7 Two wattmeters are used to measure the power in a balanced 240-V, 60-Hz, three-phase system. The indication of P_2 and P_1 are 50.6 kW and 35.1 kW, respectively. Compute the line current and the power factor of the load.

8 A delta-connected load is made up of the following impedances: $Z_{ab} = 65\underline{/30^\circ}\ \Omega$, $Z_{bc} = 70\underline{/48.9^\circ}\ \Omega$, $Z_{ca} = 20\underline{/-20^\circ}\ \Omega$. It is connected to a 240-V, 60-Hz, three-phase supply of phase sequence *abc*. Calculate the line currents.

9 A 400-V, 60-Hz, three-phase supply of phase sequence *abc* is connected to a balanced delta load of $Z = 51\underline{/60^\circ}\ \Omega$, as shown in Fig. A.12. Compute the line currents and total power.

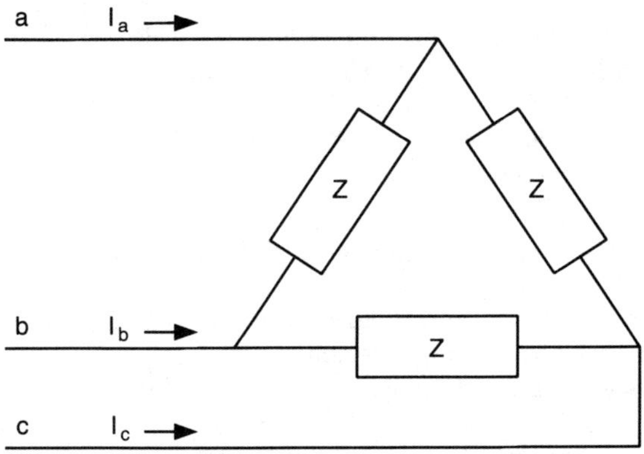

Figure A.12 Problem 9.

10 A three-phase line has an impedance of $1 + j4\ \Omega$/phase. The line feeds two balanced three-phase loads that are connected in parallel. The first load is absorbing a total of 8.3 kW and 4.6 kVAR magnetizing VARs. The second load is connected in delta and has a impedance of $10.61 - j8.6\ \Omega$/phase. The line to neutral voltage at the load end of the line is 120 V. Find the magnitude of the line voltage at the source end of the line.

Bibliography

Boylestad, Robert L.: *Introductory Circuit Analysis,* Charles E. Merrill, Columbus, Ohio, 1982.

Johnson, D. E., J. L. Hiliburn, and J. R. Johnson: *Basic Electric Circuit Analysis,* Prentice-Hall, Englewood Cliffs, N.J., 1978.

Nilson, James W.: *Electric Circuits,* Addison Wesley, Reading, Mass., 1983.

Smith, Ralph J.: *Circuit Devices and Systems,* 4th ed., John Wiley, New York, 1983.

Index